QR 105 .M46 1988

Methods in aquatic
bacteriology

D1090399

Methods in Aquatic Bacteriology

MODERN MICROBIOLOGICAL METHODS

Series Editor: Michael Goodfellow, *Department of Microbiology, University of Newcastle-upon-Tyne*

Methods in Aquatic Bacteriology
Edited by Brian Austin

0 471 91651 X

Bacterial Cell Surface Techniques (*in press*)
Edited by Ian Hancock and Ian Poxton
0 471 91041 4

Methods in Aquatic Bacteriology

Edited by

B. Austin

Department of Brewing and Biological Sciences,
Heriot-Watt University, Chambers Street,
Edinburgh EH1 1HX

A Wiley – Interscience Publication

D. HIDEN RAMSEY LIBRARY
U.N.C. AT ASHEVILLE
ASHEVILLE, N. C. 28814

JOHN WILEY & SONS

Chichester • New York • Brisbane • Toronto • Singapore

Copyright © 1988 by John Wiley & Sons Ltd.

All rights reserved.

No part of this book may be reproduced by any means, or transmitted, or translated into a machine language without the written permission of the publisher

Library of Congress Cataloging in Publication Data:

Methods in aquatic bacteriology.

  (Modern microbiological methods)
  'A Wiley–Interscience publication.'
  Includes bibliographies.
  1. Aquatic microbiology—Technique. I. Austin,
B. (Brian), 1951–   . II. Series.
QR105 .M46   1988   589.9092   87-13326

ISBN 0 471 91651 X

British Library Cataloguing in Publication Date

Methods in aquatic bacteriology —(Modern
microbiological methods).
  1. Water—Bacteriology
  I. Austin, B.   II. Series.
  589.9092   QR105

  ISBN 0 471 91651 X

Phototypesetting by Thomson Press (India) Ltd., New Delhi
Printed and bound in Great Britain by Anchor Brendon Ltd., Tiptree, Colchester.

D. HIDEN RAMSEY LIBRARY
U.N.C. AT ASHEVILLE
ASHEVILLE, N. C. 28814

Contents

PART 4: ACTIVITY

List of Contributors

DR B. AUSTIN, *Department of Brewing and Biological Sciences, Heriot-Watt University, Chambers Street, Edinburgh EH1 1HX Scotland*

DR J. H. BAKER, *Freshwater Biological Association, River Laboratory, East Stoke, Wareham, Dorset BH20 6BB, England*

DR N. S. BATTERSBY, *Water Research Centre, Medmenham Laboratory, Henley Road, Medmenham, PO Box 16, Marlow, Bucks SL7 2HD, England*

PROFESSOR C. M. BROWN, *Department of Brewing and Biological Sciences, Heriot-Watt University, Edinburgh EH1 1HX, Scotland*

PROFESSOR R. CONRAD, *Fakultät für Biologie, Universität Konstanz, Postfach 5560, D7750 Konstanz 1, FRG*

DR M. J. DAFT, *Department of Biological Sciences, University of Dundee, Dundee DD1 4HN, Scotland*

DR J. C. FRY, *Department of Applied Biology, UWIST, PO Box 13, Cardiff CF1 3XF Wales*

DR J. L. FRYER, *Department of Microbiology, Oregon State University, Corvallis, OR 97331–3804, USA*

DR R. A. HERBERT, *Department of Biological Sciences, University of Dundee, Dundee DD1 4HN, Scotland*

DR J. F. IMHOFF, *Institute für Mikrobiologie, Meckenheimer Allee 168, 5300 Bonn, FRG*

MR K. O.'CARROLL, *Institute of Offshore Engineering, Heriot-Watt University, Edinburgh, Midlothian EH14 4AS, Scotland*

PROFESSOR G. RHEINHEIMER, *Abteilung für Marine Mikrobiologie, Institut für Meereskunde an der Universität Kiel, Düsternbrooker Weg 20, D 2300 Kiel 1, FRG*

DR J. E. SANDERS, *Department of Microbiology, Oregon State University, Corvallis, OR 97331–3804, USA*

DR J. SCHNEIDER, *Abteilung für Marine Mikrobiologie, Institute für Meereskunde an der Universität Kiel, Düsternbrooker Weg 20, D 2300 Kiel 1, FRG*

DR H. SCHÜTZ, *Fraunhofer Institut für Atmosphärische Unweltforschung, Kreuzeckbahnstrasse, 19, D-8100, Garmisch-Partenkirchen, FRG*

DR J. N. WARDELL, *Department of Brewing and Biological Sciences, Heriot-Watt University, Edinburgh EH1 1HX, Scotland*

DR P. A. WEST, *Aquatic Environment Protection Division 2, Directorate of Fisheries Research, Ministry of Agriculture, Fisheries and Food, Burnham-on-Crouch, Essex CM0 8HA, England*

Series Preface

The science of microbiology owes its existence as well as its underlying principles to the talent and practical prowess of pioneers such as Leeuwenhoek, Pasteur, Koch and Beijerinck. Interest in microbiology has recently increased quite significantly given the exciting developments in genetics and molecular biology and the growth of microbial technology. There was a time when most microbiologists were acquainted with many of the techniques used in microbiology. It is, however, now becoming increasingly difficult for research workers to keep abreast of the bewildering range of techniques currently used in microbiological laboratories. This problem is compounded by the fact that scientists in any one field increasingly need to apply techniques developed in other scientific disciplines.

The series 'Modern Microbiological Methods' aims to identify specialist areas in microbiology and provide up-to-date methodological handbooks to aid microbiologists at the laboratory bench. The books will be directed primarily towards active research workers but will be structured so as to serve as an introduction to the methods within a speciality for graduate students and scientists entering microbiology from related disciplines. Protocols will not only be described but difficulties and limitations of techniques and questions of interpretation fully discussed.

In summary, this series of books is designed to help stimulate further developments in microbiology by promoting the use of new and updated methods. Both authors and the editor-in-chief will be grateful to hear from satisfied or dissatisfied users so that future books in the series can benefit from the informed comment of practitioners in the field.

MICHAEL GOODFELLOW

Preface

Significant advances have been made in our understanding of the indigenous bacterial flora of aquatic ecosystems since the pioneering work of Dr C. E. Zobell. However, apart from a book written by Rodina and published in 1972, and several subsequent specialized manuals, there is a noticeable lack of any modern comprehensive text dealing with the methods currently available for the study of aquatic bacteriology. From initial discussions with Dr M. Goodfellow of the University of Newcastle upon Tyne and Dr M. Dixon of John Wiley and Sons Ltd, a proposal was formulated for an edited text. The primary aim of the resulting book is to provide detailed information of the relevant methods. The book is primarily targeted at newcomers to the field, notably undergraduate and postgraduate students and young research workers. Thus, the volume deals with the practical as opposed to the theoretical aspect of the subject, but does not attempt to consider pollution, a subject in its own right. Methods, have, however, been presented on the isolation, enumeration, identification, and ecology of the bacterial flora of aquatic ecosystems.

The book has been divided into four sections, which deal with basic techniques, specialized environments and taxonomic groups, and the activity of aquatic bacteria. Within these sections, detailed chapters consider sampling methods, determination of biomass, isolation methods, identification, the bacterial microflora of fish, invertebrates, plants and the deep sea, specialized groups, namely phototrophs, cyanobacteria, sulphate reducers and methanogens, the assessment of bacterial activity, nitrate metabolism and attachment.

I am grateful for the co-operation of the contributors, who responded willingly to requests to prepare chapters within a tightly arranged time schedule. In addition, I am grateful to my wife, who ably assisted with the overall preparation of the book and to Mrs M. A. Dunn for providing efficient secretarial assistance.

B. Austin
Edinburgh

Dedicated to **Dr James M. Shewan** for his many
contributions to aquatic bacteriology

PART 1

Basic Techniques

Methods in Aquatic Bacteriology
Edited by B. Austin
© 1988 John Wiley & Sons Ltd.

1

Sampling Methods

R. A. Herbert

Department of Biological Sciences, University of Dundee, Dundee DD1 4HN, Scotland

1.1 INTRODUCTION

In order to investigate the roles played by micro-organisms in a particular freshwater, marine or estuarine habitat, it is often necessary to collect representative samples. Sampling thus forms an important component of an aquatic microbiology research programme and, if a meaningful and valid interpretation of the data is to be made, due evaluation of the sampling procedures to be employed must be undertaken. In planning sampling programmes, it is essential to have a clear understanding of the objectives of the work, the type of habitat to be sampled, the type of samples required and the limitation of the sampling system(s) and processing procedures used. For qualitative work, the planning of the sampling programme need not be demanding. However, when quantitative studies are being considered it is essential that standardized procedures be developed, involving a detailed layout of the sampling stations, the type and number of samples to be taken, processing procedures, etc. Under these circumstances, it is often worthwhile carrying out a pilot qualitative survey to locate the most suitable study sites, and the extra effort expended at this stage is usually well rewarded by the avoidance of costly and often time-consuming mistakes later. To facilitate selection of suitable sites, information on water depth and sediment type can be obtained from the appropriate charts or by visual examination of the shoreline. In offshore areas preliminary examinations can be made by dredging, Scuba diver or by the use of remote control cameras.

Having selected a suitable study area, the next immediate consideration must be the layout of the sampling stations. Clearly, the layout of the sampling stations will depend upon the sampling requirements and the subsequent methods of analysis. However, every attempt should be made to ensure that the sample sites selected are representative of the chosen study area. When the study area is of a uniform nature, the simplest and most effective sampling layout is in the form of a grid. However, where obvious environmental gradients exist, e.g. temperature or salinity, then the usual procedure is to lay out a transect with the individual sampling stations arranged so as to give representative samples of the different zones. In addition to selecting a suitable study area, it is essential to consider how many samples are required to give an acceptable estimate of population numbers. In practical terms, the number of samples required is frequently governed by the ability to handle the material within a reasonable period of the samples being taken. It would thus appear, as a broad generalization, that while a large number of samples may be desirable from a statistical viewpoint, the extra precision gained might not repay the additional time expended in sampling and processing effort. However, if it is required to determine, for example, whether there are significant differences in microbial population densities from one site to another or at different times of the year then a large number of samples will clearly be needed for the data to be statistically significant. A further consideration is the frequency of sampling. Where study areas are readily accessible, e.g. lakes, estuaries and inshore coastal waters, sampling may be frequent whereas in the deep sea environment, where costs are high, samples may only be obtained at infrequent intervals. Under these circumstances it is essential to make maximum use of the material obtained.

In aquatic environments the areas of most interest to microbiologists are the air–water interface, the water column, and the benthos. Each region requires specialized sampling procedures which vary in complexity depending, for example, on distance from shore and depth. For deep sea samples, equipment should be of rugged construction to withstand the high pressures encountered at great depths as well as rough handling normally encountered on board ship. The equipment should be built of inert materials preferably stainless steel or plastics to avoid bactericidal effects that can occur, for example, if copper or brass are used. In some instances it is desirable to have samplers which can be sterilized before use. A final consideration is that the selected sampling system should provide a sufficient volume for subsequent analysis.

1.2 SAMPLING METHODS

1.2.1 Surface sampling systems

The water–air interface, which comprises the top millimetre of aquatic environments, is commonly referred to as the 'microlayer' or 'neuston' and is a region of considerable biological significance (MacIntyre, 1974). Studies by Sieburth *et al.* (1976) have shown that in the marine environment there is a significant increase in dissolved organic carbon levels and heterotrophic microbial populations in the upper 150 μm surface layer compared with sub-surface waters. Thus, there is a need to study the physical, chemical and biological properties of this poorly understood region of the aquatic environment. While 'microlayers' are found in freshwater environments, few studies have been carried out and most sampling systems have been used in the marine environment. It is clearly not easy to devise a system which will efficiently sample such a thin and discrete layer of ocean waters. As pointed out by Sieburth (1979), it is essential to minimize disturbing the surface film while taking samples. A number of sampling systems have been developed and their individual advantages and disadvantages have been critically reviewed by MacIntyre (1974). Surface film samplers are basically of two types, either continuous or discontinuous systems. Continuous samplers such as those developed by Harvey (1966) have a rotating drum that picks up a continuous film of water 60–100 μm thick which can then be removed by a wiper blade into a collecting trough or bag (Fig. 1.1). However, such systems only function effectively under relatively calm conditions and are, therefore, unsuitable for open ocean work. Discontinuous samplers have proved to be more effective and practical. The earliest microlayer system developed by Garrett (1956) is still considered to be the most effective (Parker, 1978). The sampler consists of a stainless steel screen mesh which is immersed perpendicularly into the water and then withdrawn parallel to the surface. The screen sampler has a sampling efficiency of approximately 70% and retains the majority of the living matter within 250–300 μm of the surface (Sieburth, 1979). By using different screen materials different horizons of the microlayer can be sampled. Thus, Duce *et al.* (1972) were able to sample a layer approximately 150 μm thick using a 1 mm mesh Nytex screen.

Other simple samplers include a glass plate system which is dipped into

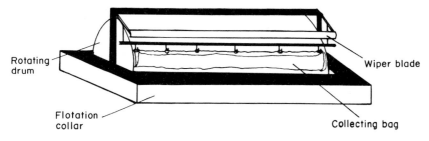

Fig. 1.1 Diagrammatic representation of continuous sampling system for collecting 'microlayer' samples.

the sea and after draining off the surplus water, the surface film is removed into a sample container using a squeegee wiper blade (Harvey and Burzell, 1972). The principal disadvantage of this technique is that the glass fails to release a major proportion of the adsorbed film (Sieburth, 1979). In addition, the sample yield is only some 20% of that obtained by the corresponding mesh screen samplers. However, thinner surface films can be obtained (20–60 μm) which may be advantageous in particular circumstances (Hatcher and Parker, 1974; Sieburth, 1979). If the principal requirement is merely to sample surface films for subsequent chemical analysis then a large filter funnel works quite successfully. Essentially the procedure involves slowly filling the funnel with surface water and allowing the contents to drain away slowly. The neuston flows out with the water leaving a thin film on the sides of the funnel. This film can then be removed by using an appropriate solvent and the chemical composition of sample determined. Various modifications of the samplers described here have been devised and each system yields different results. Since no comprehensive comparative studies have been carried out on surface samplers the individual investigator will need to determine which system is most appropriate to his/her own requirements.

1.2.1.1 Water column samplers

A wide variety of systems have been devised and described for collecting samples from the water column. The particular system used will be determined, to a large extent, by the sampling location and the information required from the material collected. Two major considerations need to be taken into account when selecting a suitable water column sampling system. Firstly, it is important that the equipment takes discrete samples at the chosen depth with the minimum degree of disturbance and contamination from other horizons. Secondly, since the range of living material in the water column is extremely diverse both in terms of species distribution and size, it is necessary to select equipment appropriate to the organisms being studied. Plankton ranges in size from 0.02 μm (femtoplankton) to 200 cm (megaplankton) and this diversity in size can be exploited to fractionate and concentrate the various biological components of the water column using nets, screens or filters with the appropriate mesh or pore size.

1.2.1.2 Net samplers

The distribution and abundance of plankton is usually determined by towing fine conical nets (Fig. 1.2) of a particular mesh size from the stern of a research vessel. The standard procedure is to attach the net to a weighted cable and tow the sampler at a given speed (e.g. 2–4 knots) and depth for a

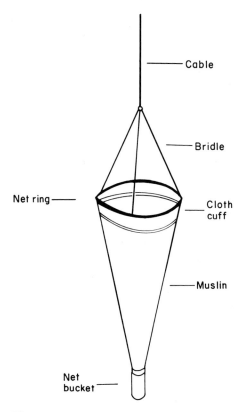

Fig. 1.2 Basic plankton net with detachable
sample bottle.

specified time. The plankton adhering to the sides of the net are washed
into the sample container, at the apex of the net, and collected. Alternatively, the plankton net may be lowered from a stationary research vessel and a
vertical haul obtained. Devices have been incorporated into more sophisticated plankton samplers which close the mouth of the net at the end of the
towing period. The simplest net closing device is one in which the towing
bridle is released by a messenger weight from the towing cable. A
secondary line secured to a bridle round the mid-point of the net collapses
the net which can then be lifted to the surface (see Sieburth, 1979). In the
Clarke-Bumpus net the throat of the sampler can be closed by a mechanical
gate which is triggered by a messenger weight (Wickstead, 1965).

Faster swimming plankton can be obtained by high-speed plankton
samplers such as the Gulf stream MK III system. This sampler comprises a
metal torpedo shaped casing which encloses the plankton net, sample

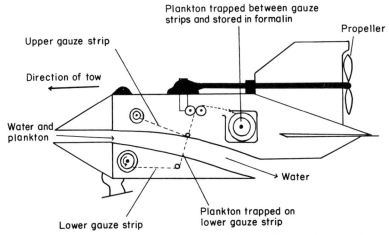

Plankton trapped between gauze
strips and stored in formalin

Propeller

Upper gauze strip

Direction of tow

Water and
plankton

Water

Plankton trapped on
lower gauze strip

Lower gauze strip

Fig. 1.3 Diagrammatic representation of a Hardy continuous flow plankton recorder.

collecting bottle, and flow meter. Since this sampler can be towed at speeds up to 11 knots, a throat restrictor is fitted to prevent too fast a current from entering the net. When there is a requirement to map plankton populations over large sea-areas, the Hardy continuous plankton recorder (Hardy, 1956) has proved extremely useful. The basic mechanism of the recorder is shown in Fig. 1.3. As water passes through the throat of the sampler, the plankton are trapped on a continuously moving gauze strip. A second gauze film 'sandwiches' the plankton which then progress into a storage chamber containing formalin. More recently, an updated version of the Hardy plankton recorder has been developed which incorporates continuous salinity and temperature recording systems (Bruce and Aitken, 1975).

While nets are widely used for collecting plankton samples, they do present considerable difficulties when carrying out quantitative surveys. The smaller phytoplankton, especially the nanoplankton, pass through standard nets and even with specially developed small mesh nets (10 μm mesh size) and slow towing speeds (0.5 knot) a large proportion of the smaller flagellates pass through. While the larger flagellates and diatoms are retained in nets with mesh sizes of 30–70 μm, problems are encountered when the nets become clogged and water by-passes the sampler. Net sampling thus grossly underestimates the total phytoplankton populations. To overcome some of these problems, Beers *et al.* (1967) developed a shipboard phytoplankton sampling system. Essentially the system involves lowering a weighted tube to the desired sampling depth and then pumping water obtained from the selected horizon through a series of stacked sieves of decreasing mesh size.

1.2.1.3 *Water samplers*

A wide variety of water sampling systems have been devised over the years. However, the fact that new systems are continuously being developed suggests that many of the existing samplers have drawbacks. Amongst aquatic microbiologists there has been much debate as to the necessity for sterile water column sampling systems or whether low-level contamination occurring as the sampler passes down the water column is acceptable. By careful attention to detail, contamination can be reduced to a minimum and certainly does not justify the criticism it has received. For example, Collins *et al.* (1973) demonstrated that there was negligible 'carry down' or cross-contamination between samples taken at 1 m depth intervals in a stratified freshwater lake. In many respects the type of metal used in the construction of samplers has caused more problems. The use of brass, which is bactericidal to many bacteria, should be avoided. For this reason the Nansen bottle has fallen into disfavour (Zobell, 1968) and has been replaced by samplers constructed from stainless steel, glass or plastic. The 'Friedinger bottle' (Fig. 1.4) is a commonly used water sampler constructed either from stainless steel or diecast metal (Buchi, Berne, Switzerland). The open sampler is lowered to the desired depth and then closed by sliding a messenger weight down the supporting cable. Closing the top and bottom lids causes the minimum of disturbance and thus the water sample taken is of a known dimension and depth. In contrast, the Ruttner bottle (Hydro-Bios Apparatebau GmbH, 23 Kiel-Holtenau, FRG), while designed on the same basic principle as the 'Friedinger' bottle, does not allow a free flow of water through the sampler. As a consequence, there is a greater risk of contamination from overlying waters. In addition, the closing mechanism induces considerable turbulence within the chosen sampling zone. More recently, plastics have been used in marine water column sampling equipment. A good example is the National Institute of Oceanography sampler (Fig. 1.5), which uses ball-type end enclosures to seal the tube once the sample has been taken. Sieburth (1979) pointed out that the only valid objection to using open-tube-type samplers of the types described is that they must pass through the surface layer with its richer microflora. To overcome this objection the Go-Flo version of the Niskin bottle (Niskin, General Oceanics, Miami, Florida) was developed. In this system the sampler remains closed as it passes through the richer surface layers. At a given depth, the end enclosures automatically open and a sample is taken before being closed by a messenger weight.

Several types of sterile water column samplers have been developed, of which the Niskin (Niskin, 1962) and J–Z systems (Zobell, 1946) are most frequently encountered. The J–Z sampler (Fig. 1.6) consists of a strong glass bottle (500 ml capacity) sealed with a rubber bung through which passes a glass tube that is connected to a short length of pressure tubing.

Fig. 1.4 The 'Friedinger' water sampler.

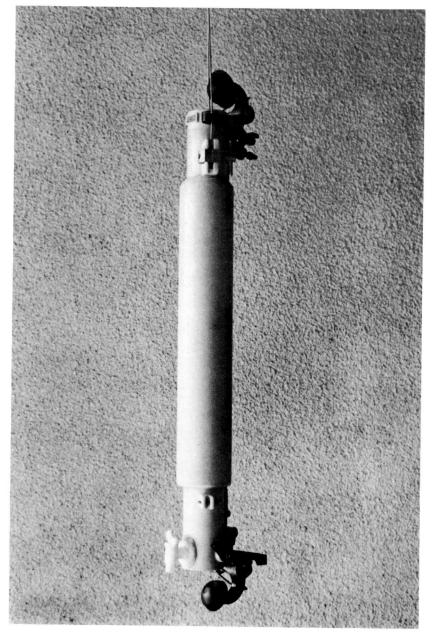

Fig. 1.5 The National Institute of Oceanography water sampler showing ball-type end enclosures.

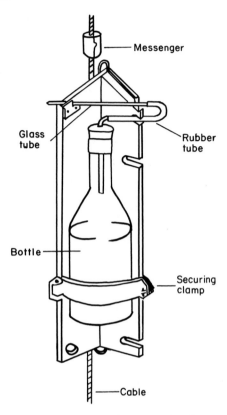

Fig. 1.6 Diagrammatic representation of the Zobell J–Z water sampler.

The end of the pressure tube terminates in a sealed small bore glass tube. After autoclaving the bung is pushed firmly into the bottle thus creating a partial vacuum as it cools. In operation, the sealed small bore glass tube is broken by a lever actuated by the messenger weight. The disadvantages of the J→Z sampler are that it cannot be used at great depth, the sample is drawn in as a 'cone' of water whose depth and dimensions cannot be precisely determined and the volume of sample is small (approximately 250 ml). To overcome the small sample volume collected by the J–Z sampler, Niskin (1962) developed a system for collecting large volumes of water aseptically. The basic sampler is shown in Fig. 1.7. The spring-loaded hinged frame is enclosed within a sterile plastic bag of approximately 3 litre capacity. When lowered to the desired depth the sampler is triggered by a messenger weight which in turn actuates a knife blade to sever a sealed tube leading to the bag. Simultaneously, the spring-loaded

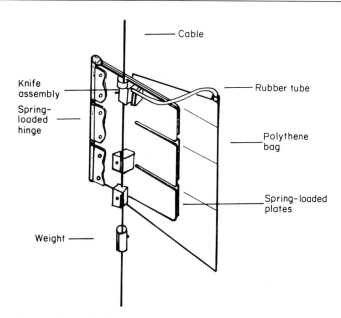

Fig. 1.7 The Niskin water sampler showing the torsion
spring assembly.

hinged frame is opened and water flows into the bag. In practice, this
sampler is difficult to operate in bad weather, leaks develop in the plastic
bag and the knife does not always sever the sealed tube. In conclusion,
non-sterile samplers are much more reliable and for all their apparent
disadvantages in respect of surface contamination appear to yield valid
results. To obtain undecompressed water samples from the deep sea
specialized water samplers such as those developed by Jannasch *et al.*
(1976) are required. These stainless steel pressure chambers are extremely
expensive but will take undecompressed water samples from depths as
great as 6000 m.

1.2.2 Bottom sampling systems

The choice of sampling equipment must necessarily take into account the
type of sample required, depth of water, type of sediment and size of
vessel. Particular problems occur when small boats are used because they
lack the capacity to handle heavy sampling equipment. Bottom sampling
systems have been extensively reviewed by Holme and McIntyre (1971),
and thus only the commonly used systems will be discussed in this section.
A major factor which dictates the type of sampler employed is the purpose
for which the sample is to be used. Microbiologists are usually interested in

the upper 5–10 cm of the sediment which is considered to be microbiologically the most active. When the samples are to be serially sectioned to determine the distribution of micro-organisms within the sediment, it is important that the surface layers should suffer the minimum disturbance and this dictates the type of sampler to be used. Sediment samplers fall into three main categories: grab samplers, gravity corers and frame mounted corers.

1.2.2.1 Grab samplers

A considerable number of grab samplers have been devised and all have deficiencies such that no individual system can be recommended for general use. The quantity of sediment obtained with a given sampler depends upon the nature of the sediment and design of the grab. Most grabs penetrate to a depth of approximately 10 cm (less on hard-packed sand) and take a sample which is semicircular in cross-section. Grab samplers commonly used cover a surface area of 0.1–0.2 m² and these are considered adequate for quantitative work. The major disadvantage of

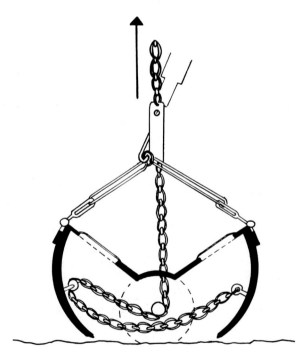

Fig. 1.8 Diagrammatic representation of a Pedersen grab sampler.

grab samplers for microbiological work is that the sediment obtained is completely mixed so that no analysis of the vertical distribution of microorganisms may be made. The Petersen grab (Fig. 1.8) is a simple design, suitable for use in sheltered waters and with soft sediments. Upon slackening the lowering cable, the release mechanism is actuated so that, with hauling, the two buckets close together and collect a sample. The van Veen grab (Fig. 1.9) is a considerable improvement on the Petersen design in that the long arms attached to the grab buckets exert a considerable leverage for closing the jaws. Models fitted with teeth are very effective for sampling firm sandy sediments although stones trapped between the interlocking teeth often allow the sample to wash out. When used in deeper water, the van Veen grab is liable to premature release during lowering of the sampler. However, for all its disadvantages, it is easy to use and is suitable for operation from small vessels in relatively shallow waters. In contrast, the Smith-McIntyre grab (Fig. 1.10) is a more suitable system for sampling continental shelf sediments. The grab is fitted within a steel framework that enables samples to be taken from a drifting ship. Only when the sampler sits squarely on the seabed are the powerful release springs triggered, forcing the buckets into the sediment. The grab buckets

Fig. 1.9 The van Veen grab sampler fitted with teeth to aid penetration into sandy sediments.

Fig. 1.10 Smith-MacIntyre grab showing the trigger plates on either side of the base.

then close as the sampler is raised on its lifting cable. The top of each bucket is fitted with a fine mesh cover to minimize resistance and downwash on descent. The disadvantage of this grab is its weight (45 kg) which makes it unsuitable for use from small vessels. However its more reliable release mechanism makes it a more suitable system for open sea work than the van Veen grab. The Shipek grab (Fig. 1.11) takes a relatively small sample (0.04 m^2) and is favoured by geologists. This system takes a sample by means of a single semicircular scoop rotating through 180°. The scoop is actuated by powerful coiled springs which are released by the momentum of a heavy weight when the sampler reaches the seabed. The Shipek grab is easy to operate although in the author's experience it does tend to 'skip' on firm sandy sediments. When deep sediment samples (up to 45 cm deep) are required, the Reineck box sampler has proved satisfactory although it is a relatively heavy and cumbersome piece of

Fig. 1.11 Shipek grab sampler showing powerful torsion spring assembly for activating sampling scoop.

equipment (750 kg). The sampler consists of a rectangular corer fitted within a steel frame (Fig. 1.12). Heavy weights attached to the sampler force the corer box into the sediments while, on raising the lifting cable, a hinged cutting arm is released which closes the bottom half of the core tube. Orange peel grab samplers, which are popular in the United States, have four curved jaws which enclose a hemisphere of sediment. The samplers are normally fitted with a canvas cover to minimize wash-out but they have never found universal acceptance.

1.2.2.2 Gravity corers

Coring systems usually have a high penetration performance and work reliably. The major difficulty is that unless a core-retaining device is fitted the sample may be lost on ascent. This is a particular problem when

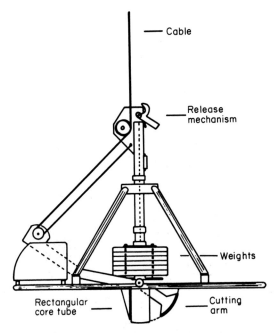

Fig. 1.12 Diagrammatic representation of Reinek box corer.

sampling sandy sediments. Where intertidal sediments are to be sampled, the simplest method of obtaining a sample is to use a perspex tube of 5 cm diameter and 20 to 30 cm long. This can be inserted gently, by hand, into the sediment until the desired depth is reached. A rubber bung is then fitted to the open end of the tube and the sampler carefully withdrawn together with the sediment sample. In subtidal areas the same system can be operated successfully by scuba divers.

An easily constructed remotely operated corer suitable for use in lakes and sheltered inshore waters is shown in Fig. 1.13. Lead weights enclose the body of the corer and the detachable nose zone ensures efficient penetration of the sediment which is then forced into the perspex lining tube. The sediment core is retained by a ring of flexible plastic which allows penetration of the sediment into the lining tube but prevents its loss upon recovery of the sampler. Internal core retainers, however, cause considerable disturbance of the sample and for microbiological work should be avoided. Gravity corers with external core retaining devices have been developed (Mills, 1961; Fenchel, 1967) and appear to function sucessfully. A major problem with gravity corers is that downwash from the corer disturbs the fine surface layers of the sediment–water interface. An

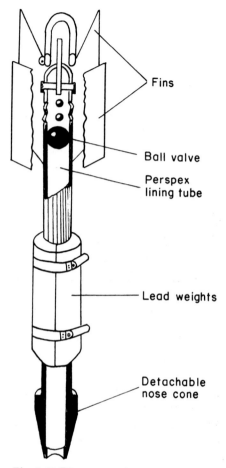

Fins

Ball valve

Perspex lining tube

Lead weights

Detachable nose cone

Fig. 1.13 Diagrammatic representation of a simple gravity core sampler (redrawn from Holme and McIntyre, 1971).

additional problem common to all coring systems but particularly apparent in gravity coring systems in the shortening (compression) of the sediment sample obtained. This shortening is principally due to wall friction (Emery and Dietz, 1941) and must be taken into account if the vertical distribution of micro-organisms within the sediment is to be undertaken. When deep sediment cores (up to 3 m long) are required the Sholkovitz gravity corer (Pedersen *et al.*, 1985) is particularly effective, although it suffers from the usual problems associated with all gravity corers, namely core compression and disturbance of the fine surface layers.

1.2.2.3 Frame mounted corers

Coring on firm sandy sediments or where it is imperative to cause the minimum disturbance of surface layers of the sediments presents particular problems. The Jenkins surface-mud sampler (Fig. 1.14) was developed in 1938 to collect sediment samples from the lakes in the English Lake District with the minimum disturbance of the surface sediment. The sampler was originally designed to sample soft silts but, with additional weights, it has been successfully used to sample more sandy sediments. The perspex sample core tube is mounted on an aluminium frame which is supported on four stainless steel legs. The sampler is lowered slowly to the lake bed and when the load on the suspending rope is removed, the release mechanism is activated and the core tube is slowly forced into the sediment under the control of a hydraulic piston. This slow penetration ensures minimum disturbance of the water-sediment surface layer. As the sampler is raised, the top and bottom sealing enclosures are slowly placed over the open ends of the core tube thereby further minimizing disturbance of the sediment sample. The operation of the Jenkins sampler and its application in limnological research are described in considerable detail by Ohnstad and Jones (1982), and are essential reading for aquatic microbiologists intending to use this type of sampler. The Craib corer (Craib, 1965) is essentially a marine version of the Jenkins surface-mud sampler and operates on the same principle of the core tube being slowly forced into the sediment under the control of a hydraulic piston. In the Craib corer, ball value end-enclosures seal the top and bottom of the core tube thereby retaining even sandy sediments effectively. The Craib corer, like the Jenkin sampler, is easy to use and provides good sediment samples for microbiological work, and can be operated from a small vessel. For deep sea sediment sampling, an uprated multi-core version has been developed (Barnett et al., 1984). This sampling system (Fig. 1.15) will take up to 12 cores simultaneously, and works on the same basic principle as the Craib corer. Core shortening still occurs with these types of sampler, but since the rate of penetration is slow the effects are minimized and the all important surface sediment layers are obtained relatively undisturbed.

1.2.3 Removal of sediment cores

Several methods have been developed for the removal of samples from standard core tubes. Holme and McIntyre (1971) recommended allowing the core to slide out under its own weight into a suitable collecting vessel rather than using a piston. However, this is not always feasible and the removal of sediment cores has to the aided either by mechanical means or by using hydraulic extrusion methods. In its simplest form the mechanical

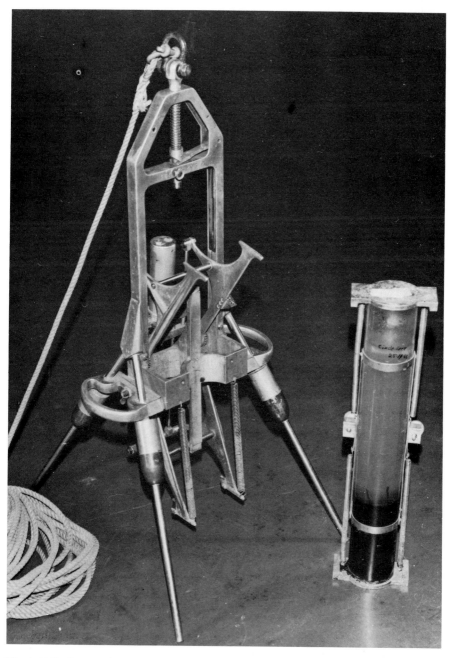

Fig. 1.14 Jenkin surface mud-sampler showing core the tube assembly and end-closures.

Fig. 1.15 Barnet multiple core sampler for taking undisturbed sediment samples at great depth.

removal of sediment samples from standard core tubes can be achieved by simply inserting a piston (steel, p.v.c, or teflon) of slightly smaller diameter than the internal diameter of the core tube and gently pushing out the sediment. A more refined mechanical system is to use a rigid frame in which the core tube can be mounted. The piston is fitted onto a fine pitch threaded rod which provides a more precisely controlled and reproducible method for driving out the sediment core. Ohnstad and Jones (1982) described the construction of this type of mechanical core-removing system specifically for use with the Jenkins surface-mud sampler, but it could be used equally effectively with other coring systems. The alternative approach is to use a hydraulically operated piston to push out the sediment core. Fenchel (1967) developed a system which allowed greater control over the release of the sediment sample from the core tube. Essentially the method involves replacing the upper seal of the core tube with a rubber bung through which passes a glass tube fitted with a tap. Opening and closing of the tap allows the sediment to slide down the tube under control. In more sophisticated systems such as that described by Ohnstad and Jones (1982), the piston is driven by control of water input and the sediment samples are extruded at the top of the core tube. These authors claim that the principal advantages of hydraulic extruding systems over the simpler mechanically operated procedures is that the sample core can be removed, if required, in one continuous operation and that the overall speed of operation is much faster. It should be stressed that all systems used to remove sediment cores from standard core tubes will induce some degree of core compression and it is inevitable that interstitial water and particles from one depth horizon will mix with those from another.

Once collected, it is important that samples should be processed as soon as possible since microbial populations can change both qualitatively and quantitatively in a relatively short time after sampling. Where possible, the sample material should be maintained under refrigerated conditions (4–5 °C) until required. Polystyrene boxes and ice packs are widely available and make suitable portable containers for holding samples at low temperatures in the absence of mechanical refrigerators.

1.3 GENERAL CONSIDERATIONS

The operation of sampling equipment presupposes that suitable boat facilities are available. Light grabs and small corers can and are frequently operated from small boats (4–5 m length) but in the cramped conditions which prevail on these vessels this is not an easy operation especially if winching facilities are not available. Larger grabs, corers and deep-sea sampling equipment can only be operated from larger vessels equipped with either A-frames or gallows from which these large, and often awkwardly shaped pieces of equipment, can be safely deployed. Rees (1971) discusses the basic requirements needed for the safe operation of sampling equipment from ships and the gauge of ropes and warps required for different operations. For hand-operated equipment, the minimum diameter of rope that should be used is 10 mm but larger diameters of 12–15 mm are much easier to work with for extended periods. To avoid problems associated with twisting of the rope used in hand-operated equipment, plaited ropes are preferable to those of traditional three-strand construction.

In conclusion, it should be apparent from the foregoing sections that while the range of sampling equipment available to the aquatic microbiologist is considerable, each system has its particular advantages and disadvantages. The individual investigator must therefore select the sampling system most appropriate to his requirements. In general terms, the simpler the sampling system the more reliable it is, and this is an important aspect which should not be overlooked when considering sampling equipment for shipboard operation.

1.4 REFERENCES

Barnett, P. R. O., Watson, J. and Connelly, D. (1984). A multiple corer for taking virtually undisturbed samples from shelf, bathyal and abyssal sediments. *Oceanologica Acta*, **7**, 399–406.

Beers, J. R., Stewart, G. and Strickland, J. D. H. (1967). A pumping system for sampling small plankton. *Journal of the Fisheries Research Board of Canada*, **24**, 1811–1818.

Bruce, R. H. and Aitken, J. (1975). The undulating oceanographic recorder—a new instrument system for sampling plankton and recording physical variables in the euphotic zone from a ship underway. *Marine Biology*, **32**, 85–97.

Collins, V. G., Jones, J. G. Hendrie, M. S., Shewan, J. M., Wynn-Williams, D. D. and Rhodes, M. E. (1973). Sampling and estimation of bacterial populations in the aquatic environment. In *Sampling—Microbiological Monitoring of Environments*, Society for Applied Bacteriology Technical Series No. 7, R. G. Board and D. W. Lovelock, (Eds), Academic Press, London and New York, pp. 77–110.

Craib, J. S. (1965). A sampler for taking short undisturbed marine cores. *Journal Conseil Permanent Internationale Exploration pour de la Mer*. **30**, 34–39.

Duce, R. A., Quinn, J. G., Olney, C. E., Piotrowicz, S. R., Ray, B. T. and Wade, T. L. (1972). Enrichment of heavy metals and organic compounds in the surface layer of Narragansett Bay. *Science*, **176**, 161–163.

Emery, K. O. and Dietz, R. S. (1941). Gravity coring instruments and mechanical sediment coring. *Bulletin of the Geological Society of America*, **52**, 1685–1714.

Fenchel, T. (1967). The ecology of the microbenthos. I. Quantitative importance of ciliates as compared with metazoans in various types of sediments. *Ophelia*, **4**, 121–137.

Garrett, W. D. (1956). Collection of oil slick forming materials from the sea surface. *Limnology and Oceanography*, **10**, 602–605.

Hardy, A. C. (1956). *The Open Sea*, Collins, London.

Harvey, G. W. (1966). Microlayer collection from the sea surface: A new method and initial results. *Limnology and Oceanogaphy*, **11**, 603–618.

Harvey, G. W. and Burzell, L. A. (1972). A single microlayer method for small samples. *Limnology and Oceanography*. **17**, 156–157.

Hatcher, R. F. and Parker, B. C. (1974). Laboratory comparisons of four surface microlayer samplers. *Limnology and Oceanography*, **19**, 162–165.

Holme, N. A. and McIntyre, A. D. (1971). *Methods for the Study of the Marine Benthos*, I. B. P. Handbook, Vol. 16. Blackwells, Oxford.

Jannasch, H. J., Wirsen, C. O. and Taylor, C. D. (1976). Undecompressed microbial populations from the deep sea. *Applied and Environmental Microbiology*, **32**, 360–367.

MacIntyre, F. (1974). The top millimetre of the ocean. *Scientific American*, **230**, 62–77.

Mills, A. A. (1961). An external core retainer. *Deep Sea Research*, **7**, 294–295.

Niskin, S. J. (1962). A water sampler for microbiological studies. *Deep Sea Research*, **9**, 501–503.

Ohnstad, F. R. and Jones, J. G. (1982). *The Jenkin Surface-Mud Sampler User Manual*, Occasional Publication No. 15, Freshwater Biological Association, The Ferry House, Ambleside, Cumbria LA 22 OLP.

Parker, B. (1978). Neuston sampling. In *A Phytoplankton Manual*, A. Sournia (Ed.), UNESCO, Paris.

Pedersen, T. F., Malcolm, S. J. and Sholkovitz, E. R. (1985). A lightweight gravity corer for undisturbed sampling of soft sediments. *Canadian Journal of Earth Sciences*, **22**, 133–135.

Rees, E. I. S. (1971). Aids and methods for benthos samplers. In *Methods for the Study of Marine Benthos*, N. A. Holme and A. D. McIntyre (Eds), Blackwells, Oxford and Edinburgh, pp. 147–155.

Sieburth, J. McN. (1979). *Sea Microbes*, Oxford University Press, Oxford.

Sieburth, J. McN., Johnson, K. M., Willis, P. J., Burney, C. M. and Lavoie, D. M. (1976). Dissolved organic carbon and living particulates in the surface microlayer of the North Atlantic. *Science*, **194**, 1415–1418.

Wickstead, J. H. (1965). *An Introduction to the Study of Tropical Plankton*, Hutchinson, London.

Zobell, C. E. (1946). *Marine Microbiology*, Chronica Botanica Co., Waltham, Massachusetts.

Zobell, C. E. (1968). Unresolved problems in marine biology. In *Marine Biology IV*, C. H. Oppenheimer (Ed.), New York Academy of Sciences, New York.

Methods in Aquatic Bacteriology
Edited by B. Austin
© 1988 John Wiley & Sons Ltd.

2

Determination of Biomass

J. C. Fry

Department of Applied Biology, UWIST, P. O. Box 13, Cardiff CF1 3XF, Wales

2.1 INTRODUCTION

Bacterial biomass may be defined as the mass of living bacteria present in a habitat or part of the habitat being investigated. To estimate biomass, the methods used must distinguish between the live bacterial cells and the other micro-organisms or non-living organic matter present. In practice, this is very difficult and several methods, although initially promising, have now slipped from general use because they have not specifically estimated bacterial biomass.

Bacteria are found in all aquatic habitats and similar techniques are used to estimate their biomass from marine, estuarine and freshwater samples. However, bacteria occupy many different niches within these broadly defined habitats, each niche presenting its own specific methodological problems. Planktonic bacteria present fewest problems, although a proportion of these apparently freely suspended bacteria are attached to particles. The biomass of bacteria in the neuston or surface film of water is also comparatively easy to measure once this rather special habitat has been sampled. Many aquatic bacteria are attached to surfaces and these populations need special approaches. Organisms attached to plants (epiphytic), animals and stones (epilithic) may have to be removed by scrubbing, scraping or other procedures. Soft sediments will often need homogenizing before estimation of the bacterial biomass of the population is possible. Such treatments result in suspensions containing large amounts of organic debris which often interfere with the assay being carried out. For these reasons, estimates of planktonic bacterial biomass are plentiful, but results from sediments or attached populations are rarer.

Many books and literature reviews contain sections on estimating bacterial biomass in aquatic habitats. However, there have been lots of changes to both the methods used and our appreciation of the relative merits of these methods in the last decade. For this reason it is better to concentrate on the more modern literature and consult older publications for mainly historical interest. Several recent reviews (Fry, 1982; Van Es and Meyer-Reil, 1982; Parkes and Taylor, 1985) discuss the relative merits of different ways of estimating biomass but give few experimental details. More practical information is given in two edited texts (Costerton and Colwell, 1979; Litchfield and Seyfried, 1979), and the excellent methodological monograph by Jones (1979) gives full details of many methods.

The methods used to estimate aquatic bacterial biomass fall into two distinct groups, i.e. direct and indirect. The direct methods aim to estimate biomass by direct microscopic observation. Most of these use a total counting procedure to enumerate bacteria and then estimate the mean cell volume in some way. The product of these two estimates is the biovolume and this must be converted to biomass by multiplication with an appropriate conversion factor. One advantage of these methods is that the organisms must be directly viewed and this keeps the scientist directly in touch with the organisms he/she is investigating. The other group of techniques may be called indirect methods; with these the scientist never sees the organisms, so they have the advantage that selectivity is eliminated. Indirect methods rely on estimating an indicator chemical that is only found in living bacteria and is rapidly degraded once the bacteria die. This chemical must also be at a known, constant ratio to the cell mass in the population. The ratio should be independent of physiological state and so growing and starving populations should have the same relative amount of the chemical being measured. Some indicator chemicals are specific to bacteria, for example lipopolysaccharide (LPS), muramic acid, bacteriochlorophyll and some of the more recently used lipid biomarkers. Others are present in bacteria and other organisms and so can only be used in habitats relatively free of eukaryotes. Adenosine triphosphate (ATP), chlorophyll *a* and phospholipids are examples. These biomass indicators must be extracted from the aquatic sample and then assayed by a variety of chemical and physiochemical procedures. Many of these procedures are adversely affected by other materials extracted from the sample along with the indicator chemical.

For the reasons outlined above, this chapter will concentrate on direct methods of biomass estimation. This is also an opportune time for a detailed consideration of these direct methods, as in the last five years they have become one of the most popular ways of estimating bacterial biomass in many aquatic habitats. Readers should be reminded, however, that no one method is best in all situations, and the cautious bacterial ecologist will use more than one technique. Much can be learnt from the differences between results obtained from several methods, with different strengths and weaknesses.

2.2 DIRECT METHODS

2.2.1 Sample preparation

Once samples have been taken they must be processed to distribute the bacteria as evenly as possible within the sample. This is essential to break up bacterial clumps and to remove organisms from particles.

Water samples normally require little pretreatment; vigorous shaking by hand 20 times or for 30 s (Jones, 1979) is all that is normally required. Samples from unpolluted marine or oligotrophic habitats contain well-separated bacteria with few attached to particles. However, nutrient-rich environments, such as ponds and salt marshes, can contain significant numbers of attached bacteria (0.99–13%; Kirchman and Mitchell, 1982). In habitats containing large amounts of suspended solids (30–1000 mg l^{-1}), such as estuarine environments, 20–100% of the bacteria may be attached to particles (Goulder *et al.*, 1981). In such habitats differential filtration is required to separate attached and planktonic bacteria. This is simply done by sticking a 3.0 μm pore size membrane filter to a perspex cylinder of appropriate diameter and inverting it in a beaker containing the water sample. A head of 22–24 cm between the water outside and inside the tube is enough to allow the particle free water and the planktonic bacteria to pass through the filter and collect inside the tube (Dodson and Thomas, 1964; Goulder, 1977).

Sediment nearly always requires treatment: homogenization or sonication can be used. A wide variety of equipment is available, for this purpose, so a few examples only will be given. Weise and Rheinheimer (1978) used sonication for 5 min at 20 kHz, 50 W for sandy sediments, and Meyer-Reil (1983) treated siltier sediments with the same sonicator for three 1-min sonications with a 30 s break between each treatment. Whatever treatment is used, samples should be cooled in ice to prevent overheating, and care taken not to use too much power and lyse the bacteria. Homogenizing must be done in a blender with the motor mounted above the blades to prevent overheating (Fry and Humphrey, 1978). Dale (1974) homogenized sediment for 5 min at 23 000 rev min^{-1} but any speed over approximately 10 000 rev min^{-1} is satisfactory. If the sediment is very thick, it will need diluting with sterile, membrane-filtered water before treatment, equal volumes of water and sediment being satisfactory. Sewage or sewage sludge samples will need homogenizing with deflocculants to stop particles from reaggregating after treatment. This may be done by incorporation of sodium pyrophosphate and cirrasol (old name Lubrol W: I.C.I. Ltd) at a final concentration of 0.01% (w/v) in the sample (Gayford and Richards, 1970).

Epiphytic bacteria should be removed from the plants. This is best done by treatment in a stomacher (Colworth Stomacher-400; A. J. Seward Ltd, London). The best technique (Fry and Humphrey, 1978; Fry *et al.*, 1985) is to put small quantities of plant material (e.g. 12–60 leaves of *Elodea canadensis*) in 50–150 ml of membrane filtered (0.22 μm porosity filters) distilled water and to treat for 5 min. The approach removes 41% of the epiphytic bacteria (Fry *et al.*, 1985).

Epilithic bacteria must be removed from stones: a 5-min scrubbing with a sterile hard-nylon tooth brush is satisfactory (Goulder, 1987). The best procedure is to put 100 ml of membrane-filtered, autoclaved water into a sterile stomacher bag; the stone is then scrubbed inside the bag to remove the epilithon, and the resulting suspension treated in a stomacher for 5 min. It is best to put another stomacher bag around the original one to prevent loss of suspension by puncturing with hard, gritty particles. This treatment gives a homogeneous epilithic suspension.

2.2.2 Enumeration by epifluorescence microscopy

Making total counts of aquatic bacteria by epifluorescence, direct counting is clearly the best method available. All intending users of the technique should read the classic papers by Jones and Simon (1975), Hobbie *et al.* (1977) and Daley (1979). In outline the method is as follows: the bacteria are stained with a fluorochrome, filtered onto a membrane filter and counted under epifluorescence illumination. Although the technique is basically straightforward, there are many important details to be considered, some of which are outlined in Table 2.1. This section attempts to describe the points which are considered to be most important.

When used as part of a biomass determination, it is important to consider whether viable or non-viable cells are counted. Although in principle, any bacteria will be stained by fluorochromes, there is mounting evidence to suggest that most of the whole, aquatic bacteria observed are viable. Van Es and Meyer-Reil (1982) have reviewed the literature, describing the proportion of active bacteria observable by direct counting procedures. These techniques depend upon autoradiography, after incubation with radioactively labelled substrates, and respiration of tetrazolium salts. They reported that up to 85% (minimum = 2.3%) of the the total number of aquatic bacteria are able to take up organic substrates or, indeed, to respire. Other work supports this contention, for example, in the case of freshwater planktonic bacteria, up to 48.7% (minimum = 18.4%) are able to divide at least once on agar (Fry and Zia, 1982), and with epilithic bacteria the maximum dividing once is 72% (Fry, unpublished data). These results strongly suggest that all aquatic bacteria observed by epifluorescence microscopy are viable and, hence, the method is a satisfactory basis for biomass estimation.

Once sampled, it is advisable to fix the bacteria immediately. This is because both the number of bacteria and their size and shape may change rapidly when stored in bottles. Increases in these variables may occur within 16 h of confinement with seawater (Ferguson *et al.*, 1984) and more rapidly with freshwater samples (Christian *et al.*, 1982). The most common fixative used is formaldehyde which has been used at final concentrations ranging from 0.2% (v/v) (Fry and Davies, 1985) to 2% (v/v) (Hobbie *et al.*, 1977). This must be added to the sample immediately after collection and should be filtered (0.2 μm pore porosity filter) before use to remove particulate debris. Such fixation has proved to be very satisfactory (Fig. 2.1). Experiments in the author's laboratory have shown that mean cell volumes of freshwater bacteria did not change until the third day of storage (Fry and Davies, 1985; Fig. 2.1a) and that, for seawater samples, volumes were constant for 10 days (Fig. 2.1b). More extensive experiments, with deep-sea sediment from the Porcupine Seabight in the NE Atlantic Ocean, showed (Fig. 2.1c) that direct counts did not change significantly during 19 weeks storage in 2% (v/v) formaldehyde. Two-way analysis of variance of cell dimensions measured during this experiment has shown that bacterial widths did not change for 10 weeks, but that lengths did increase significantly between 1 and 4 weeks of storage. The result of these changes was that mean cell volumes only increased significantly after 10 weeks, so storage for this time seemed to be acceptable. Despite these results, anyone wanting to store samples should carry out control experiments to check that microbiological changes do not occur, because samples from other habitats might well react differently.

It is important to emphasize that the reagents used for epifluorescence direct counts should be free of bacteria and particulate matter; hence membrane filtration (0.2 μm porosity filter) and autoclaving is essential; adding formalin is also a wise precaution. All water used should contain 2% (v/v) formaldehyde and should be membrane filtered (0.2 μm porosity filter), autoclaved and stored at 4 °C until used; this will be called sterile water from now on in this section. Sterile water should be used to make up all the stains and for rinsing membrane filters. Once opened, bottles of water should not be reused. Control experiments should be carried out before and after water samples are counted. Sterile water should be stained and filtered as normal, and counts made of these blank filters. Any counts obtained should be subtracted from the subsequent counts for actual water samples and if more than very few bacterially shaped objects are counted in the controls, all reagents and apparatus should be discarded and new materials used. These precautions are necessary because, despite stringent precautions, occasional contamination occurs. Contaminating bacteria are normally very large compared with aquatic organisms but their source is very hard to determine.

TABLE 2.1 Technical details of some methods used for counting aquatic bacteria by epifluorescence microscopy

Type of bacteria counted	Type of bacterial stain[1], its final concentration (mg l[-1]), contact time (min) and staining method[2]	Type of membrane filter, its pore size (μm) and filter stain used	Total volume (ml) of diluted sample filtered and filter rinsing regime used	Make of microscope and objective lens (magnification/ numerical aperture) used	Authors
Planktonic	AO, 100, 1–2, S	Polycarbonate, 0.2, irgalan black	≥2, –[3]	Zeiss or Leitz, –	Hobbie et al. (1977)
Planktonic	Euchrysine-2GNX 10, 5, S	Polycarbonate, 0.2, No. 8 Ebony black	≥6, –	Leitz, fluorite, (95×/1.32)	Jones and Simon (1975)
Planktonic	AO in 0.01 M phosphate buffer, 30, 3, M	Polycarbonate, 0.2, irgalan black	10, –	Leitz, –	Ramsay (1978)
Planktonic	AO, 200, 2–5, S	Polycarbonate, 0.2, irgalan black	1–2, 5 × 1 ml water	Zeiss, planachromat (100×/1.25)	Bjørnsen (1986)
Sediment	AO, 33, 5, M	Cellulose, 0.45, none[4]	0.1–10, filtered seawater rinse	–	Dale (1974)
Epiphytic	AO, 5, 3, S	Cellulose, 0.45, none	2.5, 2.5 ml filtered water rinse	Zeiss, planapochromat (63×/1.4)	Fry and Humphrey (1978)
Animal-associated	AO, 5, 3, whole worm[5] stained	None used	Not applicable, –	Zeiss, planapochromat (63×1.4)	Harper et al. (1981)

Planktonic	DAPI, 0.05, 5, S	Polycarbonate, 0.2, irgalan black	2, –	Zeiss, neofluor (100×/1.3)	Roberts and Sephton (1981)
Planktonic	DAPI, 5, –, S	Polycarbonate, 0.2, irgalan black	2, – 2, –	Olympus, Zeiss neofluor (100×/1.3)	Sieracki *et al.* (1985)

[1] Stains used included acridine orange (AO) and 4'6-diamidino-2-phenylindole (DAPI).
[2] S = sample stained before filtration and M = whole membrane filter stained after filtration.
[3] – = no information given by the authors.
[4] None = membrane filters prestained black by manufacturers.
[5] The worms, *Nais variabilis*, were narcotized before staining and viewing under coverslips supported by petroleum jelly.

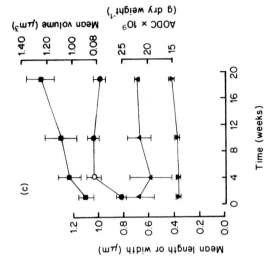

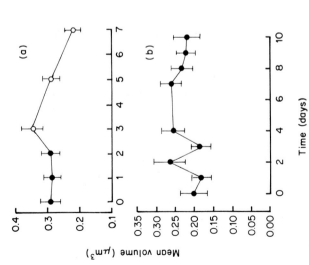

Fig. 2.1. The effect of storage in formaldehyde on cell dimensions and total numbers of aquatic bacteria. (a) Planktonic bacteria from Roath Park Lake, South Wales; volumes measured directly with phase contrast microscopy; 0.2% formaldehyde (from Fry and Davies, 1985). (b) Planktonic bacteria from Loch Etive, West Scotland; volumes measured from Kodak Tri-X pan negatives of acridine orange stained preparations; 0.2% formaldehyde. (c) Deep-sea sediment bacteria from Porcupine Seabight, NE Atlantic Ocean (Rice *et al.*, 1986); acridine orange stained preparations; total numbers ($\triangle$, $\blacktriangle$); mean cell width ($\triangledown$, $\blacktriangledown$), length ($\bigcirc$, $\bullet$) and volume ($\blacksquare$) measured from drawings of projected Kodak Tri-X pan negatives of acridine orange stained preparations. All cell dimensions were measured with the Quantimet 800 image analyser (Fry and Davies, 1985). Dimensions are mean values from one (a, b: one slide or membrane filter used per subsample; $n = 200$) or two subsamples (c: 3 filters per subsample; $n = 200$). Total counts are means of 3 replicate filters from each subsample). Vertical bars represent 95% confidence intervals (a, b) or the actual values from each subsample (c). Solid symbols indicate no significant difference from previous value, hollow symbols indicate a significant difference ($p < 0.05$) by analysis of variance (Tukey–Kramer *a posteriori* test). AODC = acridine orange direct count.

The main stains used are acridine orange and 4′,6-diamidino-2-phenylindole (DAPI). Both stains seem to work well over a wide range of concentrations (Table 2.1). It is the author's experience that a final concentration of $5 \, mg \, l^{-1}$ acridine orange, added as an appropriate volume of a $125 \, mg \, l^{-1}$ stock solution, is satisfactory with planktonic, epiphytic, epilithic and animal bacteria. This stock solution will last for many months at room temperature without apparent deterioration whether stored in the light or dark. However, many people prefer to store at 4 °C in the dark. DAPI, however, needs to be stored in the dark and will be good for at least 2 months at 0 °C (Porter and Feig, 1980; Roberts and Sephton, 1981). Other stains have also been used for direct counts such as ethidium bromide (Roser *et al.*, 1984) and euchrysine-2GNX (Jones and Simon, 1977). These have probably not proved popular because they give no colour differentiation between debris and bacteria. Acridine orange is most popular because of the striking difference between debris, which stains red, orange or yellow, and aquatic bacteria, which mainly stain green. DAPI gives a slight colour differentiation between yellowish debris and blueish-white bacteria; DAPI is also a brighter stain than acridine orange and does not fade as rapidly. The bacteria are best stained before filtration, although methods have been suggested for staining on the membrane itself (Dale, 1974; Ramsay, 1978). For acridine orange, no more than 3 min of staining is required, but for DAPI a full 5 min should be given.

Polycarbonate membrane filters of 0.2 μm pore size are the best to use. They are best because they have a perfectly flat surface on which the bacteria are collected, so the organisms all lie in one focal plane and are easy to see and count. Polycarbonate membrane filters must be stained black before use to prevent autofluorescence when viewed. This is done by staining for a minimum of 3 min in irgalan black ($2 \, g \, l^{-1}$) made up in membrane filtered 2% (v/v) acetic acid (Hobbie *et al.*, 1977). Other stains can be used, such as Dylon No. 8 Ebony black (Jones and Simon, 1975) or Lanasyn brilliant black (Fry and Davies, 1985), but these offer no particular advantage over irgalan black. Once stained the filters may be left in the staining solution all day without harm. Staining is best done in plastic petri dishes and the filters must be carefully rinsed in sterile water before use. They should then be put damp onto the base of a suitable filtration apparatus. Filters of 25 mm diameter can be used on any suitable straight-sided vacuum filtration apparatus. The author uses the filters directly on the surface of a sintered glass support, which must be kept thoroughly clean by regular boiling in detergent followed by careful rinsing in water. Some workers prefer to use a cellulose filter (0.45–1.0 μm pore size) under the polycarbonate filter, in order to spread the vacuum evenly. These precautions are taken to ensure even distribution of the organisms over the filter surface. The volume filtered also affects the distribution of bacteria. It is clear from the literature (Table 2.1) that 2 ml is the minimum total volume that should be filtered through a 25 mm filter in order to ensure even distribution. However, it is probably safer to use larger volumes (i.e. 5–10 ml). The sample may need to be diluted before filtration to ensure that the filter is not overcrowded and that counting is easy. At one time, polycarbonate filters often had hydrophobic areas on them which could be seen as large, irregular patches devoid of bacteria. This problem was cured by putting a few drops of a 0.5% (w/v) solution of surfactant (Wayfos; Hobbie *et al.*, 1977) above and below the filter before filtration, but this problem does not seem to be encountered now.

Black cellulose filters, although cheaper than polycarbonate ones, are now rarely used. The main reasons for this are that their fibrous nature makes focusing difficult and causes many bacteria to be lost in the filter matrix, giving reduced counts (Jones and Simon, 1977). If they must be used, then those made by Sartorius seem to be the best.

Rinsing the membrane filter with sterile water is essential after filtration; it is normal to use a volume approximately equal to the total sample volume filtered. Other rinses have been used, such as iso-propyl alcohol (Zimmerman, 1977), but really are unnecessary. The water rinse removes excess staining solution from the filter and enhances image contrast by reducing background fluorescence.

The filter should be left on the vacuum, until all surface water has gone, and immediately mounted. Drying the filter before mounting is unnecessary. Mounting can be done with or without a cover slip, both give good images, however, the very thin polycarbonate membrane filters stay flatter under coverslips. The author's group mount in an absolute minimum of autoclaved paraffin oil, but mounting in a suitable immersion oil is probably equally as good. A very small quantity of mountant should be spread below the membrane and a small drop put onto the filter, the coverslip then settles under its own weight to completely cover the membrane with mountant. One recent study used Citifluor (Wynn-Williams, 1985) as a mountant, which significantly reduced fading during illumination and so made counting easier.

Many suitable epifluorescence microscopes are available. Standard filter and dichroic mirror combinations are sold for most microscope systems with epifluorescence illuminators. Acridine orange needs blue light excitation (about 470 nm) and DAPI needs near ultraviolet (UV) light (about 365 nm). As DAPI needs UV light only fluorite lenses can be used with this stain, but glass or fluorite lenses are acceptable for acridine orange. Although Zeiss and Lietz microscopes have been used mostly, other makes are also suitable. Microscope systems have rarely been compared and when differences have been found they are usually small. For example, Sieracki *et al.* (1985) found that the Olympus BHT-F microscope fitted with a Zeiss Neofluor 100×/1.30 objective gave significantly higher total counts, by an average of 22%, than the other combinations tested. It is best to use objective lenses with both high magnification and high numerical aperture as this combination gets most light to the field of view and gives maximum resolution to distinguish the shape of the fluorescing bacteria to be counted. Flat field objectives make scanning large fields of view easier. The best lens used by the author's group is the Zeiss planapochromat 63×/1.4 oil objective which gives excellent resolution and even makes dividing cells easy to distinguish. Although its magnification can be increased with an intermediate magnification changer (e.g. Zeiss Optivar) it spreads light over a wider area of membrane than ×100 lenses, and makes the image less bright for a similar total magnification. For this reason, most people prefer to use a ×100 objective for routine counting. In the author's experience both the Zeiss planachromat 100×/1.25 and the Leitz Fluotar NPL 100×/1.32 are very good lenses. The Zeiss Neofluor 100×/1.3 is nearly as good, but is not a flat field objective. All these lenses are expensive, but well worth the investment if routine counting is to be done; cheap lenses invariably give poor, dimmer images. Eyepieces of sufficient quality to match the objectives should be used. The author normally uses 8× eyepieces, but 10× ones are equally good. As eyepiece magnification increases, the image appears dimmer; 16× eyepieces being the highest magnification used comfortably without resorting to intermediate magnification.

Acridine orange stained preparations should have most bacteria fluorescing green and only a few fluorescing red. In these circumstances it is easy to see bacteria attached to red-fluorescing debris or sediment. If particles are present they should not occupy more than 50% of the field of view or the free, unattached bacteria will not be seen. The background should be black, giving good contrast with the fluorescing particles. However, sometimes a pale-green haze will be seen which rapidly fades upon illumination. This is acceptable, but any other type of image should be rejected and a new filter prepared. DAPI preparations should be similar, with blueish-white bacteria and yellowish debris on a black background. However, DAPI is unsatisfactory for sediment as the bacteria and sediment particles are virtually indistinguishable.

Counting must be done randomly over a wide area of the membrane filter, and should cover the edges as well as the centre. To ensure this the filter should not be observed while the field of view is changed. This is essential to avoid stopping at the brightest areas containing most bacteria. The whole membrane can easily be covered by viewing along two transects at right angles to each other which cross in the centre of the filter. Counting should be done with a 10×10 eyepiece graticule in a focusing eyepiece. This allows the iris diaphragm on the epifluorescence illuminator to be closed down to enclose nine, four or one small square in the centre of the field of view. The number of squares selected depends on the density of bacteria on the membrane filter. Kirchman *et al.* (1982) suggested that 25 bacteria per field should be counted which reduces the coefficient of variation of the counts to below approximately 20%. However, their data are so scattered that probably 10 bacteria per field of view would also be acceptable, and this certainly makes counting easier. Because of fading, small numbers in many fields of view are much easier to count than large numbers in fewer fields. About 400 bacteria should be counted since this keeps counting errors fairly low as the bacteria are distributed on the membranes according to the Poisson distribution. Kirchman *et al.* (1982) also discussed numbers of replicate filters and subsamples which should be counted. They concluded that the optimal counting strategy consists of two subsamples with only one filter per subsample, giving an error of 8.5% of the mean. They stated that this takes about 30 min to perform.

Total numbers of bacteria per unit of original sample are easily calculated from the mean count per graticule area used, the effective filtration area of the membrane and the amount of sample actually filtered through the membrane. Jones (1979) provided a formula for those who need it. The graticule dimensions are best worked out with a slide micrometer. Enumeration by image analyser is now possible with epifluorescence microscopy (Sieracki *et al.*, 1985; Bjørnsen, 1986), but this will be discussed later.

2.2.3 Enumeration by phase contast and bright field microscopy

The most popular method of total counting bacteria, before the introduction of epifluorescence methods, was erythrosin staining on membrane filters, the bacteria being viewed by either phase contrast or bright field microscopy (Rodina, 1972; Sorokin and Overbeck, 1972). However, this is much more difficult to use than the epifluorescence methods because it is hard to distinguish bacteria from bacterial sized particles of detritus (Jones, 1979).

This problem can be overcome by viewing aquatic bacteria on agar by means of phase contrast microscopy. Under these conditions most debris is not observed as it has similar contrast to the agar, but the bacteria are clearly seen as homogeneously dark-grey objects with clear outlines and typical bacterial morphology. The method which the author has found best for planktonic bacteria is described in Fry and Zia (1982). Here, agar coated slides are prepared with thin (0.8–1.0 mm) microscope slides stored in 95% ethanol, sterilized by flaming and marked on one side with a 1 cm square. Molten agar (0.5 ml) at 65 °C is allowed to run evenly over approximately 75% of the unmarked side of the slide and solidified on a level surface at room temperature. Slides may be stored for short periods until required on supports in petri dishes containing dampened filter paper. The water sample to be counted must first be concentrated by centrifugation 2000 g for 20 min in 10 ml clean glass centrifuge tubes with tapered ends. Aliquots (9.7–9.9 ml) of the supernatant are then removed by gentle suction. This concentration procedure is quantitative if the amount of sample before and after supernatant removal is accurately determined by weighing, and the bacteria are resuspended by vigorous mixing on a vortex mixer. An aliquot (10 µl) of the concentrated sample is then spread evenly over the marked 1 cm^2 on the agar surface and allowed to dry for 10 min at room temperature. A sterile platinum wire should be used for the spreading operation to prevent scratching of the agar surface. Next the agar is covered with a coverslip, and viewed by phase contrast microscopy. When viewing, great care should be taken to focus the light on the agar surface or the best image will not be obtained. It might be necessary to use a long working distance condenser for this procedure. Once again, high quality objectives should be used, and it is personal experience that both the Zeiss Neofluor 100×/1.3 and the Zeiss Planapochromat 63×/1.4 phase contrast lenses are satisfactory. Counting should be done in a similar manner to that described for epifluorescence microscopy although a larger field of view must be counted.

This method gave excellent correlation ($r^2 = 0.999$) with the acridine orange epifluorescence method for 10 freshwater habitats (Fry and Zia, 1982; Fig. 2.2). However, it only revealed 55.9% of the bacteria as recorded by epifluorescence microscopy. The 95% confidence intervals on this percentage were very small, being 1.18% for a mean value from four replicates and 7.48% for a single, unreplicated count. As small cells are hard to see by phase-contrast microscopy, it is possible that the method misses the smallest bacteria. Despite this problem, it is probably a good method for obtaining total counts of bacteria when there is no access to an epifluorescence microscope. It is, however, technically more difficult and not nearly as well tested as the epifluorescence methods, and so cannot be fully recommended.

Morphologically distinct organisms can also be counted by transmitted light microscopy. The following technique was described by Jones (1975) for counting planktonic *Leptothrix ochracea*, and was later used for counting *Ochrobium*, *Naumanniella*, *Planctomyces* and *Metallogenium* (Jones, 1978, 1981). A known volume of water (10–30 ml) was filtered through white, cellulose membrane filters (25 mm; 0.22 or 0.45 µm pore size). These were dried under vaccum, with silica gel, at room temperature and mounted in cedarwood oil, under coverslips. Counting was done at magnifications of between 250× and 400× with bright field microscopy. Filamentous bacteria in sediments can be counted with phase contrast microscopy in

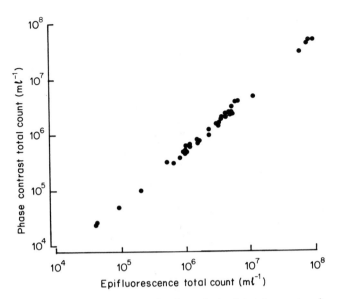

Fig. 2.2. Scatter diagram (log/log plot) of total counts of bacteria obtained by acridine orange epifluorescence and phase contrast counting of bacteria spread onto agar films on microscope slides. The data have been previously summarized by Fry and Zia (1982) and consist of four samples from nine different freshwater sites and one set of samples from comminuted sewage.

wet mounts, but the method underestimates the total bacterial count, by as much as 100–1000×. This is best obtained with acridine orange on membrane filters (Godinho-Orlandi and Jones, 1981).

Hossell and Baker (1979) have described an excellent technique for counting total numbers of epiphytic bacteria directly on the leaves of submerged aquatic plants Initially whole leaves are stained in filtered phenolic-aniline-blue (20 ml glacial acetic acid; 80 ml distilled water; 3.75 g phenol; 0.05 g aniline blue) for 1–2 min and mounted in a little of the stain solution. Counting is by bright field microscopy; the bacteria appear dark blue and can readily be distinguished from other material by their shape. Some aquatic plants, such as *Ranunculus*, absorb the stain and need pretreatment to prevent stain absorption. Pretreatment is as follows: 40% formaldehyde for 30 min, distilled water rinse, fresh eosin yellowish (0.2 g l^{-} for 1 h, and a distilled water rinse.

2.2.4 Estimating sizes of aquatic bacteria

To estimate biomass from direct counts the mean cell volume must be determined. Some of the reported values of average volume obtained since

1977 have been included in Table 2.2. This shows that the estimates vary a great deal. Some of this variation is due to methodology, but most is probably due to size variation between habitats. For example, with the acridine orange direct counting methods, estimated volumes varied from $0.05 \, \mu m^3$ to $0.31 \, \mu m^3$, this being a range of over $6\times$. In one study using scanning electron microscopy, the observed range was nearly $9\times$ (Krambeck *et al.*, 1981). So biomass cannot be estimated from a single, universally applied cell size. The mean cell volume must be estimated in each study, and because we know so little about what controls size changes in aquatic bacterial populations, mean volume should be measured in every sample.

It now seems clear that the very harsh preparative procedures used in electron microscopy result in severe shrinkage (Fuhrman, 1981; Fry and Davies, 1985). So, electron microscopy should not be used to estimate mean cell volume unless only internal comparisons are required. For this reason, neither scanning nor transmission electron microscopy will be considered further.

Aquatic bacteria are known to change size rapidly when put into bottles (Ferguson *et al.*, 1984), so rapid fixing of all samples with 0.2–2% formaldehyde is essential. The effect of confinement and fixing on size changes have already been discussed and so need no further comment.

Most methods for estimating cell volumes first determine the width and length of individual cells, then calculate the volume of each cell and finally the mean cell volume. There are two sizing techniques which depend on direct visual estimates from either epifluorescence or phase contrast images, and these will be considered first.

A simple eyepiece graticule or a moving-hairline micrometer eyepiece may be used to directly measure lengths and widths of bacteria. Although the latter is far superior to the former, even this device seems to overestimate widths by 65% and so the mean cell volumes obtained are 47–73% larger than estimates by other methods (Fry and Davies, 1985). So eyepiece graticules should not be used.

The size class method has been used by many scientists to estimate average volumes. The method involves dividing the population of bacteria seen by microscopy into a number of size and shape categories, which can be easily recognized, while total counting is performed. Then the number in each category or size class is recorded during the total count. The mean volume within each size class will already have been estimated and so the total and then the average volumes of the bacteria counted can be calculated. Several different schemes for these size and shape categories have been used and three examples are given (Table 2.3). Most have used about 5–7 size classes but Zimmerman (1977) and Meyer-Reil (1983) used 10 and only three, respectively. It is impossible to be sure about the merits or, indeed, problems with this method because no comparisons between this and other approaches have been published.

TABLE 2.2 Some estimates of the mean cell volume from different aquatic habitats

Type of bacteria	Habitat	Method used for size estimation[1]	Microscope methodology[2]	Range of mean cell volumes (μm³)	Authors
Planktonic (marine)	Kiel Fjord, Germany	Size classes (10)	AODC	0.06	Zimmerman (1977)
	Coastal Water, North Carolina, USA	Size classes (5)	AODC	0.053	Ferguson et al. (1984)
	Gulf of Mexico	Size classes (5)	AODC	0.078–0.096	Palumbo et al. (1984)
	Coastal Water, South Carolina, USA	Measurements from enlarged photographs	{ AODC / SEM	0.081–0.145 / 0.036–0.058	Fuhrman (1981)
	Baltic Sea	Projected negatives, digitizer	AODC	0.05–0.1	Hagström (1984)
	Sargasso Sea and Narragansett Bay, Rhode Island, USA	Image analysis, direct	DAPIDC	0.16–0.19	Sieracki et al. (1985)
	Roskidle Fjord, Denmark	Image analysis, direct	AODC	0.08–0.115	Bjørnsen (1986)
	Coastal Water, Georgia, USA	—	AODC	0.12–0.17	Newell and Fallon (1982)
	Saline, Lake Grevelingen, The Netherlands	—	AODC	0.05	Laanbroek and Verplanke (1986)
Planktonic (estuarine)	Newport River estuary North Carolina, USA	—	SEM	0.047	Bowden (1977)
	Newport River estuary	Size classes (5)	AODC	0.072–0.096	Palumbo et al. (1984)

Planktonic (freshwater)				
12 Finnish lakes	Eyepiece micrometer	Phase contrast of erythrosin stained bacteria on membrane filters	0.041–0.241	Salonen (1977)
8 Oligotrophic lakes, Norway	Eyepiece micrometer	AODC	0.076–0.176	Hessen (1985)
9 Freshwaters of mixed types, South Wales, UK	Projected negatives	Phase contrast on agar films	0.32 –0.77	Fry and Zia (1982)
Lake, near Plön, Germany	Projected negatives, digitizer	SEM	0.015–0.130	Krambeck et al. (1981)
Lake Plussee, Germany	Projected negatives, digitizer	SEM	0.026–0.048	Krambeck (1984)
Llanishen reservoir, South Wales, UK	Mixed image analysis and projected negative methods	AODC / Phase contrast on agar films / SEM	0.083–0.146 / 0.087–0.196	Fry and Davies (1985)
Barber Pond, Rhode Island, USA	Image analysis, direct	DAPIDC	0.031 / 0.08	Sieracki et al. (1985)
2 Lakes, Denmark	Image analysis, direct	AODC	0.188–0.310	Bjørnsen (1986)
Mill Beck, Yorkshire Wolds, UK	Image analysis of negatives	AODC	0.22	
4 Wolds streams, Yorkshire, UK / 4 Galloway sreams Scotland, UK	—	TEM	0.07 –0.11 / 0.04 –0.05	Rimes and Goulder (1986)

(continued overleaf)

TABLE 2.2 (*contd.*)

Type of bacteria	Habitat	Method used for size estimation[1]	Microscope methodology[2]	Range of mean cell volumes (μm^3)	Authors
Sediment	Coastal, Georgia, USA	—	AODC	0.23 –0.35	Newell and Fallon (1982)
	Deep Sea, Porcupine *Sea Bight*, NE Atlantic Ocean	Image analysis of drawings from projected negatives	AODC	0.088–0.98	Rice *et al.* (1986)
Epiphytic	*Zostera marina* Woods Hole, USA	Eyepiece micrometer	AODC	0.31 –0.49	Kirchman *et al.* (1984)
Epilithic	5 Mountain streams, South Wales, UK	Image analysis of negatives	AODC	0.10 –0.40	Fry, Kemmy and Taylor (unpublished data)

[1] — = Methodological details not given by authors. The number of size classes used are given in brackets.
[2] AODC: acridine orange direct counting methods; SEM: scanning electron microscopy; DAPIDC: 4'6-diamidino-2-phenylindole direct counting method; TEM: transmission electron microscopy.

TABLE 2.3 Examples of size and shape categories used for determination of mean cell volumes for planktonic bacteria

Authors	Cell shape	Diameter (μm)	Length (μm)	Mean volume (μm³)
Zimmerman	Cocci	0.3	<0.4	0.014
(1977)	Cocci	0.5	0.4 –0.6	0.065
	Cocci	0.7	0.6 –0.8	0.180
	Rods	0.4	<0.6	0.023
	Rods	0.8	0.6 –1.0	0.095
	Rods	1.2	1.0 –1.4	0.185
	Rods	1.7	1.4 –2.0	0.292
	Rods	2.4	2.0 –2.8	0.508
Palumbo *et al.*	Coccoid rods	<0.6	<0.6	0.065
(1984)	Cocci	0.6 –1.2	0.6 –1.2	0.320
	Rods	0.38–0.44	0.6 –1.2	0.111
	Rods	0.44–0.5	1.2 –1.8	0.254
	Rods	0.54–0.63	1.8 –3.0	0.574
Turley and	Mini (Coccoid)	0.22–0.57	0.90–1.13	0.038
Lochte	Mini (Vibrioid)	0.11–0.22	0.90–1.13	0.038
(1985)	Cocci	0.57–0.90	0.57–0.90	0.20
	Rods	0.22–0.45	1.01–1.24	0.11
	Large rods	0.45–0.68	1.13–2.03	0.45
	Spirillage	0.22–0.34	1.70–2.83	0.16
	Dividing cells	0.57–0.90	–	0.48

Photographic methods have proved popular for size estimation (Table 2.2). The best of these are probably as good as the direct image analyser methods, although they take considerably longer to perform. Photography has two major advantages. Firstly, a permanent record is obtained and, secondly, photographs are quick to take, can be stored for subsequent processing and so make more time available while the total counts are being done. It also saves time, if the same microscopy method is used for the total counts and cell sizing. Most of what is stated about the photographic methods applies equally to epifluorescence, phase contrast or bright field microscopy.

A wide range of photographic films are available. Colour films might appear preferable for acridine orange stained images but, in practice, a high-contrast black and white negative film will give good monochrome separation of green and red. In this connection, Kodak Tri-X pan is excellent, in so far as it is fast (400 ASA), contrasty and exposure times are short (about 10 s for most acridine orange epifluorescence images). Many sophisticated, purpose-built, automatic photomicrography devices are available from microscope manufacturers. These are not really essential because an automatic, aperture-priority, 35 mm camera fitted to a trinocular head will give equally good results. The camera should have a sensitive exposure meter and be of the type that measures light levels constantly during exposure; this takes account of the fading of epifluorescence images with time. The camera should also have automatic exposure for up to 2 min. As an example, the Olympus OM-2 camera fulfils all these criteria. A motor drive is not necessary in so far as manual film winding causes no problems. The effect of errors in exposure are likely to be small (Fig. 2.3). With Tri-X pan film and photographs of planktonic bacteria, there were no significant differences in estimated volume over an exposure range of four f-stops (Fry, unpublished results). Kodak Ecktachrome 400 colour slide film has less latitude.

Calibration of the photomicrographs is done most appropriately from photographs of a slide micrometer with 10 μm graduations. Fluorescent latex beads with known diameters might seem suitable, however, it has been found recently that their estimated volume varied greatly with exposure time (Fig. 2.3). With these beads, automatic exposures were very overexposed and gave volumes over four times greater than the stated volume. Even the apparently best exposed negatives (-3 f-stops) gave a mean volume 80% greater than stated. These results suggest that caution should be exercised when using latex beads for calibrating microscope images.

Once photographs have been taken, size data must be obtained from them. This can be done either directly from the negative or from positive prints made from the negatives. Working directly with the negatives saves time and has proved popular (Table 2.2). Negatives can be used for direct estimates of cell length and width, or with an image analyser.

The negatives need to be enlarged for direct estimates. This can be done with either a projector fitted with a carrier for rolls of film or with an enlarger. Enlargers give dimmer images and so projectors are often preferred. Projected negatives from acridine orange stained bacteria are easy to work with, giving an image of dark-grey bacteria against a light background. It has been found (Fry and Davies, 1985) that varying the magnification of the projected image produces significant variation in the measured sizes, hence an image 70 cm × 48 cm from a whole 35 mm negative was recommended. Lengths and widths of bacteria may then be measured with a transparent, plastic ruler or a micrometer gauge.

Another method of measurement that has been found to be satisfactory is to project the negative onto a white sheet of paper and to draw around the entire outline of the bacteria. With practice this can be done quickly and provides a permanent record of the measured cells. An image size of about 50 cm × 35 cm has proved satisfactory for this approach.

A quicker and perhaps a less subjective way of measuring bacteria, either directly from microscope images or from photographs, is to use an image analyser or other computer assisted system. The least expensive

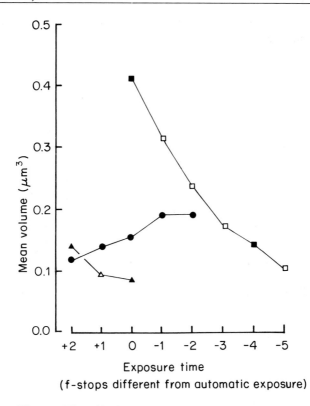

Fig. 2.3 The effect of exposure time on volume determinations from photographs. Except for the latex beads, the measurements were made with the Quantimet 800 image analyser as described by Fry and Davies (1985). Planktonic bacteria from Llanishen Reservoir, South Wales, photographed with Kodak Tri-X pan negative film (●) and Kodak Ektachrome 400 colour transparency film (△, ▲; data plotted from Table 3, Fry and Davies, 1985). Fluorescent latex beads mounted on a microscope slide under a coverslip, photographed with Kodak Tri-X pan film, volumes of these beads (□, ■) were calculated from area measurements ($n = 50$) assuming them to be spherical (volume $= 4/3 \sqrt{A^3/\pi}$, where $A =$ area). The stated diameter of the latex beads was $0.57\,\mu m$, so the volume was $0.097\,\mu m^3$. Solid symbols indicate no significant difference from previous value; hollow symbols indicate a significant difference ($p < 0.05$) by analysis of variance (Tukey-Kramer *a posteriori* test).

approach is to use a graphics tablet or digitizer (Krambeck *et al.* 1981; Krambeck, 1984; Hagström, 1984). This method involves projection of the image directly onto the graphics tablet or digitizer field. The lengths and widths of bacteria are recorded by marking coordinates, two for each length and width, for each organism. The computer receiving the information from the graphics tablet must be preprogrammed to accept it, and to calculate the length and width of each bacteria. Such programs may either be purchased commercially or specifically written by the scientist.

Television-based image analysis systems are more sophisticated versions of computer-based systems, and can be used to measure bacteria, drawings, projected negatives (Fry and Davies, 1985) or directly on epifluorescence images (Sieracki *et al.*, 1985; Bjørnsen, 1986). When used directly with epifluorescence images they can also make total counts of bacteria. So in principle, it could be used to estimate the bacterial biomass in a sample automatically, without recourse to manual counting or taking measurements. Although these methods are in their infancy, it is probable that they will be used more and more until they become the most popular. Readers completely unfamiliar with the principles of image analysis will find the general descriptions of Bradbury (1977, 1979, 1981) very useful.

So far only three image analysers have been reported to have been used for bacterial volume estimation. These are the Quantimet 800 (Cambridge Instruments; Fry and Davies, 1985), the Artek 810 (Artek Systems; Sieracki *et al.*, 1985) and the IBAS system (Zeiss/Kontron; Bjørnsen, 1986). The Quantimet 800 cannot detect weak epifluorescence images and so should only be used with photographs; but the other two systems have been used directly on microscope images. All these systems are constructed different-ly so a personal account of the necessary components of an ideal image analyser used for estimating the biovolume of aquatic bacteria will be given. This should help the would-be purchaser to buy wisely.

The basic configuration of an ideal image analyser is outlined in Fig. 2.4. All the components, apart from the television camera, will inevitably be driven by software that is programmed into the computer which will not be hard-wired like older systems (e.g. Quantimet 800). The camera must be sensitive enough to give a good video image from the acridine orange or DAPI stained microscope image, and this should be checked with the microscope system to be used before purchase. It is better to work on a stored image as this overcomes fading problems, allows plenty of time to manipulate images during practice sessions, minimizes electronic image disturbance and allows the use of recorded images on video-tapes. A full range of image enhancement facilities allows the hazy outlines of bacteria to be sharpened and uneven backgrounds to be made more homogeneous. In the author's limited experience with the IBAS system, image enhance-ment is of tremendous value. Once maximally improved, the video image

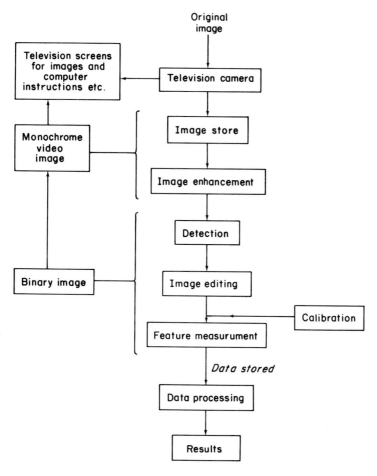

Fig. 2.4 Configuration of the main components of an image analyser suitable for direct counting, estimating mean cell volume and total biovolume of aquatic bacteria.

has to be turned into a binary image. This is done by selecting grey levels in the monochrome video image which will be detected or highlighted for future measurement. The binary image is made of small squares (or pixels) which are either detected or not detected; hence bacteria are seen as highlighted groups of pixels against the undetected background. Image editing may be either interactive or automatic. Interactive editing is essential for removal of detected debris from the binary image. The ability to switch quickly between the video and binary image while editing is very valuable. Automatic editing can be used to remove small groups of detected pixels, which are clearly not bacteria, or to separate bacteria which are touching in the binary image. Once the binary image is edited, each detected bacterium or feature can be measured; this aspect of image analysis will be discussed later. The basic data from the measurements should be stored on a floppy disc for future use. It should then be fully analysed to give mean numbers per field, distributions, means and variances of lengths, widths and volumes, and the final bacterial biovolume in the sample. It is best if this analysis can be done within the image analysers computer, but data-transfer to another machine for the calculations is perhaps an acceptable alternative.

Image analysers measure a variety of basic parameters for each detected bacterium or feature. The most useful ones are given below and, with others, are described in more detail elsewhere (Bradbury, 1977).

Area: this is the total number of pixels in the detected feature.

Perimeter: this is an approximation of the outside edge length of the detected feature, usually measured from the diagonals or edges of the detected pixels along the outside edge of the feature.

Feret lengths: these are the caliper lengths or the length which would be measured between parallel lines just touching the feature at fixed orientations (e.g. 0°, 45°, 90° and 135° from vertical for the Quantimet 800).

Convex perimeter: the length of an imaginary string pulled taught around the feature; this is shorter than the true perimeter for curved rods or spiral bacteria.

These measurements are approximations to the true values, due to the differences between the video and binary images. Thus, it is important to arrange the microscope and image analyser magnifications so that the binary images of the bacteria do not contain too few pixels. From the published data it appears that the systems used by Sieracki *et al.* (1985), Fry and Davies (1985) and Bjørnsen (1986) would give about 10, 80 and 105 pixels, respectively, for a bacterium of about $0.15\,\mu m^3$. As mean cell volumes as low as $0.05\,\mu m^3$ have been found using acridine orange methods (Table 2.2), the configuration used by Sieracki *et al.* (1985) seems barely adequate. The other two systems are probably satisfactory as Fry and Davies (1985) reported that, with a sample giving a mean cell volume of about $0.15\,\mu m^3$, the smallest, i.e. 2% of cells, averaged 17 pixels and bacteria smaller than 12 pixels were exceedingly rare.

The basic measurements from the image analyser are not directly useful so they must be converted to useful dimensions. Some of the methods described earlier yield length and width data. To convert lengths and widths to volumes, a model shape has to be assumed, and aquatic bacteria are most usually considered to be straight-sided rods with hemispherical ends. The formula for this model is:

$$v = (d^2 \pi/4)\,(l - d) + \pi d^3/6 \tag{1}$$

where the cell dimensions are $v =$ volume, $d =$ width, $l =$ length. This formula works equally well for cocci and rods, as for cocci $l - d$ becomes zero and formula for a sphere alone is left. A simplified form of this formula is sometimes used (Fuhrman, 1981; Krambeck *et al.*, 1981), i.e.:

$$v = (\pi/4)d^2(l - d/3) \tag{2}$$

These formulae cannot be used directly to estimate volume with an image anlyser as absolute lengths and widths cannot be measured directly. Using the hemispherical ends model a pair of simultaneous equations may be written to define the area (A) and perimeter (P) of the two-dimensional silhouette of a bacterium seen by the image analyser in terms of l and d. These equations are:

$$A = (l - d)\, d + \pi\, (d/2)^2 \tag{3}$$

$$P = (l - d)\, 2 + \pi\, d \tag{4}$$

They can then be solved for d and l to give:

$$d = (P - \sqrt{P^2 - 4\pi A})/\pi \tag{5}$$

$$l = P/2 + (1 + \pi/2)d \tag{6}$$

These equations allow calculation of d and l directly from the area and perimeter data provided by the image analyser and this is the approach used by Fry and Davies (1985). Others have used simpler models to calculate volumes from image analyser data. Sieracki *et al.* (1985) used only the area data and calculated volumes assuming either a sphere:

$$v = 4/3 \sqrt{A^3/\pi} \tag{7}$$

or a prolate spheroid with a length/width ratio of 2:

$$v = 0.94 \sqrt{A^3/\pi} \tag{8}$$

Bjørnsen (1986) used area and convex perimeter (C) measurements to calculate volume:

$$v = 8.5A^{2.5}C^{-2} \tag{9}$$

and claims that the equation gives $< 10\%$ error for spheres and rods with length/width ratios of up to 5.

The effectiveness of these different calculation methods can easily be compared using diagrams of model cells. Figure 2.5 is a set of computer generated drawings, with approximately equal widths, of a coccus, a rod with a length/width ratio of 2, a shallow vibrio, a deep vibrio and a spirillum. These drawings have been used to estimate cell dimensions using the equations discussed above and the results from a Quantimet 800 image analyser are tabulated in Table 2.4. From these results it is clear that equations 5 and 6 give very good estimates of length and width, from the

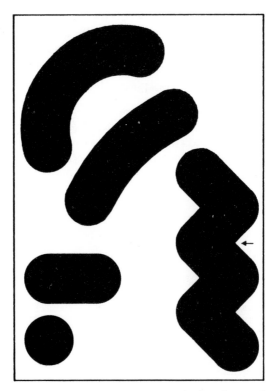

Fig. 2.5 Computer generated drawings of some typical shapes of aquatic bacteria. All the shapes have the same nominal diameter, except the spirillum which has a true diameter of about 1.2 nominal diameters at the point marked with an arrow. The short side of the rectangle containing the shapes is 5.15 nominal diameters long.

image analyser measured area and perimeter, for all the shapes. The poorest estimates are for the coccus; the reasons for this are unclear but the author has consistently noted this phenomenon, even with other spherical silhouettes. The volumes calculated from these measurements using equation 1 are also very close to the calculated volumes. The most inaccurate volume is that for the spirillum, but for this shape the real average width is greater than the 1.0 assumed for the calculated volume, because at the sharp angles of the drawing (Fig. 2.5, arrowed) the width is 1.2 diameters. All the other methods of estimating volume are considerably worse, with the greatest errors for the vibrios and spirillum. The spherical volume based solely on area (equation 7) is clearly very bad. Although in

TABLE 2.4 Dimensions of computer drawings[1] representing five silhouettes of differently shaped bacteria measured with a Quantimet 800 image analyser

Dimensions and units[2]	Coccus	Rod	Shallow vibrio	Deep vibrio	Spirillum
			Cell shape		
Measured central length, dia	1.0	2.0	3.5	4.5	5.5
Calculated volume[3], dia[3]	0.524	1.31	2.49	3.27	4.06
Image analyser measurements					
Area, dia[2]	0.846	1.88	3.39	4.21	5.65
Perimeter, dia	3.31	5.29	8.25	9.99	12.3
Convex perimeter, dia	3.29	5.18	8.12	9.15	10.6
Maximum feret length, dia	1.09	2.00	3.57	3.85	4.61
Dimensions (dia, dia[3]) calculated from image analyser measurements, equation number[4] used in brackets					
Spherical volume (7)	0.585	1.94	4.70	6.50	10.1
Spheroid volume (8)	0.413	1.37	3.31	4.58	7.12
Bjørnsen's volume (9)	0.517	1.54	2.73	3.69	5.74
Hemispherical ends volume (1)	0.518	1.41	2.62	3.21	4.61
Hemispherical ends width (5)	0.872	1.02	1.02	1.00	1.06
Hemispherical ends length (6)	1.16	2.06	3.54	4.42	5.54

[1] Fig. 2.5 shows the computer drawings used here.
[2] dia = diameter: all the cell shapes have the same nominal diameter of 1.0.
[3] Uses the measured central length, diameter = 1.0 and assumes a rod with hemispherical ends (equation 1)[4].
[4] See text for equations.

the author's experience spirilli are rare in aquatic samples, curved rods are often very common (Watson *et al.*, 1977 give 32% for one sample). Therefore it is important to use a system of volume estimation which takes into account the true length of curved rods down the central axis. It is also clear from Table 2.4 that the maximum feret diameter gives a good estimate of length for straight rods and cocci but, as expected, underestimates the length of curved rods and spirilli.

Typical distributions of length, width and volume (Fig. 2.6) show that most aquatic bacteria are small, and that the distributions are highly skewed. These distributions are typical for aquatic bacteria, as many similar distributions have been reported (Fuhrman, 1981; Maeda and Taga, 1983; Sieracki *et al.*, 1985). They also showed that small numbers of rather large bacteria are present. Fry and Davies (1985) found that in one freshwater sample, 3.7% of the bacteria accounted for 23.6% of the biovolume. This means that it is very important to measure enough cells to include these rare large bacteria; 200 cells would seem adequate for this purpose. It is also clear (Fig. 2.6) that the cell dimensions are logarithmically-normally distributed in the population. Thus to use

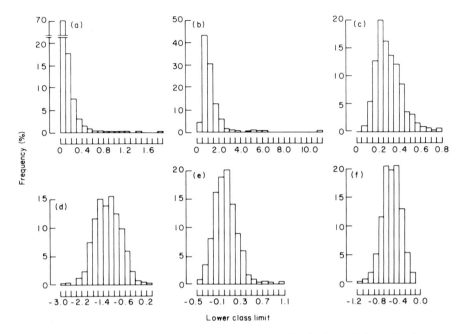

Fig. 2.6 Frequency distributions of dimensions of planktonic bacteria from a reservoir ($n = 808$). The dimensions are (a) volume, μm^3; (B) length, μm; (c) width, μm; (d) $\log_{10}$ volume, μm^3; (e) $\log_{10}$ length, μm; (f) $\log_{10}$ width, μm (from Fry and Davies, 1985, reproduced by permission of Blackwell Scientific Publications,Ltd.).

analysis of variance to compare mean values in different samples without replication, the values should be logarithmically transformed. If mean values from replicate samples are used the central limit theorem ensures that untransformed data can be used.

The author has found that planktonic bacteria, suspensions of epiphytic and epilithic bacteria are all amenable to the methods of size estimation using the image analyser recommended here. However, sediment samples present real difficulties (Rice *et al.*, 1986). The problems arise because most sediment bacteria are attached to particles, and the particles vary greatly in grey-level when examined with the image analyser. Thus although Tri-X pan negatives of acridine orange preparations allow resolution of the darker (green) bacteria from lighter (orange) sediment particles only very few bacteria can be detected at one time. The rest of the image remains either undetected or severely overdetected. An alternative method, which is a lot quicker than using direct image analysis on such difficult photographs, is to project the negatives with an enlarger and to draw round the bacteria. The author uses a red pen to draw them and to cross out any errors; the correctly drawn outlines can then be filled in with a dark felt

TABLE 2.5 Differences in apparent cell dimensions with two different methods of measuring deep sea (2000 m) sediment bacteria and planktonic bacteria from a reservoir with the Quantimet 800 image analyser from Kodak Tri-X negatives taken after staining with acridine orange

Type of sample (number of bacteria measured)	Method[1] of image analysis	Volume[2] (μm^3)	Width[2] (μm)	Length[2] (μm)
Sediment (167)	Projection	0.12***	0.40***	0.92
	Direct	0.09	0.29	1.33***
Sediment (108)	Projection	0.15***	0.44***	1.12
	Direct	0.11	0.30	1.37**
Sediment (49)	Projection	0.13***	0.49***	0.83
	Direct	0.09	0.34	1.09***
Water (579)	Projection	0.07	0.30	0.92
	Direct	0.10**	0.30	1.17***

[1] Projection = negatives projected, outlines of bacteria drawn in red pen, filled in with dark pen and sizes estimated on image analyser (Fry and Davies, 1985). Direct = sizes estimated directly on image analyser (Fry and Davies, 1985).
[2] Value significantly larger at $p = 0.01$ (**), 0.001 (***).

pen. The drawings are crowded together on one page of plain paper and can be rapidly measured with the image analyser. Detection can be adjusted so that only the dark pen is detected; the red pen remains undetected. Comparisons of these two methods for both deep-sea sediment samples and a water sample are given in Table 2.5. These results show that, although volumes of sediment bacteria were consistently overestimated by an average of 42%, those from water were underestimated by 33%. These differences were due to differential errors in estimating widths and lengths. Consequently, the projection method seems just acceptable, but only when direct image analysis is impractical.

To summarize, it is recommended that mean cell volumes of aquatic bacteria should preferentially be estimated by image analyser directly on epifluorescence microscope images of acridine orange stained bacteria. Areas and perimeters of individual cells should be used to estimate widths, lengths and hence volumes (equations 1, 5 and 6). When this is not possible, Tri-X pan negatives of the epifluorescence images should be projected and the lengths and widths obtained by direct measurement or with a graphics tablet or digitiser.

2.2.5 Conversion of biovolume to biomass

Once biovolume has been estimated it must be converted to biomass, by multiplication by a conversion factor. One of the most popular conversion factors, which has been used in the aquatic literature, is that devised by Watson *et al.* (1977) of $121 \, fg \, C \, \mu m^{-3}$. This was devised from older literature values for bacterial density (1.1 g wet weight cm^{-3}), a dry weight/wet weight ratio (0.22) and a carbon/dry weight ratio (0.50); the conversion factor being simply the product of these values, suitably adjusted for units. Others have used similar empirical conversion factors calculated from other constants, ranging from 87 to 143 fg C μm^{-3} (Hagström, 1984). Rather than use these older literature values, it is probably better to estimate a conversion factor specifically for naturally occurring bacteria and several authors have attempted to do this with rather variable results (Table 2.6). Krambeck et al. (1981) pointed out that from data for Escherichia coli, presented by Watson et al. (1977), it is possible to calculate a conversion factor of 130 fg C μm^{-3}. This value was very similar to the empirically derived factor and had little variability (range = 126 − 132 fg C μm^{-3}). However, from the work of Van Veen and Paul (1979), it is possible to calculate another factor, averaging 365 fg C μm^{-3}. This is nearly three times higher than the factor of Watson et al. (1977). This large difference has lead others to determine conversion factors experimentally, and these attempts have produced even greater discrepancies (Table 2.6).

There are two main ways that a suitable conversion factor could be calculated from all these results. Firstly, new, and perhaps better, values for density, and the relevant ratios, could be calculated to give a suitable conversion and, secondly, a value from the directly obtained factors (Table 2.6) could be used. The average density from the two relevant papers is 1.11 g wet weight cm^{-3}, which is so close to the older values that it seems suitable. The main problem with the older values for dry weight/wet weight ratio was the removal of intercellular water from centrifuged cell pellets. Bakken and Olsen (1983) obtained their value for this ratio (0.33) by a suction technique, but admitted that it probably represented a minimum estimate. Bratbak and Dundas (1984) used a better method to account for intercellular water by using [^{14}C]-inulin. So their average value of 0.438 is probably the best to use. There is little difference between the two recent estimates of the carbon/dry weight ratio so the average value of 0.446 seems acceptable. Using these figures a conversion factor of 217 fg C μm^{-3} results, which is naturally very similar to the 220 fg C μm^{-3} recommended by Bratbak and Dundas (1984) on whose work the crucial dry weight/wet weight ratio is based. The directly estimated conversion factors, given by the authors who have calculated them from cell volume and carbon data, range from 130 fg C μm^{-3} to the figure of 560 fg C μm^{-3} recommended by Bratbak (1985). This range is over four times and it is hard to understand how such experimental discrepancies can exist, although Bjørnsen (1986) argued that the figure of 560 g C μm^{-3} is theoretically impossible. Until work can be done to examine these differences, it is probably best to consider the average figure which is 395 fg C μm^{-3}. Although this factor is

TABLE 2.6 Some recently estimated factors which have been used for converting biovolume to biomass.

Test bacteria used	Density (g wet weight cm⁻³)	Dry weight/wet weight ratio	Cell C/dry weight ratio	Conversion factor (fgC μm⁻³)	Authors
Escherichia coli	—	—	—	130	Watson et al. (1977)[1,2]
Arthrobacter globiformis	—	—	0.369	374	Van Veen and Paul (1979)[3]
Enterobacter aerogenes	—	—	0.428	355	
9 Soil isolates	1.09	0.33	—	—	Bakken and Olsen (1983)[4]
Bacillus subtilis	1.13	0.515	0.488	284	Bratback and Dundas (1984)
Escherichia coli	1.09	0.314	0.480	164	
Pseudomonas putida	1.12	0.484	0.463	251	
Pseudomonas putida	—	—	—	478	Bratbak (1985)[1]
Mixed estuarine bacteria	—	—	—	650	
Mixed lacustrine bacteria	—	—	—	307	Bjørnsen (1986)[1]
Mixed estuarine bacteria	—	—	—	409	

[1] Authors estimated conversion factors directly from mean cell volumes and carbon.
[2] Value calculated from authors' results.
[3] Conversion factors calculated from authors' values for density, estimated as g dry weight cm⁻³ (*A. globiformis* = 0.85; *E. aerogenes* = 1.01) and cell C/dry weight ratios. Authors do not recommend a conversion factor.
[4] Authors do not estimate cell carbon so a carbon conversion factor cannot be calculated.

1.8 times greater than that obtained from the best estimates of density and the ratios, there seems no logical basis on which to choose between them. Consequently, it is recommended that a conversion factor of $308\,\mathrm{fg}\,C\,\mu\mathrm{m}^{-3}$ is used. This figure is midway between the best estimates obtained by the two approaches outlined above, and, being midway in the range of published estimates, will minimize errors in biomass calculations until further research can be done.

2.3 INDIRECT METHODS

2.3.1 Adenosine triphosphate

The use of adenosine triphosphate (ATP) as a biomass indicator in aquatic habitats was first suggested 20 years ago (Holm-Hansen and Booth, 1966) and it has been extensively used since then. Karl (1980) has written an extensive review of this technique, which is very comprehensive and should be read by everyone who uses the approach. Despite its popularity, ATP is an indicator of total biomass and not a specific bacterial biomass indicator. In most samples it is relatively easy to remove interference from large organisms. For example, water samples can be filtered with 200–400 μm nylon mesh to remove zooplankton, and sediment samples can be homogenized to disrupt macrofauna and meiofauna, which might contribute up to 90% of the sediment ATP (Sikora et al., 1977). It is more difficult to remove eukaryotic algae especially from water samples. Attempts to separate ATP from algae and bacteria have been numerous. One early suggestion was to filter the water through a 10 μm nylon mesh (Rudd and Hamilton, 1973). This has proved totally unsuccessful as many algae can squeeze through the pores of the mesh. For example, Euglena spp., which are approximately 50 μm wide, easily penetrate the mesh and hardly any are retained. This problem results in the <10 μm ATP concentrations often being more related to algal biomass than bacterial biomass. An example of this is seen in Fig. 2.7 which is taken from a study carried out in 1975 (Fry and Humphrey, unpublished data). There was strong correlation between <10 μm ATP and chlorophyll a concentrations ($r = 0.696$, $p < 0.001$), but no correlation between the ATP concentration and the acridine orange direct count of bacteria ($r = 0.04$). More recently, filtration experiments with 3 μm and 1 μm pore size polycarbonate membrane filters has shown that 1 μm filters separate algae and bacteria quite well (Riemann, 1978). However, this technique will be unsatisfactory in waters where a high proportion of the bacteria are attached to particles. Because of these problems ATP can only be considered as a good bacterial

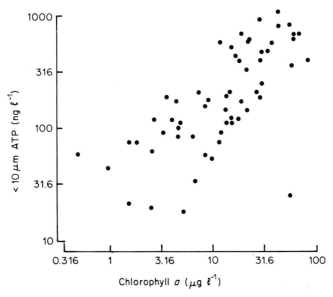

Fig. 2.7 Scatter diagram (log/log plot) of the $< 10\,\mu m$ particulate adenosine triphosphate (ATP) and the particulate chlorophyll *a*. The data come from the water of a small pond containing a large stand of mixed aquatic macrophytes and was collected over 18 months in 1975–76. Points are means of triplicate determinations. The ATP was extracted with boiling 0.02 M Tris buffer for 5 min and the chlorophyll *a* with cold methanol (4°C) for 24 h.

biomass indicator in habitats devoid of algae, such as in the deep sea below the photic zone (Watson and Hobbie, 1979).

Sampling natural habitats must be done carefully if it is intended to estimate biomass with ATP. As the physiological state of bacteria controls the amount of ATP in the cells and bacterial populations change rapidly when enclosed, samples must be analysed speedily and not kept for long periods. Once in the laboratory, the samples must be processed as soon as possible. Sediment samples will need dilution to about 10^{-3} (Jones and Simon, 1977) which reduces interference with the ATP assay. Conceivably, this interference is caused by inhibitors or guanosine triphosphate (GTP) which is often abundant in sediment (Karl, 1980). The next step is filtration of an appropriate volume through a $0.2\,\mu m$ cellulose membrane filter. Although many people have suggested that filtration reduces the amount of ATP measured, it is essential to separate bacteria in the sample from dissolved ATP in the water, which is now thought to contribute significantly to the total amount of ATP in water. For this reason the direct injection techniques suggested by some should not be used.

After filtration, extraction of the ATP from the bacteria should be done. This must follow immediately to prevent breakdown of the ATP in any damaged cells by ATP degrading enzymes. Many extraction procedures have been suggested and compared (Karl, 1980). Most methods fall into one of five groups: boiling buffer solutions, inorganic acids, organic acids, organic solvents and inorganic bases. No single method can be universally recommended because different research groups have obtained contrasting results, and it is probable that the best extraction method is dependent on the sample characteristics. However, alkaline extractions should be avoided and buffers are often unsatisfactory with sediment samples. One of the most popular methods uses boiling Tris (tris-hydroxymethylaminomethane hydrochloride) buffer, and this will be described as an example.

The membrane filter is inverted into 5 ml of boiling 0.02 M Tris buffer (pH 7.8) and extracted at 100 °C for 5 min. The buffer must be above 92 °C to ensure degradation of the ATPases released from the broken bacteria. This is best achieved in a boiling water bath on a metal gauze support grid. Fifty millilitre wide-necked conical flasks, covered with foil caps to minimize evaporation, are suitable for 47 mm filters. The buffer is decanted off into a 10 ml measuring cylinder, and the filter rinsed with a further 3 ml of buffer at 100 °C for 3 min. The rinse is mixed with the buffer initially used for extraction, the total volume measured, and the extracted sample can then be stored for up to a month at -20 °C for the ATP assay.

The amount of ATP in the extracted samples is then measured. The most popular method is to assay ATP with the enzyme luciferase. This involves a light emission reaction:

$$\text{ATP} + \text{reduced luciferin} + O_2 \xrightarrow{\text{luciferase}}$$
$$\text{AMP} + \text{inorganic pyrophosphate} + CO_2 + \text{product} + \text{light}$$

Crude extracts of firefly luciferase (Sigma) are quite satisfactory. The extract is prepared by reconstituting the crude enzyme with 0.02 M Tris buffer (pH 7.8), and storing it overnight at 4 °C. After storage, it is centrifuged at 6000 g for 10 min to remove particulate debris, and equilibrated in the dark at room temperature for 1 h. Some workers recommend the use of purified enzymes to decrease the limit of detection of the assay from about $1-10 \times 10^{-3}$ μg ATP ml^{-1} with crude enzyme to $1-10 \times 10^{-6}$ μg ATP ml^{-1} with pure enzyme (Deming et al., 1979), or with extra luciferin (Jones and Simon, 1977). These enhancements are rarely necessary as large volumes can be concentrated at the filtration stage.

The assay is carried out in a photometer by injecting about 0.1 ml of sample into 0.2 ml of enzyme preparation. The typical light emission is best recorded on a chart recorder and the peak height can be measured from the

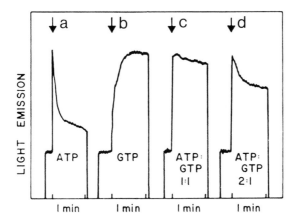

Fig. 2.8 Recorder tracings of light emission obtained when adenosine triphosphate (a, ATP), guanosine triphosphate (b, GTP) or mixtures of the two nucleotides (c, 1 ATP:1 GTP; d, 2 ATP: 1GTP) were added to crude luciferin–luciferase enzyme reagents. The arrows indicate the point of injection (from Stevenson *et al.*, 1979; copyright ASTM, reprinted with permission).

trace, a typical light emission curve is shown in Fig. 2.8. It can also been seen from this figure that GTP present in the sample can also give light emission and changes the shape of the recorder tracing. It is best to measure peak height rather than an integration of light emission over time because GTP does not affect peak height but it does affect subsequent light levels. In fact, the ratio of initial peak height to the integrated light output can be used as a measure of the amount of GTP in the sample (Karl, 1979). Interference with the light reaction by inhibitors may occur in some samples, particularly in sediments. For this reason it is best to examine each recorder tracing, to check for a typically shaped curve, and to add internal standards of pure ATP at the extraction stage in some samples. The amount of ATP may be calculated from a calibration curve, normally prepared between 1×10^{-3} and $100 \,\mu g$ ATP ml^{-1}, when crude enzyme is used. A wide variety of photometers are on sale today. A simple machine seems to be quite satisfactory, but more complex instruments may handle large numbers of samples automatically.

One very simple extraction procedure has been recommended (Deming *et al.*, 1979) which speeds up the assay considerably. This uses 0.1 N nitric acid which extracts ATP immediately from the filters and inactivates the ATPases. No neutralization after extraction is required if the crude enzyme extract is made up in 0.25 M Tris buffer (pH 8.2) and the sample is diluted $20 \times$ with water before injecting 0.1 ml quantities into the photometer tube.

There has been much discussion of the correct conversion factor to use for converting ATP concentrations to carbon biomass. Although C/ATP ratios of between 28 and 1972 have been obtained, a ratio of 250 seems a good figure to use (Karl, 1980) and should prove acceptable in most cases.

2.3.2 Phospholipids

All cell membranes contain phospholipids, which are quantitatively extractable from the cells. They constitute approximately 50% of eukaryotic lipids and 98% of bacterial membrane lipids (White *et al.*, 1979a; White, 1983), but are not present as storage lipids. The phospholipid content of sediment is highly correlated with the extractable ATP content ($r^2 = 0.94$, White *et al.*, 1979b; $r^2 = 0.71$, White *et al.*, 1979c). The extraction efficiency from sediment micro-organisms is good, for example, Parkes and Taylor (1985) have reported 98% recovery of phospholipids from bacteria added to sediment. Lipid phosphates are degraded fairly readily on death; their half-life in sediment being 2–12 days (White *et al.*, 1979c). For these reasons, phospholipids make a good general biomass indicator, very similar in character to ATP. A conversion factor of 50 μmoles lipid phosphate/g dry weight has been recommended (White *et al.*, 1979a).

The principle of the assay is to extract the lipids with chloroform and methanol, digest them with perchloric acid, and measure the released phosphate colorimetrically. Full details of the procedure are discussed in White *et al.* (1979c), and a summary will be given here.

Approximately 5–20 g of fresh, frozen or frozen-lyophilized sediment should be extracted with 30 ml 50 mM phosphate buffer (pH 7.4), 75 ml anhydrous methanol and 37.5 ml chloroform for 2 h. The sediment must be well mixed in this first stage which should be a single phase extraction with no separation of the components. If the sediment is wet, the volume of phosphate buffer used should be reduced by the amount of water added with the sediment, as easily determined by centrifugation at 8000 g for 30 min. A further 37.5 ml chloroform and 37.5 ml water is subsequently added, and the mixture agitated before being allowed to separate for 24 h. The upper methanol/water layer is removed by suction and the lower chloroform layer decanted through a filter paper from above the sediment. This gives about 80% recovery of the chloroform. An aliquot of the chloroform is then evaporated at 40 °C in a stream of air or nitrogen, and the resulting dried residue digested with 1.5 ml 35% perchloric acid at 190 °C for 2 h. The chloroform must be totally removed to prevent explosion. To the cooled extract is added 2.4 ml molybdate reagent (4.4 g l^{-1} ammonium molybdate; 14 ml l^{-1} concentrated H_2SO_4) and 2.4 ml of 12:1 diluted ANSA reagent (30 g sodium bisulphite, 2 g sodium sulphite, 0.5 g 1-amino-2-naphthol-4-sulphonic acid, 200 ml water). This reaction mixture is heated in a boiling water bath for 7–10 min, and the phosphate measured by absorbance at 830 nm.

This assay gives a minimum sensitivity of about 10^{-9} mole lipid phosphate, equivalent to about 2×10^9 bacteria of *E. coli* size, which is satisfactory for sediment. Greater sensitivity (10^{-11} mole, about 10^6 *E. coli*, 10^{-12} mole, 5×10^5 *E. coli*) can be obtained by using gas liquid chromatography (GLC) to measure glycerol or palmitic acid (White, 1983). Such sensitivity might just allow biomass estimation in nutrient rich waters.

2.3.3 Chlorophylls

Both chlorophyll *a* and the bacteriochlorophylls may be used as biomass indicators. Chlorophyll *a* is present in both eukaryotic algae and cyanobacteria, so it should only be used as an indicator of cyanobacterial biomass if eukaryotic algae are absent, and this situation only arises in some microbial mats. So chlorophyll *a* is a general biomass indicator for oxygenic photosynthetic micro-organisms. Bacteriochlorophylls are confined to prokaryotes, and so are specific indicators for the presence of photosynthetic bacteria. Several detailed papers have discussed the relative merits of measuring chlorophyll *a* concentrations with acetone and methanol extraction techniques (Marker, 1972). Simple, practical accounts have also been written (Jones, 1979; Parsons *et al.*, 1984). Measuring the bacteriochlorophylls is similar in principle (Jones, 1979) but has been less extensively discussed in the literature. However, a recent paper (Stal *et al.*, 1984) gives an excellent, highly practical account of how to measure both chlorophyll *a* and bacteriochlorophylls when present in mixtures. Chlorophyll *a* may be measured either colorimetrically or by fluorimetry. As colorimetric methods are used more commonly, only these will be described here. A good account of the more sensitive, fluorimetric approach may be found by combining information given by Jones (1979) and Parsons *et al.* (1984).

Firstly, the chlorophylls must be extracted either directly from samples or after concentration on a filter. Glass fibre filters (47 mm), such as Whatman GF/C or GF/F, are normally used but 0.2 μm or 0.45 μm cellulose membrane filters are also suitable. One litre of water is usually the maximum volume that can be filtered before particulate debris clogs the filters. Boiling or cold methanol and cold 90% acetone are the most popular solvents. The efficiency of extraction is variable and may be increased by grinding or sonicating the filter in the cold solvent. Often cold solvents need 18 h extraction in the dark, while extraction in boiling methanol takes only a few seconds. If mixtures of pigments are expected then a two-phase extraction with *n*-hexane (13 ml) and a methanol–NaCl mixture (13 ml) (Table 2.7) partitions the pigments between the two phases, making measurement easier. The quantity of pigment is then determined spec-

TABLE 2.7   Absorption maxima (nm), absorption coefficients (g^{-1} 1 cm) and other factors needed to measure concentration of chlorophylls in various solvents[1].

| Type of Chlorophyll | Values[2] for the following solvents | | | | % Chlorophyll in methanol/ NaCl after hexane extraction |
	90% Acetone	Methanol	Methanol/ NaCl[3]	Hexane	
Chlorophyll a	664 89.7	665 74.9	665 89.4	660 119.0	10
Bacteriochlorophyll a	772 92.3	770 84.1	770 88.1	768 149.5	38
c	654 98.0	668 86.0	670 90.2	664 51.2	53
d	662 92.6	657 82.3	658 86.4	650 89.9	66
e	—	659 82.3	659 94.8	645 76.3	79

[1] Values obtained from Jones (1979), Stal et al. (1984) and Korthals and Steenbergen (1985).
[2] Values are absorption maxima, absorption coefficients.
[3] 10 ml methanol, 3 ml 0.05% NaCl (w/v) in water.

trophotometrically at fixed wavelengths (Table 2.7), which give maximum absorption at the secondary peak. This is specific for the chlorophyll measured. To allow for other coloured compounds extracted with the chlorophyll, measurements at a secondary wavelength where no chlorophyll absorption is expected should be made. This is normally 750 nm for chlorophyll a or 820–850 nm for bacteriochlorophylls. Pieces of filter often get into the solvent and should be removed by centrifugation before measuring the absorbance. To estimate the amount of chlorophyll present the absorption coefficient (Table 2.7) should be used in the following formula:

$$(\text{Chlorophyll}), \mu g\, l^{-1} = \frac{V_e E \times 10^3}{V_s A l}$$

where V_e = volume of solvent, ml; V_s = volume of samples, l; E = absorbance in solvent; A = absorption coefficient, g^{-1} 1 cm; l = light path of curvette, cm. If the two-phase extraction is used allowance should be made for the partitioning of the pigments between the solvents (Table 2.7).

The amount of chlorophyll present in cells is very dependent on the physiological state. So it is normal to express results in terms of the pigments, and not as biomass, hence chlorophylls are used more for comparison rather than to obtain absolute biomass estimates. Despite these problems, a value of 40 has been suggested as the chlorophyll a/carbon ratio for algal cells.

The methods outlined above provide a simple way of estimating phototroph biomass. It is better to take into account the phaeophytins which are natural degradation products of the chlorophylls. This makes the methods and calculations slightly more difficult and full details of these can be found elsewhere (Marker 1972; Jones, 1979; Parsons *et al.*, 1984; Stal *et al.*, 1984).

2.3.4 Lipopolysaccharide

Lipopolysaccharide (LPS) is a component of all Gram-negative bacterial cell walls. So it is only a good biomass indicator for aquatic bacteria when the community contains few, if any, Gram-positive bacteria. It is relevant to emphasize that it has been claimed that 80–95% of prokaryotes in seawater are Gram negative (Watson *et al.*, 1977), but polluted waters and sediments contain a much higher proportion of Gram-positive bacteria (Moriarty, 1980). In seawater samples, one study (Watson *et al.*, 1977) observed very good correlation between LPS concentration and the total number of bacteria ($r^2 = 0.899$, $n = 188$), but another investigation found distinct differences in the vertical distribution of these two variables at other marine sites (Maeda and Taga, 1979). Thus, it seems unlikely that LPS should be used as a biomass indicator in anything but clean, unpolluted waters where it could be quite convenient (Watson and Hobbie, 1979).

LPS is most commonly assayed on whole water samples by adding an aqueous extract from the blood cells of the horseshoe crab (*Limulus polypherus*) to give a turbid suspension. Watson *et al.* (1977) originally proposed a turbidometric assay. However, a colorimetric assay, using a complex amino acid/*p*-nitroanalide reagent, has also been suggested. This uses extracts from the Japanese horseshoe crab (*Tachypleus tridentatus*; Maeda and Taga, 1979). The latter assay seems less sensitive (calibration curve, 0.2–1 ng ml^{-1}) than the former (calibration curve, 1–100 pg ml^{-1}), which has also been made more sensitive by using 10 μM EDTA and 0.9% NaCl in the reaction mixture (Coates, 1977). The tubidometric assay is said to detect 500 bacteria per sample (Jones, 1979); this is clearly much more sensitive than a GLC method (Parker *et al.*, 1982) which only detects 10^6–10^7 bacteria (White, 1983).

The turbidometric LPS assay is straightforward but great care must be taken to make sure all reagents, water and glassware are completely free of LPS (Watson and Hobbie, 1979). *Limulus* amoebocyte lysate is reconstituted in 5 ml of LPS-free seawater, prepared by autoclaving 1 l of seawater with 100 g activated charcoal and filtering it through a 0.2 μm membrane filter. A sample (1 ml) is then added to the reconstituted lysate (0.2 ml), gently mixed, incubated at 37 °C for 1 h, and the absorbance read at 360 nm. Calibration curves to estimate the LPS concentration from turbidity must be prepared every day, as batch to batch variability of the lysate is large. The concentration of LPS is multiplied by the conversion factor of 6.35 to estimate the amount of bacterial carbon present.

2.3.5 Muramic acid

The cell walls of all prokaryotes contain muramic acid in their mucopeptides. However, Gram-positive bacteria contain much more muramic acid than Gram-negative bacteria, and cyanobacteria may possess up to 500 times more than is present in other bacteria (King and White, 1977). Also the muramic acid present in these three groups of prokaryotes is different (White, 1983). This means that using muramic acid as a bacterial biomass indicator is rather difficult. Despite these problems Moriarty (1980) has suggested it as a suitable method for sediments. Conversion factors of 6.4–12 µg muramic acid mg^{-1} bacterial carbon have been suggested when few Gram-positive bacteria are present (Jones, 1979). Alternatively, if large numbers of Gram-positive bacteria are present, a formula may be used which allows for the proportion of these bacteria in the sediment (Moriarty, 1977).

Several methods for determining muramic acid in sediment have been suggested, but none are sensitive enough for water samples. One method depends on an enzymic conversion to lactate (Moriarty, 1977; 1980); another is a complex chemical analysis (King and White, 1977), and a third uses acid hydrolysis, thin layer chromatography and GLC (Fazio *et al.*, 1979). Because these methods are all quite complex and because of the other problems with muramic acid discussed earlier, it cannot be recommended for routine use in aquatic habitats, and so will not be described further.

2.3.6 Lipids and fatty acids

There is a wide variety of lipids and fatty acids present in micro-organisms. Recently, the large body of work published, mainly by Dr D. C. White and his colleagues (White, 1983) has shown that some of these compounds can be used as biomass indicators for specific groups of bacteria and eukaryotes (Table 2.8). Thus a 'signature' or 'fingerprint' of various organisms may be obtained. These compounds are detected by solvent extraction followed by GLC or high performance liquid chromatography (HPLC). Although the methodology is complex, the techniques show a lot of promise. The sensitivity is poor, and so they are only suitable for use in biofilms or sediments. At present these methods are in their infancy and so conversion factors are not available. These compounds are used to indicate the presence of biomass rather than to quantify it. Many more of these 'signature' compounds are probably still to be discovered. It is also likely that individual groups of genera or even individual genera will eventually be quantifiable. Recent work has shown that methane-oxidizing bacteria (Nichols *et al.*, 1985) and *Desulfobulbus* spp. might be identifiable with this

TABLE 2.8 Some fatty acids and lipids that are potential biomass indicators for specific types of organisms in aquatic sediments (data from White (1983) and Parkes and Taylor (1985)).

Group of organisms	Biomass indicator
Eukaryotes	Polyenoic fatty acids (>18 C-atoms)
Microfauna	Monohydroxy steroids
	Polyenoic fatty acids, with linolenic acid unsaturation patterns
Filamentous bacteria, cyanobacteria and fungi	Linolenic acid
Anaerobic bacteria	Plasmalogens
Anaerobic fermentative bacteria	Phosphosphingolipids
Sulphate reducing bacteria	Ester-linked iso-3-hydroxy fatty acids
	Anteiso-branched 3-hydroxy fatty acids
Methanogens	Phantanyl glycerol ether phospholipids

type of technique (Parkes and Calder, 1985). The complexity of the methodology and the novelty of the approach precludes its further description.

Acknowledgements

I would like to thank all my undergraduate students, postgraduate students and research staff whose observations have helped to formulate my ideas about the methods described in this review. I specifically thank Dr C. M. Turley for providing the fluorescent latex beads, Miss H. J. Taylor for the practical work with these beads and Miss F. Kemmy, Dr N. C. B. Humphrey and Dr A. R. Davies for the use of their unpublished results. Also I thank Mr D. Thomas for preparing the computer drawings of the bacterial silhouettes.

2.4 REFERENCES

Bakken, L. R. and Olsen, R. A. (1983). Buoyant densities and dry-matter contents of microorganisms: conversion of a measured biovolume to biomass. *Applied and Environmental Microbiology*, **45**, 1188–1195.

Bjørnsen, P. K. (1986). Automatic determination of bacterioplankton biomass by image analysis. *Applied and Environmental Microbiology*, **51**, 1199–1204.

Bowden, W. B. (1977). Comparison of two direct-count techniques for enumerating aquatic bacteria. *Applied and Environmental Microbiology*, **33**, 1229–1232.

Bradbury, S. (1977). Quantitative image analysis. In *Analytical and Quantitative Microscopy*, G. A. Meek and H. Y. Elder (Eds), Cambridge University Press, Cambridge, pp. 91–116.

Bradbury, S. (1979). Microscopical image analysis: problems and approaches. *Journal of Microscopy*, **115**, 137–150.

Bradbury, S. (1981). Automatic image analysers and their use in anatomy. In *Eleventh International Congress of Anatomy, Part B: Advances in the Morphology of Cells and Tissues*, M. A. Galina (Ed.), Alan R. Liss, New York, pp. 129–150.

Bratbak, G. (1985). Bacterial biovolume and biomass estimations. *Applied and Environmental Microbiology*, **49**, 1488–1493.

Bratbak, G. and Dundas, I. (1984). Bacterial dry matter content and biomass estimations. *Applied and Environmental Microbiology*, **48**, 755–757.

Christian, R. R., Hanson, R. B. and Newell, S. Y. (1982). Comparison of methods for measurement of bacterial growth rates in mixed batch cultures. *Applied and Environmental Microbiology*, **43**, 1160–1165.

Coates, D. A. (1977). Enhancement of the sensitivity of the *Limulus* assay for the detection of Gram-negative bacteria. *Journal of Applied Bacteriology*, **42**, 445–449.

Costerton, J. W. and Colwell, R. R. (1979). *Native Aquatic Bacteria: Enumeration, Activity and Ecology*, ASTM STP 695, American Society for Testing and Materials, Philadelphia.

Dale, D. G. (1974). Bacteria in intertidal sediments: factors related to their distribution. *Limnology and Oceanography*, **19**, 509–518.

Daley, R. J. (1979). Direct epifluorescence enumeration of native aquatic bacteria uses, limitations, and comparative accuracy. In *Native Aquatic Bacteria Enumeration, Activity and Ecology*, ASTM STP 695, J. W. Costerton and R. R. Colwell (Eds), American Society for Testing and Materials, Philadelphia, pp. 29–45.

Deming, J. W., Picciolo, G. L. and Chappelle, E. W. (1979). Important factors in adenosine triphosphate determinations using firefly luciferase: applicability of the assay to studies of native aquatic bacteria. In *Native Aquatic Bacteria: Enumeration, Activity and Ecology*, ASTM STP 695, J. W. Costerton and R. R. Colwell (Eds), American Society for Testing and Materials, Philadelphia, pp. 89–98.

Dodson, A. N. and Thomas, W. H. (1964). Concentrating plankton in a gentle fashion. *Limnology and Oceanography*, **9**, 455–456.

Fazio, S. D., Mayberry, W. R. and White, D. C. (1979). Muramic acid assay in sediments. *Applied and Environmental Microbiology*, **38**, 349–350.

Ferguson, R. L., Buckley, E. N. and Palumbo, A. V. (1984). Response of marine bacterioplankton to differential filtration and confinements. *Applied and Environmental Microbiology*, **47**, 49–55.

Fry, J. C. (1982). The analysis of microbial interactions and communities. In *Microbial Interactions and Communities*, Vol. 1 A. T. Bull and J. H. Slater (Eds), Academic Press, London, pp. 103–152.

Fry, J. C. and Davies, A. R. (1985). An assessment of methods for measuring volumes of planktonic bacteria, with particular reference to television image analysis. *Journal of Applied Bacteriology*, **58**, 105–112.

Fry, J. C., Goulder, R. and Rimes, C. (1985). A note on the efficiency of stomaching for the quantitative removal of epiphytic bacteria from submerged aquatic plants. *Journal of Applied Bacteriology*, **58**, 113–115.

Fry, J. C. and Humphrey, N. C. B. (1978). Techniques for the study of bacteria epiphytic on aquatic macrophytes. In *Techniques for the Study of Mixed Populations*, D. W. Lovelock and R. Davies (Eds), Academic Press, London, pp. 1–29.

Fry, J. C. and Zia, T. (1982). Viability of heterotrophic bacteria in freshwater. *Journal of General Microbiology*, **128**, 2841–2850.

Fuhrman, J. A. (1981). Influence of method on the apparent size distribution of bacterioplankton cells: epifluorescence microscopy compared to scanning electron microscopy. *Marine Ecology Progress Series*, **5**, 103–106.

Gayford, C. G. and Richards, J. P. (1970). Isolation and enumeration of aerobic heterotrophic bacteria in activated sludge. *Journal of Applied Bacteriology*, **33**, 342–350.

Godinho-Orlandi, M. J. L. and Jones, J. G. (1981). Filamentous bacteria in sediments of lakes of differing degress of enrichment. *Journal of General Microbiology*, **123**, 81–90.

Goulder, R. (1977). Attached and free bacteria in an estuary with abundant suspended solids. *Journal of Applied Bacteriology*, **43**, 399–405.

Goulder, R. (1987). Evaluation of the saturation approach to measurement of V_{max} for glucose mineralization by epilithic freshwater bacteria. *Letters in Applied Microbiology*, **4**, 29–32.

Goulder, R., Bent, E. J. and Boak, A. C. (1981) Attachment to suspended solids as a strategy of estuarine bacteria. In *Feeding and Survival Strategies of Estuarine Organisms*, N. V. Jones, and W. J. Wolff (Eds), Plenum Press, New York, pp. 1–15.

Hagström, A. (1984). Aquatic bacteria: Measurements and significance of growth. In *Current Perspectives in Microbial Ecology*, M. J. Klug and C. A. Reddy (Eds), American Society for Microbiology, Washington DC, pp. 495–501.

Harper, R. M., Fry, J. C. and Learner, M. A. (1981). A bacteriological investigation to elucidate the feeding biology of *Nais variabilis* (Oligochaeta: Naididae). *Freshwater Biology*, **11**, 227–236.

Hessen, D. O. (1985). The relation between bacterial carbon and dissolved compounds in oligotrophic lakes. *FEMS Microbiology Ecology*, **31**, 215–223.

Hobbie, J. E., Daley, R. J. and Jasper, S. (1977). Use of nucleopore filters for counting bacteria by fluorescence microscopy. *Applied and Environmental Microbiology*, **33**, 1225–1228.

Holm-Hansen, O. and Booth, C. R. (1966). The measurement of adenosine triphosphate in the ocean and its ecological significance. *Limnology and Oceanography*, **11**, 510–519.

Hossell, J. C. and Baker, J. H. (1979). A note on the enumeration of epiphytic bacteria by microscopic methods with particular reference to two freshwater plants. *Journal of Applied Bacteriology*, **46**, 87–92.

Jones, J. G. (1975). Some observations on the occurrence of the iron bacterium *Leptothrix ochracea* in fresh water, including reference to large experimental enclosures. *Journal of Applied Bacteriology*, **39**, 63–72.

Jones, J. G. (1978). The distribution of some freshwater planktonic bacteria in two stratified eutrophic lakes. *Freshwater Biology*, **8**, 127–140.

Jones, J. G. (1979). *A. Guide to Methods for Estimating Microbial Numbers and Biomass in Fresh Water*, Freshwater Biological Association, Windermere.

Jones, J. G. (1981). The population ecology of iron bacteria (genus, *Ochrobium*) in a stratified eutrophic lake. *Journal of General Microbiology*, **125**, 85–93.

Jones, J. G. and Simon, B. M. (1975). An investigation of errors in direct counts of aquatic bacteria by epifluorescence microscopy, with reference to a new method for dyeing membrane filters. *Journal of Applied Bacteriology*, **39**, 317–329.

Jones, J. G. and Simon, B. M. (1977). Increased sensitivity in the measurement of ATP in freshwater samples with comment on the adverse effect of membrane filtration. *Freshwater Biology*, **7**, 253–260.

Karl, D. M. (1979). Adenosine triphosphate and guanosine triphosphate determinations in intertidal sediments. In *Methodology for Biomass Determinations and*

Microbial Activities in Sediments, ASTM STP 673 C. D. Litchfield and P. L. Seyfried (Eds), American Society for Testing and Materials, Philadelphia, pp. 5–20.

Karl, D. M. (1980). Cellular nucleotide measurements and applications in microbial ecology. *Microbiological Reviews*, **44**, 739–796.

King, J. D. and White, D. C. (1977). Muramic acid as a measure of microbial biomass in estuarine and marine samples. *Applied and Environmental Microbiology*, **33**, 777–783.

Kirchman, D. L., Mazzella, L., Alberte, R. S. and Mitchell, R. (1984). Epiphytic bacterial production on *Zostera marina*. *Marine Ecology Progress Series*, **15**, 117–123.

Kirchman, D. L. and Mitchell, R. (1982). Contribution of particle-bound bacteria to total microheterotrophic activity in five ponds and two marshes. *Applied and Environmental Microbiology*, **43**, 200–209.

Kirchman, D. L., Sigda, J., Kapulscinski, R. and Mitchell, R. (1982). Statistical analysis of the direct count method for enumerating bacteria, *Applied Environmental Microbiology*, **44**, 376–382.

Korthals, H. J. and Steenbergen, C. L. M. (1985). Separation and quantification of pigments from natural phototrophic microbial populations. *FEMS Microbiology Ecology*, **31**, 177–185.

Krambeck, C. (1984). Diurnal responses of microbial activity and biomass in aquatic ecosystems. In *Current Perspectives in Microbial Ecology*, M. J. Klug and C. A. Reddy (Eds), American Society for Microbiology, Washington D. C., pp. 502–508.

Krambeck, C., Krambeck, H. and Overbeck, J. (1981). Microcomputer-assisted biomass determinations of plankton bacteria on scanning electron micrographs. *Applied and Environmental Microbiology*, **42**, 142–149.

Laanbroek, H. J. and Verplanke, J. C. (1986). Seasonal changes in percentages of attached bacteria enumerated in a tidal and a stagnant coastal basin: relation to bacterioplankton productivity. *FEMS Microbiology Ecology*, **38**, 87–98.

Litchfield, C. D. and Seyfried, P. L. (1979). *Methodology for Biomass Determinations and Microbial Activities in Sediments*, ASTM STP 673, American Society for Testing and Materials, Philadelphia.

Maeda, M. and Taga, N. (1979). Chromogenic assay method of lipopolysaccharide (LPS) for evaluating bacterial standing crop in seawater. *Journal of Applied Bacteriology*, **47**, 175–182.

Maeda, M. and Taga, N. (1983). Comparisons of cell size of bacteria from four marine localities. *La Mer*, **21**, 201–204.

Marker, A. F. (1972). The use of acetone and methanol in the estimation of chlorophyll in the presence of phaeophytin. *Freshwater Biology*, **2**, 361–385.

Meyer-Reil, L. A. (1983). Benthic response to sedimentation events during autumn to spring at a shallow water station in the Western Kiel Bight. II. Analysis of benthic bacterial populations. *Marine Biology*, **77**, 247–256.

Moriarty, D. J. W. (1977). Improved method using muramic acid to estimate biomass of bacteria in sediments. *Oecologia*, **26**, 317–323.

Moriarty, D. J. W. (1980). Measurement of bacterial biomass in sandy sediments. In *Biogeochemistry of Ancient and Modern Environments*, P. A. Trudinger, M. R. Walter and B. J. Ralph (Eds), Australian Academy of Science, Canberra, pp. 131–137.

Newell, S. Y. and Fallon, R. D. (1982). Bacterial productivity in the water column and sediments of the Georgia (USA) coastal zone: estimates via direct counting and parallel measurement of thymidine incorporation. *Microbial Ecology*, **8**, 33–46.

Nichols, P. D. , Smith, G. A., Antworth, C. P., Hanson, R. S. and White, D. C. (1985). Phospholipid and lipolysaccharide normal and hydroxy fatty acids as potential signatures for methane-oxidising bacteria. *FEMS Microbiology Ecology*, **31**, 327–335.

Palumbo, A. V., Ferguson, R. L. and Rublee, P. A. (1984). Size of suspended bacterial cells and association of heterotrophic activity with size fractions of particles in estuarine and coastal waters. *Applied and Environmental Microbiology*, **48**, 157–164.

Parker, J. H., Smith, G. A., Fredrickson, H. L., Vestal, J. R. and White, D. C. (1982). Sensitive assay based on hydroxy fatty acids from lipopolysaccharide lipid A for Gram-negative bacteria in sediments. *Applied and Environmental Microbiology*, **44**, 1170–1177.

Parkes, R. J. and Calder, A. G. (1985). The cellular fatty acids of three strains of *Desulfobulbus*, a propionate-utilising sulphate-reducing bacterium. *FEMS Microbiology Ecology*, **31**, 361–363.

Parkes, R. J. and Taylor, J. (1985). Characterization of microbial populations in polluted marine sediments. *Journal of Applied Bacteriology Symposium Supplement*, **59**, 155S–173S.

Parsons, T. R., Maita, Y. and Lalli, C. M. (1984). *A Manual of Chemical and Biological Methods for Seawater Analysis*, Pergamon Press, Oxford.

Porter, K. G. and Feig, Y. S. (1980). The use of DAPI for identifying and counting aquatic microflora, *Limnology and Oceanography*, **25**, 943–948.

Ramsay, A. J. (1978). Direct counts of bacteria by a modified acridine orange method in relation to their heterotrophic activity. *New Zealand Journal of Marine and Freshwater Research*, **12**, 265–269.

Rice, A. L., Billett, D. S. M., Fry, J., John, A. W. G., Lampitt, R. S., Mantoura, R. F. C. and Morris, R. J. (1986). Seasonal deposition of phytodetritus to the deep-sea floor. *Proceedings of the Royal Society Edinburgh*, **88B**, 265–279.

Riemann, B. (1978). Differentiation between heterotrophic and photosynthetic plankton by size fractionation, glucose uptake, ATP and chlorophyll content. *Oikos*, **31**, 358–367.

Rimes, C. A., and Goulder, R. (1986). Suspended bacteria in calcareous and acid headstreams: abundance, heterotrophic activity and downstream change. *Freshwater Biology*, **16**, 633–651.

Roberts, R. D. and Sephton, L. M. (1981). The enumeration of aquatic bacteria using DAPI. *Journal of the Limnology Society South Africa*, **7**, 72–74.

Rodina, A. G. (1972). *Methods in Aquatic Microbiology*, University Park Press, Baltimore.

Roser, D., Nedwell, D. B. and Gordon, A. (1984). A note on plotless methods for estimating bacterial cell densities. *Journal of Applied Bacteriology*, **56**, 343–347.

Rudd, J. W. M. and Hamilton, R. D. (1973). Measurement of adenosine triphosphate (ATP) in two precambrian shield lakes of northwestern Ontario. *Journal of the Fisheries Research Board of Canada*, **30**, 1537–1546.

Salonen, K. (1977). The estimation of bacterioplankton numbers and biomass by phase contrast microscopy. *Annales de Botanica Fennici*, **14**, 25–28.

Sieracki, M. E., Johnson, P. W., and Sieburth, J. M. (1985). Detection, enumeration, and sizing of planktonic bacteria by image-analyzed epifluorescence microscopy. *Applied and Environmental Microbiology*, **49**, 799–810.

Sikora, J. P., Sikora, W. B., Enkenbrecker, C. W. and Coull, B. C. (1977). Significance of ATP, carbon and caloric content of meiobenthic nematodes in partitioning benthic biomass. *Marine Biology*, **44**, 7–14.

Sorokin, Y. I. and Overbeck, J. (1972). Direct microscopic counting of microorganisms. In *Techniques for the Assessment of Microbial Production and Decomposition in Fresh Waters, IBP Handbook No. 23*, Y. I. Sorokin and H. Kadota (Eds), Blackwell Scientific Publications, Oxford, pp. 44–47.

Stal, L. J., van Gemerden, H. and Krumbein, W. E. (1984). The simultaneous assay of chlorophyll and bacteriochlorophyll in natural microbial communities. *Journal of Microbiological Methods*, **2**, 295–306.

Stevenson, L. H., Chrzanowski, T. H. and Erkenbrecher, C. W. (1979). The adenosine triphosphate assay: conceptions and misconceptions. In *Native Aquatic Bacteria: Enumeration, Activity and Ecology*, ASTM STP 695 J. W. Costerton and R. R. Colwell (Eds), American Society for Testing and Materials, Philadelphia, pp. 99–116.

Turley, C. M. and Lochte, K. (1985). Direct measurement of bacterial productivity in stratified waters close to a front in the Irish Sea. *Marine Ecology Progress Series*, **23**, 209–219.

Van Es, F. B. and Meyer-Reil, L. A. (1982). Biomass and metabolic activity of heterotrophic marine bacteria. *Advances in Microbial Ecology*, **6**, 111–170.

Van Veen, J. A. and Paul, E. A. (1979). Conversion of biovolume measurements of soil organisms, grown under various moisture tensions to biomass and their nutrient content. *Applied and Environmental Microbiology*, **37**, 686–692.

Watson, S. W. and Hobbie, J. E., (1979). Measurement of bacterial biomass as lipopolysaccharide. In *Native Aquatic Bacteria: Enumeration, Activity, and Ecology*, ASTM STP 695, J. W. Costerton and R. R. Colwell (Eds), American Society for Testing and Materials, Philadelphia, pp. 82–88.

Watson, S. W., Novitsky, T. J., Quinby, H. L. and Valois, F. W. (1977). Determination of bacterial number and biomass in the marine environment. *Applied and Environmental Microbiology*, **33**, 940–946.

Weise, W. and Rheinheimer, G. (1978). Scanning electron microscopy and epifluorescence investigation of bacterial colonization of marine sand sediments. *Microbial Ecology*, **4**, 175–188.

White, D. C. (1983). Analysis of microorganisms in terms of quantity and activity in natural environments. In *Microbes in their Natural Environments*, J. H. Slater, R. Whittenbury and J. W. T. Wimpenny (Eds), Cambridge University Press, Cambridge, pp. 37–66.

White, D. C., Bobbie, R. J., Herron, J. S., King, J. D. and Morrison, S. J. (1979a). Biochemical measurements of microbial mass and activity from environmental samples. In *Native Aquatic Bacteria: Enumeration, Activity and Ecology*, ASTM STP 695 J. W. Costerton and R. R. Colwell (Eds), American Society for Testing and Materials, Philadelphia, pp. 69–81.

White, D. C., Bobbie, R. J., King, J. D., Nickels, J. and Amoe, P. (1979b). Lipid analysis of sediments for microbial biomass and community structure. In *Methodology for Biomass Determinations and Microbial Activities in Sediments*, ASTM STP 673, C. D. Litchfield and P. L. Seyfried (Eds), American Society for Testing and Materials, Philadelphia, pp. 87–103.

White, D. C., Davis, W. M., Nickels, J. S., King, J. D. and Bobbie, R. J. (1979c). Determination of the sedimentary microbial biomass by extractible lipid phosphate. *Oecologia*, **40**, 51–62.

Wynn-Williams, D. D. (1985). Photofading retardant for epifluorescence microscopy in soil micro-ecological studies. *Soil Biology and Biochemistry*, **17**, 739–746.

Zimmerman, R. (1977). Estimation of bacterial number and biomass by epifluorescence microscopy and scanning electron microscopy. In *Microbial Ecology of a Brackish Water Environment*, G. Rheinheimer (Ed.), Springer-Verlag, New York, pp. 103–120.

Methods in Aquatic Bacteriology
Edited By B. Austin
© 1988 John Wiley & Sons Ltd.

3

Isolation Methods

J. Schneider and G. Rheinheimer

Abteilung für Marine Mikrobiologie, Institut für Meereskunde an der Universität Kiel, Düsternbrooker Weg 20, D 2300 Kiel 1, FRG

3.1 INTRODUCTION

The aim of enrichment and isolation procedures in microbiology is to obtain pure cultures of micro-organisms mainly for biochemical, taxonomic and autecological studies.

The knowledge of a few basic operations generally is sufficient with most bacteria, and the success with 'difficult' species depends to a great extent on the choice of the appropriate cultivation medium rather than on sophisticated procedures. A student unfamiliar with microbiology should always keep in mind, however, that the maintenance of aseptic conditions at all steps of an operation is indispensable not only to reach the desired aim, but to avoid serious injuries to health by the unexpected presence of pathogens in the samples.

For isolation, it is necessary:

1. to separate cells from neighbouring cells which is conveniently done on or in solid substrates (i.e. media containing agar);
2. to stimulate the organism to multiply, and thus, to form a colony.

It may be necessary to apply enrichment procedures (often in liquid media) to select the organism of interest. Thence, the second step will be purification and recovery of the organism in large numbers for subsequent study.

3.2 BASIC EQUIPMENT

The laboratory should be equipped with washable benches, and dust and draught should be avoided. There is an absolute requirement for water, gas and electricity. As to equipment, it is necessary to have access to autoclaves, incubators, refrigerators and refrigerated incubators (for storage and for the cultivation of psychrophilic bacteria), safety-cabinets or laminar air-flow hoods (to alleviate potential problems of contamination). Detailed information on these items is given by Philips (1981). A well-equipped laboratory will have good supplies of assorted glassware such as screw-cap tubes, flasks (250 ml, 1 l and 5 l capacity), pipettes (1 and 10 ml), glass or plastic Petri dishes and bent glass rods for spreading samples on solid media. In addition, anaerobic jars for the cultivation of anaerobes would be useful.

3.3 BASIC TECHNIQUES

3.3.1 Aseptic techniques

As a word of caution, it is imperative to adopt stringent aseptic methods to ensure the recovery of native aquatic bacteria rather than chance contaminants. In particular, it is a wise precaution to verify the sterility of stored media immediately before use to ensure the absence of micro-colonies. It is also sound policy to discard bottles of media, of which some has been previously used.

3.3.2 Sterilization of media and equipment

Three methods may be applied to sterilize glassware and media, namely: wet heat, i.e. autoclaving at 115–121 °C for up to 30 min; dry heat, 160° or 180 °C for 2 h or 30 min, respectively; and filtration (membrane filters of $\leqslant 0.3\,\mu m$ pore size).

3.3.3 Culture vessels

For cultivation of aquatic bacteria, the following types of vessel are most suitable: Petri dishes, test tubes and Erlenmeyer flasks (100, 300 ml or larger capacity). As a word of caution, it is important to keep any cotton wool plugs in a dry condition. Although screw-caps are more effective in terms of maintaining sterility, the aeration of the cultures may be retarded, therefore it is often advantageous to use cotton wool plugs. With Petri dishes, a common problem is water of condensation, which accumulates in the inner part of the lid. This generally occurs when the agar is poured at temperatures higher than 50 °C. To control this problem, the agar plates should be dried overnight at 45 °C, prior to use. If the water is not removed it may cause colonies to mix, giving rise to a more- or less continuous layer of bacteria rather than well-isolated colonies.

3.4 ENRICHMENT CULTURES

In nature, bacteria inevitably occur as mixed communities. The bacteria within such populations will be in various stages of activity from dormancy to rapid metabolism. As much research depends upon recovery of certain preselected taxa, it is sometimes inevitable that the desired organism has to be enriched in order to ensure its subsequent growth on laboratory media. This enrichment can be achieved by changing conditions in favour of the desired organism. The principal means of enrichment involve altering the nutritional and physiological conditions, and physically separating the organisms from mixed populations (Gerhardt *et al.*, 1981).

In terms of metabolic activity, the reader is reminded that bacteria may be divided into three groups, namely:

1. *Photoautotrophs.* These produce organic material by using the energy derived from sunlight to transform CO_2 into simple organic molecules. The process utilizes water and some mineral salts.
2. *Chemolithotrophs.* These derive the energy for life-giving processes from chemical oxidations (e.g. ammonia to nitrite and nitrate, as in the case of the nitrifying bacteria).
3. *Heterotrophs (saprophytes).* This group comprises a diverse array of genera, which depend completely on the presence of complex organic compounds, e.g. polysaccharides, proteins and lipids, for their existence.

3.4.1 Enrichment by direct plating

With heterotrophs, it is usually sufficient to use simple media, containing a nitrogen and a carbon source. One example contains:

Bacteriological peptone	5.0 g
Yeast extract	1.0 g
Agar (for a solid medium)	15.0 g
Water to 1 l	

For freshwater bacteria, distilled or tap water is used. However, with marine bacteria, seawater is usually supplied, either undiluted (mean salinity 32–35°/oo s) or diluted with tap water (750 ml seawater plus 250 ml tap water) depending on the salinity at the sampling location. The pH will need to be adjusted before sterilizing (121 °C/15 min) to 6.9 (for freshwater bacteria) or 7.0–7.2 (for marine bacteria).

In the case of lipolytic bacteria, which maintain their metabolism principally by the decomposition of fats (some *Pseudomonas* spp. fall into this category), the medium described by Smibert and Krieg (1981) may be used. Essentially to the standard recipe (as described above $CaCl_2 \cdot 6H_2O$

(0.001% w/v) and Tween 80 (1.0% v/v) are added. Tween 80 (oleic ester) needs to be sterilized separately by autoclaving (121 °C/20 min). Thence, the sterile Tween 80 is added to the molten cooled basal medium at 45–50 °C. The medium is shaken thoroughly and dispensed into Petri dishes. Aquatic samples are diluted (if appropriate) and streaked out onto the solidified agar medium. Colonies of lipolytic bacteria develop an opaque halo around them because of the deposition of insoluble calcium salts. It should be mentioned that Tween 40 (palmitic acid ester) or Tween 60 (stearic acid ester) may be substituted for Tween 80. It is also possible to incorporate fats or plant oils into the basal medium in combination with an indicator, such as nile blue, or gentian blue. In this case, the presence of lipolytic bacteria is indicated by a blue ring around the colonies (Reichardt, 1978). The isolation of pure colonies is subsequently performed by repeated streaking and re-streaking onto fresh medium.

3.4.2 Membrane filtration

A known volume of the sample may be passed through a sterile filter, which is carefully removed (e.g. by sterile forceps) and placed onto the surface of an agar medium. The bacteria may also be concentrated by centrifugation or by initial filtration of the sample. The micro-organisms adhering to the filter are resuspended in the liquid which serves as a start for the inoculation of spread or pour plates.

A more difficult procedure involves the recovery of some of the nutritionally fastidious bacteria. However, as a general rule, it is argued that such bacteria comprise a minor component of the microbial population. The precise cultural conditions which need to be adopted vary with the special demands of the organism in question. For example, by adopting low temperatures psychrophilic bacteria are favoured, sulphate-reducing bacteria are enhanced by use of anaerobic conditions with a medium containing sulphate and an organic compound. Similarly, chitin decomposers are enriched by including fine particles of chitin in a medium, which is devoid of other carbohydrates. From the great number of methods designed specifically for the enrichment of specialist groups, only those for the more frequently occurring taxa will be considered further. For more detailed information see Gerhardt et al. (1981), Norris and Ribbons (1969) and Starr et al. (1981).

3.4.3 Psychrophilic bacteria

Psychrophiles are mainly found in colder water, especially springs or deep seawater. For such bacteria, it is important to maintain the sample at a constant low temperature approximating to that of the in situ conditions.

This can be done in Dewar vessels or cooling boxes. For enrichment purposes, prepare spread plates on a standard medium and incubate at a temperature of $\leq 0°$ to $5-10°C$ for at least 3 weeks.

3.4.4 Decomposers of cellulose, chitin and agar

Representatives of *Cytophaga, Sporocytophaga, Flexibacter, Microscilla,* and *Lysobacter* are important decomposers of complex polymers, which arrive in water from leaves and other plant material, dead insects and (in marine environments) from algae. A main characteristic of these organisms is their gliding movement. Moreover, they are strict aerobes, growing at mesophilic temperatures, and occur in freshwater as well as in marine habitats (Reichenbach and Dworkin, 1981).

3.4.4.1 Recovery of cellulose decomposers

To recover freshwater organisms, a strip of filter paper is immersed in the following medium so that a part of the paper remains above the surface of the solution:

$(NH_4)_2SO_4$	1.0 g
K_2HPO_4	1.0 g
(autoclaved separately)	
$MgSO_4 \cdot 7H_2O$	0.2 g
$CaCl_2 \cdot 2H_2O$	0.1 g
$FeCl_3$	0.02 g
Distilled water to 1 l	
pH 7.0 to 7.5 (after sterilization)	

For the enrichment of marine cellulose decomposers, the following medium is recommended:

$(NH_4)_2SO_4$	1.0 g
(autoclaved separately)	
Agar	15.0 g
Natural seawater to 1 l	
pH, not adjusted	

The tubes are inoculated with water, mud or plant debris and incubated. After some weeks, translucent shiny white or yellow colonies indicate the development of cellulose decomposers.

To obtain pure cultures, prepare a cellulose overlay-agar as follows:

Powdered cellulose (0.4% w/v) is added to the medium described above. Two-layer plates are prepared by overlaying the medium *without* cellulose (after it has solidified) with a thin layer of the cellulose-containing medium.

On this medium a little material from the enrichment broth is streaked out and incubated for some weeks. Reichenbach and Dworkin (1981) recommended inoculating small pieces of sterile filter paper placed onto the two-layered plate. This favours the separation of cellulose decomposers from other bacteria. However, in order to suppress the growth of fungi, the addition of cycloheximide (2.5 mg per 100 ml; Brockman, 1967) is advisable.

The somewhat tedious purification of cellulose decomposers is executed by repeated streaking on solid media. To overcome the problem of slow growth, it is recommended to add low concentrations of glucose (0.1% w/v) to the medium. For further information refer to Reichenbach and Dworkin (1981).

3.4.4.2 *Chitin decomposers*

A suitable enrichment has been described (Reichenbach and Dworkin, 1981):

$(NH_4)_2SO_4$	2.0 g
KH_2PO_4	0.7 g
Na_2HPO_4	1.1 g
$MgSO_4 \cdot 7H_2O$	0.2 g
$FeSO_4$	trace
$MnSO_4$	trace
Chitin, precipitated	3.0 g
Agar	15.0 g
Water to 1 l	

The chitin precipitate is prepared as follows:

40 g of finely powdered chitin (coarse material is not suitable) is dissolved in cold concentrated HCl (this takes between 30 and 60 min). The chitin is then precipitated by pouring the solution slowly into 2 l of water at approximately 8 °C. The precipitate is dialysed against running water, the pH adjusted to 7.0 (with KOH) and autoclaved (this is the stock solution). The precipitate must not be dried.

3.4.4.3 *Agar decomposers*

Representatives of this group of bacteria are found in marine habitats, and may be enriched on solid media containing agar and seawater, and low concentrations of nutrients. Purification is performed by repeated subculturing. Reichenbach and Dworkin (1981) recommended the following medium:

Agar	1.5% (w/v)
$(NH_4)_2SO_4$	0.1% (w/v)
(autoclaved separately)	
Glucose	0.2% (w/v)
(autoclaved separately)	
Natural seawater	
pH, not adjusted	

The presence of agar decomposers is indicated by depressions or cavities in the surface of the medium.

3.4.5 Nitrogen-fixing bacteria (Azotobacteriaceae)

Although most representatives of this group are inhabitants of both soil and water, some species (i.e. *Azotobacter agilis* and *A. insignis*) have so far been isolated only from water. All members are aerobic, heterotrophic bacteria, which are characterized by the ability to fix atmospheric nitrogen when living in an environment which is poor in nitrogen compounds (Becking, 1981).

For enrichment, 100 ml volumes of sample are added to a medium containing

Glucose	2.0 g
$CaCO_3$	2.0 g
K_2HPO_4	0.1 g
$MgSO_4 \cdot 7H_2O$	0.05 g

The inoculated medium is incubated in small volumes in Erlenmeyer flasks (300 ml capacity) for several days at 27 °C. A 'scum'-like growth develops at the surface; a little of this culture is transferred to a solid medium, prepared as follows:

Glucose	20.0 g
K_2HPO_4	0.8 g
KH_2PO_4	0.2 g
$MgSO_4 \cdot 7H_2O$	0.5 g
$FeCl_3 \cdot 6H_2O$	0.1 g
$CaCO_3$	20.0 g
$Na_2MoO_4 \cdot 2H_2O$	0.05 g
Agar	20.0 g
Distilled water to 1 l	
pH 7.4–7.6	

The presence of dark pigmented colonies is indicative of *A. chroococcum*. Further purification needs to be carried out by repeated streaking on fresh media. Confusion may arise, however, from the pleomorphic nature of *A. chroococcum* strains (Becking, 1981).

3.4.6 Prosthecate bacteria (*Hyphomicrobium* spp., *Pedomicrobium* spp. and *Hyphomonas* spp.)

The members of this group are characterized by the presence of hypha-like structures (prosthecae), at the end of which a flagellated motile cell develops which separates from the 'mother' organism. Another feature of these bacteria is the ability to use C_1-compounds, and advantage may be taken of this ability for enrichment purposes. According to Moore and Hirsch (1972), the following medium is suitable for the enrichment of *Hyphomicrobium* spp.:

KH_2PO_4	1.36 g
Na_2HPO_4	1.13 g
$(NH_4)_2SO_4$	0.50 g
$MgSO_4 \cdot 7H_2O$	0.20 g
$CaCl_2 \cdot 2H_2O$	3.0 mg
$FeSO_4 \cdot 7H_2O$	2.0 mg
$MnSO_4 \cdot 4H_2O$	0.88 mg
$Na_2MoO_4 \cdot 2H_2O$	1.00 mg
Methylamine	5 ml
(67.5%; filter sterilized)	
Distilled water to 1 l	
pH 7.2–7.5 (after sterilization)	

For the preparation of a solid medium, it is recommended to add 1.5% (w/v) Nobel-agar (Difco).

The enrichment process may be achieved as follows:

Glass slides are exposed to a water sample for a predetermined period. A slime layer develops on the surface of the glass and this slime is transferred to the solid medium with incubation at 30 °C. However, growth is very slow. A more rapid (and selective) enrichment is possible when the cultivation is performed anaerobically in the presence of nitrate and methanol (Attwood and Harder, 1972; see also Starr *et al.*, 1981). *Hyphomicrobium* spp. and related bacteria are purified by repeated streaking of samples from enrichment cultures onto solid medium.

3.4.7 Denitrifying bacteria

Representatives of many different taxa are capable of denitrifying activity. Generally, they are facultatively anaerobic organisms which may oxidize organic compounds by transferring electrons to nitrites and nitrates (these may be reduced to N_2). Denitrifying bacteria are found in all types of soils, in muds within freshwater and brackish water, and in water of lakes, rivers and the sea.

The enrichment of *Thiobacillus denitrificans* may be described here as an example. This organism, which is also a sulphur oxidizer, may be enriched on the medium of Taylor *et al.* (1971). This comprises:

Solution 1

KNO_3	2.0 g
NH_4Cl	1.0 g
KH_2PO_4	2.0 g
$NaHCO_3$	2.0 g
$MgSO_4 \cdot 7H_2O$	0.8 g
$Na_2S_2O_3 \cdot 5H_2O$	5.0 g
Distilled water to 1 l	
(Anaerobic medium)	

Solution 2 (trace metals)

Na_2-EDTA	50.0 g
$ZnSO_4 \cdot 7H_2O$	2.2 g
$CaCl_2 \cdot 2H_2O$	7.34 g
$MnCl_2 \cdot 4H_2O$	2.5 g
$CoCl_2 \cdot 6H_2O$	0.5 g
$(NH_4)_6Mo_7O_{24} \cdot 4H_2O$	0.5 g
$FeSO_4 \cdot 7H_2O$	5.0 g
$CuSO_4 \cdot 5H_2O$	0.2 g
NaOH	11.0 g
Distilled water to 1 l	

For the preparation of solution 2, EDTA is dissolved, the pH adjusted to 6.0, and then the salts are dissolved one by one (keeping the pH at 6.0). Then the pH is adjusted to 4.0, and the solution stored in small volumes at 4 °C, until required. Enrichment is carried out anaerobically, yet purification is done aerobically on the medium devoid of KNO_3 (Starr *et al.*, 1981).

3.4.8 Sulphate-reducing bacteria (*Desulfovibrio* spp., *Desulfotomaculum* spp. and *Desulfomonas* spp.)

These bacteria occur mainly in the mud of freshwater and saline habitats. Sulphate-reducing bacteria are especially plentiful in saline waters (estuaries, beach ponds, etc.) and also in polluted freshwater lakes and ponds where sulphate and organic material is available. It is also possible, however, to obtain them by enrichment from the overlying water. They are characterized by the reduction of sulphate, involving the transfer of hydrogen from organic compounds, with the final production of hydrogen sulphide. The organisms are strict anaerobes and require a redox potential of 0 to 100 mV in the medium. The media for enrichment and isolation are described in Chapter 11.

3.4.9 Sheathed bacteria (*Sphaerotilus* spp. and *Leptothrix* spp.)

The filamentous morphology of these bacteria results from long chains of single cells held together by a sheath, which is often encrusted by $Fe(OH)_3$ or MnO_2. Both groups inhabit freshwater. *Sphaerotilus natans* occurs in slowly running, heavily polluted rivers and estuaries or in activated sludge, whereas *Leptothrix* spp. prefers unpolluted iron and manganese-containing water. These bacteria may easily be recognized by their flocculant aggregates either adhering to submerged stones and twigs in ponds and ditches, or drifting freely in the water.

The enrichment procedures differ a little between the two genera (Mulder and Deinema, 1981).

Enrichment of *Sphaerotilus natans* is performed conveniently on a simple solid medium:

Meat extract	0.5 g
Agar	15.0 g
Tap water to 1 l	

The low nutrient content of the medium is important to reduce the development of accompanying heterotrophic bacteria. Tufts (which should be washed and blended before processing) are distributed on agar plates (with well-dried surfaces). Small samples of activated sludge are directly spread onto the agar. Within 2 to 6 days of incubation at 20–25 °C, colonies of *Sphaerotilus natans* may be identified by their radial growth which somewhat resembles that of moulds.

In case *Sphaerotilus* spp. occurs only sparsely in activated sludge or non-polluted waters, the organism may be enriched in a first step by incubation of straw, cut into small pieces and inoculated with water or mud of the location in question (Mulder and Deinema, 1981). Further enrichment is performed on agar plates, as described above.

The enrichment of *Leptothrix* spp. may be a little easier, insofar as the samples are drawn from non-polluted waters etc.; competition by other bacteria is, therefore, less important in most cases. An enrichment medium for *Leptothrix* spp. (Mulder and Deinema, 1981) is as follows:

| $Fe(NH_4)_2(SO_4)_2$ | 0.15 g |
| Beef extract (Difco) | 1.0 g |

Yeast extract (Difco)	$0.075\,g$
Sodium citrate	$0.15\,g$
Vitamin B_{12}	$5 \times 10^{-6}\,g$
Agar	$7.5\,g$
Distilled water to 1 l	

After sterilization (121 °C/15 min) and cooling, plates are poured and allowed to set. A hole is then cut into the centre of the medium using a sterile cork borer. Washed tufts of *Leptothrix* spp. are distributed on the surface of the agar, and finally a few drops of a concentrated solution of $MnCO_3$ are pipetted into the hole. By this procedure, turbidity in the plates is avoided, and the resultant colonies of *Leptothrix* spp. may be recognized more reliably.

3.4.10 Iron bacteria (*Gallionella* spp. and *Siderocapsa* spp.)

Apart from filamentous (sheathed) bacteria which form ferrous or manganese incrustations on or in cell sheaths, there exists another group of bacteria apparently metabolizing iron or manganese compounds. *Gallionella* and *Siderocapsa* spp. are thus encountered in waters containing iron or manganese. *Gallionella* spp. prefers clean habitats with minimal organic contents whereas *Siderocapsa* spp. occurs in locations with higher quantities of organic nutrients, and often under micro-aerophilic conditions. Both genera are not restricted to freshwater (ponds, ditches, especially drainages of water and iron works) but have also been isolated from seawater and thermal springs. The presence of these bacteria is easily recognized in nature by the brown coloration of the water.

For the enrichment of *Gallionella* spp., and especially *Siderocapsa* spp., a series of media have been developed (see Hanert, 1981).

3.4.11 Marine luminescent bacteria (*Photobacterium* spp. and others)

In marine habitats, luminescent bacteria may be regularly recovered as free-living, symbiotic or parasitic in shrimps and insects. Dominant taxa include *Photobacterium phosphoreum* and *P. leiognathi* and several *Vibrio* spp. From freshwater, only a very few species have been reported.

The following medium is appropriate for the isolation of *Photobacterium* spp. (Dunlop, personal communication):

Glyceroi	3 ml
Peptone	5.0 g
Yeast extract	3.0 g
Agar	15.0 g
Aged seawater	750 ml

Distilled water   250 ml
pH 7.8 (adjusted before
    sterilization; 121 °C/15 min)

Plates are poured, and the agar surfaces carefully dried, as luminescence of some species is difficult to observe when the colonies spread too much, or colonies of non-luminescent bacteria come in contact with them. The volume of sample used depends on the source. For example with coastal water, 0.1–0.3 ml may be sufficient whereas larger volumes are required with water from the open sea. In fact, up to 300 ml may need to be filtered through nitrocellulose filters (*ca.* 0.3 μm pore size), which subsequently are placed onto the agar surface.

It is important for successful enrichment and isolation that the agar surface is not too crowded with colonies, as luminescence may be inhibited. In contrast, a sparse population density is preferred. To reduce the growth of undesired bacteria, it is recommended to incubate at approximately 10 °C; temperatures of >15–20 °C should be avoided as there are indications that the luminescent bacteria may be inhibited.

Another simple way to enrich for luminescent bacteria is to place a piece of fish into a glass dish filled with a shallow layer of seawater. After incubation at temperatures between 5 and 10 °C for 2–10 days luminescent colonies can be seen on those parts of the fish that are exposed to the air.

Detailed information about the isolation procedure is given by Baumann and Baumann (1981). However, it should be noted that luminescent colonies will only be recognized in the dark.

3.4.12 Photosynthetic sulphur-bacteria *(Chromatium* spp. and *Chlorobium* spp.)

This group is able to perform photosynthesis under anaerobic conditions in the presence of hydrogen sulphide. Representatives of these taxa occur in freshwater, in sulphur springs, and are especially abundant in brackish or seawater ponds, where massive development may occur. Isolation methods are dealt with in Chapter 9.

3.4.13 Nitrifying bacteria (Nitrobacteraceae)

This group comprises strictly aerobic chemolithotrophic bacteria, which occur in soil as well as in freshwater and marine habitats. They have also been isolated from sewage. The nitrifiers are able to obtain energy by the oxidation of reduced nitrogen compounds, the process of which is divided into two steps performed by two distinct groups of organisms:

1. The oxidation of ammonia to nitrite is carried out by *Nitrosomonas* spp. and *Nitrosococcus* spp.

2. The oxidation of nitrite to nitrate is done by *Nitrobacter* spp. and *Nitrococcus* spp.

Enrichment is often complicated by the fact that nitrifying bacteria excrete organic compounds which stimulate the growth of undesired heterotrophic bacteria. These will quickly overgrow any nitrifiers. Enrichment and purification may therefore be time-consuming.

Enrichment for ammonia-oxidizing bacteria (Nitrosomonas *spp.*). The following simple medium is well suited for most purposes (Drews, 1974). It should be mentioned that more complicated media have been described by Watson *et al.* (1981)

$(NH_4)_2SO_4$	1.0 g
K_2HPO_4	0.5 g
NaCl	2.0 g
$MgSO_4 \cdot 7H_2O$	0.2 g
$FeSO_4 \cdot 7H_2O$	0.05 g
$CaCO_3$	6.0 g
Phenol red (0.5%)	0.01 g
Distilled water to 1 l	

(For marine species a seawater (S = 35‰)–distilled water mixture = 400:600 or 600:400 is used. To prepare a solid medium, agar (15.0 g) is added. The pH is adjusted to 7.6.

Volumes (60 ml) of the solution are added to Erlenmeyer flasks (300 ml capacity) and sterilized by autoclaving (121 °C/15 min). These flasks are subsequently inoculated with 5 ml volumes of water or a spatula tip full of mud, and incubated at 28 °C in the dark for several weeks to a month.

The formation of nitrite is indicated by the appearance of acid which turns the colour of the indicator from orange to yellow. Spot tests for nitrate formation should be carried out weekly (for reagents and procedure see Watson *et al.*, 1981; Gerhardt *et al.*, 1981). Nitrite will not be found if both ammonia and nitrite oxidizers are also present; in this case the presence of nitrate should be examined (see below).

Watson *et al.* (1981) recommended using a dilution series in a liquid medium for enrichment and isolation as well as to separate different species of nitrifying bacteria. Generally, isolation can be performed on solidified medium as described above (with a trace of cresol red instead of phenol red added). After the appropriate incubation time, colonies of acid-producing bacteria become fairly red. From these colonies (preferably from the smallest) subcultures are prepared as a dilution series.

To ascertain if the cultures are free from heterotrophic micro-organisms, several liquid media which contain different organic substrates should be employed. Dense cultures of ammonia oxidizers, assumed to be pure, are then examined visually under a microscope (high power objective; phase contrast). If the preparations show only one morphological type of organism then it is highly likely that the outcome has been successful.

Enrichment for nitrite-oxidizing bacteria (Nitrobacter *spp.*). The following medium has proved reliable and convenient. However, it should be emphasized that some species are suppressed by the relative high concentration of nitrite in this recipe (whereas *N. winogradskyi* is enhanced! Watson *et al.*, 1981).

$NaNO_2$	2.0 g
NaCl	0.5 g
$MgSO_4 \cdot 7H_2O$	0.05 g
KH_2PO_4	0.150 g
$CaCO_3$	7 mg
$(NH_4)_6Mo_7O_{24} \cdot 4H_2O$	0.05 mg
$FeSO_4 \cdot 7H_2O$	0.15 mg
Distilled water to 1 l	
pH 8.9 before sterilizing by	
autoclaving (121 °C/15 min).	

For marine species, the distilled water is replaced by a mixture of seawater (S = 35‰): distilled water = 700:300 or 300:700, depending on the salinity at the sampling site.

The liquid medium is inoculated with water or mud dilutions, and incubated at 25–30 °C in the dark. Regularly, the decrease in nitrite content within the cultivation vessels is tested. From the highest dilutions showing nitrite consumption, fresh media are inoculated (another dilution series).

Pure cultures are prepared by continued dilution procedures (which may take months). Isolates, apparently free of impurities, are tested several times as follows:

1. for the ability to use ammonia or nitrite;
2. uniformity in respect of cell and colony morphology;
3. for the absence of heterotrophic bacteria, as determined by streaking and re-streaking on nutrient rich media.

3.4.14 Sulphur oxidizing bacteria (*Thiobacillus* spp. and *Thiospira* spp.)

The genera *Thiobacillus* and *Thiospira* comprise several species which require specialized, yet different, media for isolation. A common feature of all members of this group is their ability to oxidize inorganic sulphur compounds in order to gain energy, and nearly all can grow autotrophically.

The habitat, from which *Thiobacillus* can be isolated, is mud containing hydrogen sulphide (and iron); although sulphur springs and mine effluents are also possible sources.

The main problem with the isolation of autotrophic bacteria is that heterotrophic contaminants may remain in contact with the chemoautotrophs (Taylor *et al.*, 1971). Another problem is that several species of thiobacilli may often occur in the same habitat, and it is notoriously difficult to separate them if growth requirements are similar. For enrichment anaerobically the following recipe is recommended (Taylor *et al.*, 1971; Kuenen and Tuovinen, 1981):

KNO_3	2.0 g
NH_4Cl	1.0 g
KH_2PO_4	2.0 g
$NaHCO_3$	2.0 g
$MgSO_4 \cdot 7H_2O$	0.8 g
$Na_2S_2O_3 \cdot 5H_2O$	5.0 g
Trace metals (see below)	1 ml
Distilled water to 1 l	

(For brackish water bacteria use a seawater (S = 35‰): distilled water mixture = 300:700.) pH 6.8–7.0 (after sterilization). For the preparation of an aerobic medium, KNO_3 is deleted.

The trace metal solution contains:

Na_2-EDTA	50.0 g
$ZnSO_4 \cdot 7H_2O$	2.2 g
$CaCl_2 \cdot 2H_2O$	7.34 g
$MnCl_2 \cdot 4H_2O$	2.5 g
$CoCl_2 \cdot 6H_2O$	0.5 g
$(NH_4)_6Mo_7O_{24} \cdot 4H_2O$	0.5 g
$FeSO_4 \cdot 7H_2O$	5.0 g
$CuSO_4 \cdot 5H_2O$	0.2 g
NaOH	11.0 g
(adjust to pH 6)	

For storage, adjust to pH 4; for use, the pH is readjusted to 6.

As with nitrifying bacteria, enrichment is best performed with a repeated dilution series. Purification is usually carried out by spreading onto a solid medium (without KNO_3) (add 15.0 g agar to the formula given above). The purity test is performed in the liquid anaerobic medium, where the development of N_2 must be proved.

3.4.15 Filamentous sulphur bacteria (Beggiatoaceae)

These bacteria (*Beggiatoa* spp. and *Thioploca* spp.) are common in freshwater and brackish habitats, where a certain amount of hydrogen sulphide is available, and the conditions are not strictly anoxic, e.g. shallow waters over mud where there is abundant organic material such as leaves. The organisms have also been found in sulphur springs. Typically, filaments are capable of gliding movement.

Enrichment: Hay is cut into small pieces, extracted three times in boiling water and suspended in tap water, in Erlenmeyer flasks. The medium is inoculated with rotted plant material or mud and incubated at 25 °C for 10 days. Growth is recognized by the presence of a white layer of filaments on the glass wall and by a distinct smell of hydrogen sulphide. Small portions of the filaments are washed in sterile water and subsequently distributed on a beef extract agar (0.2 g beef extract; 1.0 g agar; 100 ml tap water), and incubated at 28 °C. Filaments, which are clearly growing out from the source material, are cut out with a sterile scalpel and placed onto fresh media. This procedure is continued until the filaments are free of other bacteria, and colonies of undesired micro-organisms are not observed on the agar (Wiessner, 1981).

3.5 ISOLATION (PURE CULTURES)

Enrichment methods may not generate pure cultures. The fact that micro-organisms form, in many cases, slimy capsules, which may stick to other (different) organisms, will necessitate purification stages. The following methods of obtaining pure cultures are practicable for all important groups of bacteria:

1. Streaking (spreading) onto solid media.
2. Incorporation into molten solid media (pour plates).
3. Dilution in a liquid medium.
4. Agar shake-cultivation (or 'roll-tubes').
5. Isolation of single cells.

Streaking (spreading) on solid media. In common with other areas of microbiology, well-isolated colonies may be obtained by streaking material onto the surface of solid media, with an inoculation-loop (Fig. 3.1). A typical example of this method may be seen in Fig. 3.2. It is also possible to

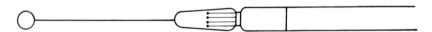

Fig. 3.1. Inoculation-loop: the wire should have a length of 12 cm, and the loop, which must be formed in a completely closed and regular circle, should have a diameter of 2–4 mm.

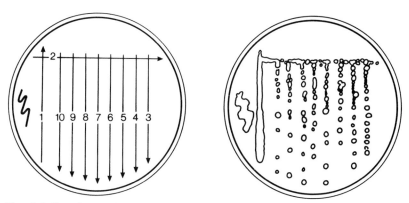

Fig. 3.2 Streaking of a small portion of inoculum on a solid agar surface with an inoculation loop. The bacteria are stripped off near the edge and then a straight stroke is made with the loop without damaging the agar surface. After flaming the loop, a second stroke is made, crossing the first one in a perpendicular direction. After repeated flaming, further strokes are made, parallel to the second one. Well isolated colonies will result as shown by the figure on the right.

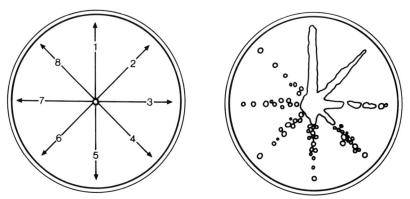

Fig. 3.3 Spreading of an inoculum by a glass spatula (see Fig. 3.4): a small portion of a suspension of bacteria is placed in the centre of the agar surface with the wire-loop, which is then well distributed by radiating strokes with the sterile spatula.

spread the material (after inoculation into the centre of the agar medium) by means of a bent glass rod (Fig. 3.3), also called a Drygalski spatula (Fig. 3.4). However, care should be taken not to tear the agar surface. The inoculated plates are incubated at room temperature, or, for faster development, in an incubator at *ca.* 25 °C. Lower incubation temperatures (i.e. 10 to 15 °C or less) are necessary when psychrophilic bacteria (from cool springs or marine habitats) are under investigation.

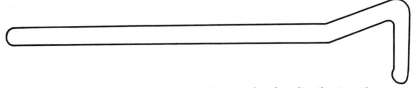

Fig. 3.4 Bent glass-rod (Drygalski spatula) for distributing the inoculum in a more spreaded way than with the loop.

After 2 or more days, the inoculated plates are scrutinized with a lens or low-power objective of a microscope. Well-isolated colonies are picked with a flamed needle and subcultured. Generally, a pure culture will be obtained after a few subculturing stages. If the colonies are very small, it is recommended that the material be suspended in sterilized water before restreaking (Drews, 1974). Inoculated plates should be observed for several weeks (drying out of the agar will set the limit), because some species develop rather slowly. However, in these cases, attention must be paid to contaminants which may develop especially at the edges of the agar.

A *purity test* is indispensable. An isolate will be regarded as 'pure' if: (1) the same type of colony develops following subculturing and (2) the cells appear to be morphologically uniform (Gram stained smears; high power objective with oil immersion and phase contrast lenses). One should take note that, for example, some bacteria form filaments as well as individual cells, and secondly in cultures (such as *Vibrio*) not all individual cells will appear to be motile.

Pour plates. A small amount (*ca.* 0.1 ml) of the sample (water or sediment) is poured into a sterile dish. Then *ca.* 20 ml of the molten agar medium (kept at *ca.* 48 °C in a water bath) is added, with thorough mixing.

Dilution in liquid media. To obtain pure cultures of bacteria which grow poorly on solid media, it is recommended to prepare a dilution series in a liquid medium (in screw-cap tubes) by stepwise dilution, i.e. Most Probable Numbers method. The appropriate concentration of bacteria is reached when turbidity is seen in only a few tubes of the last dilution step. In all probability, such turbidity has developed from a single cell (or at least from very few cells). Unfortunately, there are several disadvantages which must be taken into account, i.e. the method is more laborious than the two mentioned previously; furthermore, the separation of single cells might be more difficult than on solid media, and it is hardly to be expected that slow-growing bacteria will be enriched by this method.

Agar-shake culture ('roll-tubes', etc.). In some cases (e.g. sulphate reducing bacteria, some purple bacteria and other anaerobes), it is necessary to reduce or inhibit access of oxygen to the cells. Facultatively anaerobic bacteria often develop in the pour-plate method (perhaps even with the addition of a reducing substance, such as Na-thioglycolate). With strict anaerobes, the agar-shake or the roll-tube method is quite reliable. The agar-shake technique is as follows:

A series of sterile screw-cap tubes are filled with an appropriate medium to as far as 2 cm below the rim. These tubes are immediately placed in a water bath at 42 °C in order to keep the agar molten. These tubes are then inoculated as for a dilution series, i.e. 1 ml of the sample is placed in the first tube, which is thoroughly shaken. One millilitre of this dilution is then pipetted into a second tube, and so forth. The inoculated and shaken tubes should be cooled at once with tap water.

After several days of incubation of the tightly closed tubes the first colonies will be observed. With a sterile Pasteur pipette, a well-isolated colony is picked and some material transferred to fresh medium and the shake procedure repeated. However, it is more convenient to utilize sterilized glass tubes (*ca.* 2 cm wide) open at both ends, which are closed with rubber bungs. The procedure is carried out as described above. When colonies have developed, the agar column can easily be pushed out of the tube and placed in a sterile dish. With a sterile scalpel , slices are cut in order to expose appropriate colonies.

The cultivation of strict anaerobes is also possible on agar plates, if they are incubated anaerobically.

Isolation of single cells. The most reliable way to obtain a truly pure culture of a bacterial isolate is of course to start from a single cell. It is, however, the small size of most bacteria which makes this task difficult. Most procedures require both patience and skill. A review about the methods available is given by Johnstone (1969). A simple method is described here:

A suspension (preferably an enrichment culture) is diluted such that small droplets (2 mm diam.) will contain only one cell. With a sterile capillary, several droplets are distributed on sterile coverslips. These coverslips are then placed upside down over the depressions of cavity slides (the margin is sealed with vaseline). Under the microscope (phase contrast) those droplets which contain only one cell are searched for. With a sterile capillary, the drop is transferred to liquid medium and incubated (some authors recommend that the droplets should be sucked up with small pieces of sterile filter paper, which are then put as a whole into medium). The disadvantage of this procedure lies in the uncertainty about whether or not one cell is actually transferred. It is possible that additional bacteria, such as those hidden in the dark edge of the droplet (formed by light refraction), may be overlooked.

Johnstone (1969) described a reliable procedure based on relatively simple equipment. The principle of the method is to pick up single cells distributed on the surface of an agar gel with a sterile needle moved mechanically under microscopical control. The disadvantage of this procedure may be that no highpower (oil immersion) objectives can be used.

3.6 PRESERVATION OF CULTURES

An important, but often neglected part of microbiology, is the preservation of stock cultures. Two methods are recommended: (a) subculturing and (b) drying. With these methods, preservation is possible usually only for a few weeks or months, except with spore formers.

Subculturing. This method is widely used because of its simpicity. Effectively, there is a periodic transfer of the organism to a fresh solid medium, which should contain only a minimal content of nutrients. The medium may be prepared as agar slants, which are inoculated on the surface, or as 'deep cultures' by stabbing the inoculum deep into the agar (this is especially suitable for the preservation of facultative anaerobes or micro-aerophilic bacteria). The tubes are then stored in a refrigerator at about 5–8 °C. However, this method has serious disadvantage insofar as contamination and genetic variation may ensue. Furthermore, there is the perpetual risk of loosing cultures, and even of mislabelling at each transfer. In addition, much storage room is required.

Drying. This method is especially useful with spore-forming micro-organisms. Sterilized soil or paper is used in most cases. In addition, Năvecke and Tepper (1979) described a method using skimmed-milk powder.
   Quite simply, soil or sand is sterilized in small portions (in screw-cap tubes) by autoclaving for several hours on 2 successive days. One millilitre of a suspension of the organisms (or spores) is inoculated and the content of the tubes dried in a vacuum-desiccator (containing silica gel) with the caps loosely closed. Subsequently, the caps are closed tightly and the tubes stored in another evacuated desiccator, which may be kept in a refrigerator at 5–8 °C.
   It is convenient to employ sterile paper discs or strips, soaked with a suspension of the bacterium. The discs are dried and stored, as described above, in screw-cap tubes. For revival, it is convenient to put several discs (inoculated with the same organism) into one tube. When necessary, one disc is removed with sterile forceps and the rest stored again. Precautions must be observed not to contaminate the stock cultures.

The method generally applied to preserve stock cultures for long periods of time (for many years) is freeze-drying (lyophilization). The principle of this reliable and convenient procedure is to remove the water from the frozen bacterial suspension by sublimation under vacuum. Freeze-drying has been proven to be very reliable, with many different bacteria. It is also economic as only small vials are required. Although there is a need for rather expensive equipment, reliable results are obtained. For further information see Gherna (1981) and Nävecke and Tepper (1979).

Finally, it must be stressed that every care applied to any preservation procedure is useless if some basic principles are overlooked. Thus, it is indispensable to label the vials with indelible ink with a successive series of numbers. The organisms must be recorded carefully, with reference to their number, together with all relevant data concerning their taxonomy, origin, date of isolation, special properties, the name of the person who isolated the organism and the date of preparation of the stock culture. It may seem ironic that many failures with stock cultures can be attributed to careless keeping of the data files.

3.7 REFERENCES

Attwood, M. M. and Harder, W. (1972). A rapid and specific enrichment procedure for *Hyphomicrobium* spp. *Antonie van Leeuwenhoek Journal of Microbiology and Serology*, **38**, 369–378.

Baumann, P. and Baumann, L. (1981). The marine Gram-negative Eubacteria: genera *Photobacterium*, *Beneckea*, *Alteromonas*, *Pseudomonas*, and *Alcaligenes*. In *The Procaryotes*, vol. 2, M. P. Starr *et al.* (Eds), Springer–Verlag, Berlin, Heidelberg, New York, pp. 1302–1331.

Becking, J.-H. (1981). The family Azotobacteriaceae. In *The Procaryotes*, vol. 1, M. P. Starr *et al.* (Eds), Springer–Verlag, Berlin, Heidelberg, New York, pp. 795–817.

Brockmann, E. R. (1967). Fruiting myxobacteria from the South Carolina coast. *Journal of Bacteriology*, **94**, 1253–1254.

Drews, G. (1974). *Mikrobiologisches Praktikum*, Springer–Verlag, Berlin, Heidelberg, New York, pp. 230.

Gerhardt, P., Murray, R. G. E., Costilow, R. N., Nester, E. W., Wood, W. A., Krieg, N. R. and Phillips, G. B. (1981). *Manual of Methods for General Bacteriology*, American Society for Microbiology, Washington DC, pp. 524.

Gherna, R. L. (1981). Preservation. In *Manual of Methods for General Bacteriology*, P. Gerhardt *et al.* (Eds), American Society for Microbiology, Washington DC, pp. 208–217.

Hanert, H. H. (1981). The genus *Gallionella*. In *The Procaryotes*, vol. 1, M. P. Starr *et al.* (Eds), Springer–Verlag, Berlin, Heidelberg, New York, pp. 509–515.

Johnstone, K. J. (1969). The isolation and cultivation of single organisms. In *Methods in Microbiology*, vol. 1, Norris, J. R. and D. W. Ribbons (Eds), Academic Press, London, New York, pp. 455–471.

Kuenen, G. and Tuovinen, O. H. (1981). The genera *Thiobacillus* and *Thiomicrospira*.

In *The Procaryotes*, vol. 1, M. P. Starr *et al.* (Eds), Springer–Verlag, Berlin, Heidelberg, New York, pp. 1023–1036.

Moore, R. L. and Hirsch, P. (1972). Deoxyribonucleic acid base sequence homologies of some budding and prosthecate bacteria. *Journal of Bacteriology*, **110**, 256–261.

Mulder, E. G. and Deinema, M. H. (1981). The sheathed bacteria. In *The Procaryotes*, vol. 1, M. P. Starr *et al.* (Eds), Springer–Verlag, Berlin, Heidelberg, New York, pp. 425–440.

Nävecke, R. and Tepper, K.-P. (1979). *Einführung in die mikrobiologischen Arbeitsmethoden*, Gustav Fischer, Stuttgart, New York, pp. 43–51.

Norris, J. R. and Ribbons, D. W. (1969). *Methods in Microbiology*, vols. 1–13, Academic Press, London, New York.

Phillips, G. B. (1981). Laboratory safety. In *Manual of Methods for General Bacteriology*, P. Gerhardt *et al.* (Eds), American Society for Microbiology, Washington DC, pp. 475–503.

Reichenbach, H. and Dworkin, M. (1981). The order Cytophagales (with addenda on the genera *Herpetosiphon, Saprospira* and *Flexithrix*). In *The Procaryotes*, vol. 1, M. P. Starr *et al.* (Eds), Springer–Verlag, Berlin, Heidelberg, New York, pp. 356–379.

Reichardt, W. (1978). *Einführung in die Methoden der Gewässermikrobiologie*, Gustav Fischer, Stuttgart, New York, pp. 250.

Smibert, R. M. and Krieg, N. R. (1981). General characterization. In *Manual Methods for General Bacteriology*, P. Gerhardt *et al.* (Eds), American Society for Microbiology, Washington DC, pp. 418.

Starr, M. P., Stolp, H., Trüper, H. G., Balows, A. and Schlegel, H. G. (1981). *The Procaryotes: A Handbook on Habitats, Isolation and Identification of Bacteria*, vols 1 and 2, Springer–Verlag, Berlin, Heidelberg, New York.

Taylor, B., Hoare, D. S. and Hoare, S. L. (1971). *Thiobacillus denitrificans* as an obligate chemolithotroph. Isolation and growth studies. *Archiv für Mikrobiologie*, **78**, 193–204.

Watson, S. W., Valois, F. W. and Waterbury, J. B. (1981). The family Nitrobacteraceae. In *The Procaryotes*, vol. 1, M. P. Starr *et al.* (Eds), Springer–Verlag, Berlin, Heidelberg, New York, pp. 380–389.

Wiessner, W. (1981). The Family Beggiatoaceae. In *The Procaryotes*, vol. 1, M. P. Starr *et al.* (Eds), Springer–Verlag, Berlin, Heidelberg, New York, pp. 380–389.

Methods in Aquatic Bacteriology
Edited by B. Austin
© 1988 John Wiley & Sons Ltd.

4

Identification

B. Austin

*Department of Brewing and Biological Sciences, Heriot-Watt University,
Edinburgh, EH1 1HX Scotland*

4.1 INTRODUCTION

Identification is the end result of the taxonomic process. Essentially, organisms can only be identified if classification has been previously carried out, in a sound and thorough manner. As a general rule, if organisms represent undescribed taxa (taxonomic groups), then fresh isolates can not be identified. Unfortunately, most aquatic microbiologists are not interested in taxonomy and, for that matter, taxonomists are seldom interested in the aquatic environment. This is a pity because progress can only ensue if community structures are known. Sometimes, novel organisms are encountered that are distinct from, but resemble, recognized taxa. Then, a favoured ploy is to label such organisms as 'atypical' or 'presumptive.'

Identification is a comparative process by which unknowns are examined and compared to established taxa. Conventionally, the first step is to obtain pure cultures, followed by an onslaught of phenotypic tests (e.g. morphological, biochemical and physiological tests), and thence a comparison with existing diagnostic schemes. Cowan (1974) described some viewpoints for identifying medical bacteria. His discussion is relevant to all aspects of identification and, consequently, it is appropriate to consider the points here. Essentially, one strategy is that an organism may be examined for a wide battery of tests. (This is a blunderbuss approach.) Thence, when the data are available, use is made of standard texts such as *Bergey's Manual of Systematic Bacteriology* (Krieg, 1984; Sneath, 1986). Alas, the value of such texts for the identification of aquatic bacteria is sometimes suspect. In any case, the logic of the approach is that it may be envisaged that the battery of test results will permit ready and 'rapid' identification of the isolate. Unfortunately, not only is the approach time-consuming and costly in resources, but some so-called important tests are usually forgotten. Another view involves a stepwise approach in which diagnostic schemes are followed progressively. Here, the responses to an initial series of tests allow the scientist to proceed with the next series, and continue along the avenue until an answer is obtained. An obvious drawback is that this approach is time-consuming. However, there will be (hopefully) many occasions when a scientist will suspect the identity of an organism. Therefore, specialist schemes, suitable for the group(s) in question, may be consulted. Such occasions will include the recovery of cultures on selective media, such as thiosulphate citrate bile salt sucrose agar in the case of marine vibrios. A fundamental problem, however, is a general lack of good identification schemes for the wide range of bacteria found in the aquatic environment. Nevertheless, in the future, it is envisaged that specific taxa may be recognized/identified among mixed populations by means of sensitive serological or molecular genetic techniques.

Oblique reference has been made to the use of phenotypic tests, which are involved in identification. Unfortunately, many of these are classical tests designed for medical bacteriology, such as the Voges Proskauer reaction and methyl red test. However, modern tests such as the determination of the presence or absence of specific subcellular components, e.g. lipids, are slowly gaining more widespread use. With the classical tests, the reproducibility of results may be highly suspect. In sugar fermentation reactions, the presence of plasmid DNA may give rise to a positive response in organisms not normally associated with such ability. Here, it should be emphasized that many marine bacteria are packed with plasmids, the expression of which may interfere with the identification process. Moreover, freshly recovered environmental isolates often behave very differently to cultures (of the same taxon) maintained under laboratory conditions for prolonged periods of time. As an oversimplification, stored cultures may lose enzymic ability, which may hinder identification. Within such constraints, it is timely to consider the alternatives available for the identification of aquatic bacteria.

4.2 DICHOTOMOUS KEYS

Historically, dichotomous keys (diagnostic keys) were the first schemes available for bacterial identification. Such keys may be found in the older literature, including the earlier editions of *Bergey's Manual of Determinative Bacteriology*. With dichotomous keys, identification occurs in a stepwise fashion. Here, the scientist proceeds along progressively branching routes depending upon positive or negative responses to tests. For example, if a test generates a positive reaction, then a given route is followed, but if a

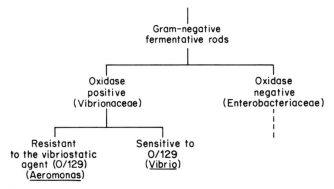

Fig. 4.1 An example of a dichotomous key for the delineation of *Aeromonas* and *Vibrio* (based on Austin and Priest, 1986).

negative result is obtained then an alternative pathway is used (Fig. 4.1). The principal drawback is that an incorrect result, or an isolate with an aberrant feature, will send the scientist along the wrong branch of the key. Thus, misidentification will occur. Another problem is that dichotomous keys have been formulated with a minimum of data. Consequently, dissimilar organisms could be identified into the same taxon, solely on the basis of responses to a few so-called 'key' tests. For these reasons, dichotomous keys have lost favour to diagnostic tables. In conclusion, dichotomous keys are not recommended for the identification of aquatic bacteria.

4.3 DIAGNOSTIC TABLES

An end result of comprehensive taxonomic studies, such as typified by numerical taxonomy procedures, should be the construction of diagnostic tables, which permit the ready identification of fresh isolates. Such tables should lack the disadvantage of dichotomous keys, because they are polythetic (refers to groups with high numbers of common characters; the membership of such groups does not require the presence or absence of particular characters) and the presence of a few aberrant results is less likely to adversely effect the outcome of identification. However, there has been a general lack of comprehensive taxonomic studies of aquatic bacteria. Consequently, diagnostic tables have been described for a restricted number of groups, such as some aerobic heterotrophs from the coastal marine environment (Table 4.1) and from freshwater (Table 4.2). With diagnostic tables, it is recognized that not all members of a taxon will

TABLE 4.1 A diagnostic table for the identification of aerobic heterotrophic bacteria from a coastal marine environment (modified from Austin, 1982)

Taxon	Gram stain	Motility	Fermentative (F) Oxidative (O)	Catalase production	Oxidase production	Sensitivity to O/129	Arginine dihydrolase	Methyl red test	Indole production	Phenylalanine deaminase	Casein degradation	Chitin degradation	DNA degradation	Tween 20 degradation	Growth at 37°C	Growth in 0% (w/v) NaCl	Growth in 4% (w/v) NaCl	Utilization of maltose	Utilization of mannose
Vibrio sp.	−	+	F	+	+	−	+	+	D	−	+	+	+	+	−	−	−	−	−
Flavobacterium sp.	−	−	O	−	+	+	+	D	−	−	−	−	+	+	−	−	−	+	D
Cytophaga-Flexibacter	−	+	O	+	+	D	−	−	−	−	−	+	+	−	−	−	+	−	−
Caulobacter sp.	−	+	O	+	+	−	+	+	D	−	D	+	+	+	−	−	−	+	D
Hyphomicrobium vulgare	−	+	O	+	+	+	−	−	−	−	−	−	−	+	−	−	+	+	−
Hyphomonas polymorpha	−	+	O	+	+	D	−	+	−	−	−	−	+	+	−	−	−	+	D
Hyphomicrobium neptunium	−	+	O	+	+	D	+	+	−	−	+	−	+	+	−	−	+	−	−
Prosthecomicrobium sp.	−	+	O	D	+	−	−	−	D	−	−	D	D	−	−	−	−	−	−
Asticcacaulis sp.	−	−	O	+	D	−	−	−	−	−	−	−	−	−	−	−	+	−	−
Vibrio nereis	−	D	F	+	+	−	−	−	−	+	−	−	+	+	−	−	+	+	−
V. fischeri	−	+	F	+	+	−	−	−	−	−	−	+	−	+	−	−	+	+	+
V. costicola	−	+	F	D	+	−	−	−	−	−	+	−	+	+	D	D	+	+	+
Photobacterium phosphoreum	−	D	F	+	D	−	−	−	−	D	+	−	+	+	−	−	+	−	−
V. campbellii	−	D	F	+	−	−	−	−	−	+	−	−	+	+	−	−	−	−	−
V. harveyi	−	+	F	+	−	−	−	−	−	−	−	+	−	−	−	−	−	−	−
Photobacterium angustum	−	+	F	+	+	−	+	−	−	−	−	+	−	−	−	−	+	−	−
Photobacterium logei	−	−	F	−	+	−	−	−	−	−	+	+	+	−	+	+	+	+	+
Alteromonas haloplanktis I	−	−	O	+	+	−	−	−	+	−	+	−	+	+	−	−	+	+	−
A. haloplanktis II	−	D	O	+	+	+	+	+	+	−	+	D	+	+	D	−	+	+	D
Alcaligenes faecalis	−	+	O	+	+	−	+	−	−	−	D	−	−	+	+	−	+	−	−
Janthinobacterium lividum	−	+	O	+	+	−	−	−	−	−	−	−	+	+	−	−	+	−	−
Erwinia herbicola	−	+	F	+	−	−	−	−	−	+	−	−	+	−	−	−	−	−	+
Escherichia coli	−	D	F	+	−	+	+	+	−	−	−	−	+	+	+	−	+	−	−

TABLE 4.1 (*contd.*)

Taxon	Gram stain	Motility	Fermentative (F) Oxidative (O)	Catalase production	Oxidase production	Sensitivity to O/129	Arginine dihydrolase	Methyl red test	Indole production	Phenylalanine deaminase	Casein degradation	Chitin degradation	DNA degradation	Tween 20 degradation	Growth at 37°C	Growth in 0% (w/v) NaCl	Growth in 4% (w/v) NaCl	Utilization of maltose	Utilization of mannose
Serratia liquefaciens	−	+	F	+	−	−	−	−	−	−	+	−	−	+	−	+	−	−	−
S. marinorubra	−	+	F	+	−	−	−	−	−	−	−	−	+	+	+	−	D	D	+
Pseudomonas fluorescens	−	+	O	+	+	−	+	−	−	−	−	D	−	−	D	−	+	+	D
Pseudomonas multivorans	−	+	O	D	+	−	+	+	−	−	−	−	+	+	−	−	+	+	−
Pseudomonas sp.	−	+	O	D	D	−	−	−	−	−	−	−	+	−	D	−	−	−	−
Pseudomonas marina	−	−	O	+	+	−	+	−	−	−	+	−	−	+	−	+	+	−	−
Acinetobacter calcoaceticus	−	−	O	+	−	−	−	−	−	−	−	−	−	+	+	−	−	−	−
coryneform	+	−	F	+	−	−	−	−	−	−	−	−	+	+	−	−	+	+	−
coryneform	+	−	F	D	−	+	+	+	−	−	−	−	+	+	−	−	−	−	−
coryneform	+	+	F	+	+	+	−	+	+	−	−	−	+	+	−	−	−	−	−
Bacillus firmus	+	+	F	+	+	−	−	−	−	−	+	+	+	+	−	−	+	−	−
B. cereus	+	+	F	+	+	D	D	+	D	−	−	−	−	−	−	−	−	−	−

+, − and D = ⩾80%, ⩽20%, and 21–79% positive responses, respectively.

give uniform test results. For example, the presence of positive or negative reactions to tests indicate that most, but not necessarily all, isolates of a given taxon correspond to the results. This author considers that positive and negative responses correspond to ⩾80% and ⩽20% of positive results, respectively. Results, labelled as 'D' or 'V', indicate a variable level of positive responses, i.e. in the range of 21–79%. Diagnostic tables, which contain a matrix of test results for a wide range of taxa, may be found in *Bergey's Manual of Systematic Bacteriology* (Krieg, 1984; Sneath, 1986). In all cases, the unknown isolate is examined for the tests indicated in the diagnostic table. Then the data are examined against the profile for each taxon included in the diagnostic table. Obviously, for large tables, this will be a laborious process. However, there is scope for automation, such as

TABLE 4.2 Diagnostic table for freshwater bacteria (after Allen et al., 1983)

Taxon	Fluorescein production	Gram-positive	Motility	Oxidase production	Arginine dihydrolase	Ornithine decarboxylase	Citrate utilization	Gluconate oxidation	Indole production	Methyl red test	Voges–Proskauer reaction	Phosphatase production	Growth at 4°C	Growth at 37°C	Growth on TCBS agar	Growth on MacConkey agar	Growth on 4% (w/v) NaCl	Growth on 6% (w/v) NaCl	Aesculin hydrolysis	Casein degradation	Gelatin hydrolysis	Lecithin degradation	RNA degradation	Starch degradation	Tween 20 degradation	L(−)-Arabinose utilization	Maltose utilization	Sodium formate utilization
coryneforms	−	D	−	−	−	−	+	D	D	−	−	+	+	−	−	+	+	+	D	−	−	−	+	−	D	−	−	−
coryneforms	−	+	+	−	−	−	−	−	−	−	−	D	D	−	+	+	+	+	D	−	D	−	D	+	−	+	+	+
Micrococcus roseus	−	+	−	−	−	−	−	−	−	−	−	D	−	+	+	+	+	+	−	−	−	−	−	+	−	−	+	D
Micrococcus varians	−	+	−	−	−	−	−	−	−	−	−	−	−	−	−	+	D	−	−	−	−	D	−	−	D	−	−	−
Arthrobacter sp.	−	D	+	−	−	−	−	−	−	−	−	D	+	−	−	−	+	−	+	−	−	−	−	−	+	+	+	−
Listeria sp.	−	D	D	D	D	−	D	−	−	−	−	D	+	+	D	+	+	D	+	−	D	−	D	+	−	+	+	−
Acinetobacter calcoaceticus	−	−	−	+	−	−	−	−	−	−	−	−	D	D	−	+	+	D	−	+	−	D	−	−	+	−	D	−
Acinetobacter sp.	−	−	+	+	−	−	+	−	D	−	−	+	−	−	−	+	+	+	−	+	−	−	−	+	+	−	−	−
Alcaligenes denitrificans	D	−	+	+	−	−	−	−	−	−	−	D	D	+	D	D	−	−	−	−	D	−	−	−	D	−	D	−
Alcaligenes faecalis	−	−	+	+	−	−	−	−	−	−	−	D	D	−	−	D	+	−	+	−	−	−	−	+	D	+	+	−
Alcaligenes sp.	−	−	+	+	−	−	−	−	−	−	−	+	D	D	−	D	+	−	−	−	−	−	D	+	+	−	+	−
Moraxella sp.	−	−	+	+	−	−	−	−	−	−	−	−	−	−	−	−	−	+	−	−	−	−	D	+	−	−	−	−
Alcaligenes piechaudii	−	−	+	+	−	−	−	−	−	+	−	−	D	−	−	D	+	+	−	−	−	−	−	−	D	−	+	−
Agrobacterium sp.	−	−	+	+	−	−	−	−	−	−	−	+	D	−	−	D	+	+	−	−	−	D	−	+	D	D	−	−
Agrobacterium sp.	−	−	+	+	−	−	−	−	−	−	−	+	−	+	−	D	+	+	−	−	−	D	D	+	−	+	+	−
Bordetella bronchiseptica	−	−	D	+	−	−	−	−	−	−	−	−	−	+	−	+	+	−	+	−	+	−	+	−	+	−	−	−

Bacillus firmus	−	+	−	−	−	−	−	−	−	D	+	+	+	−	−	−
Staphylococcus sp.	+	+	D	−	D	+	−	+	+	+	+	D	−	−	+	+
Staphylococcus epidermidis	+	−	−	−	−	D	−	−	−	−	+	−	−	D	−	−
Bacillus pumilus	+	+	−	+	+	+	+	−	+	+	+	+	D	+	−	−
Bacillus cereus	+	D	D	+	−	D	D	D	−	+	+	+	+	D	−	+
Bacillus megaterium	+	+	−	+	−	+	+	+	+	+	+	+	D	+	−	+
Escherichia coli	−	−	−	+	−	D	+	−	D	−	−	−	+	D	+	+
Enterobacter aerogenes	−	+	D	D	D	+	−	+	+	+	D	D	+	−	D	+
Hafnia alvei	−	D	D	+	−	+	−	−	D	D	+	+	D	D	−	+
Serratia sp.	−	−	D	+	−	+	D	+	+	+	+	+	+	−	−	−
Klebsiella sp.	+	+	+	+	+	−	+	+	D	−	D	+	D	+	+	+
Erwinia stewartii	−	−	−	−	−	+	+	−	+	−	−	+	−	+	+	D
Flavobacterium sp.	+	−	−	+	D	−	+	−	−	+	+	+	+	D	D	+
Flavobacterium sp.	+	−	−	D	D	−	−	−	−	−	−	+	+	−	−	−
Pseudomonas fluorescens	−	+	D	+	−	−	−	D	−	+	+	+	−	D	−	−
Aeromonas hydrophila	+	+	+	+	D	D	−	+	+	+	+	+	−	D	D	D
Vibrio fluvialis	D	+	+	+	+	+	+	+	+	+	+	+	−	+	+	D
Flexibacter-Cytophaga	−	−	−	−	−	−	−	−	−	−	−	−	−	−	−	−
Aeromonas salmonicida	−	−	−	−	−	+	−	−	+	+	−	−	−	−	−	−
Aeromonas sp.	−	−	−	−	−	+	+	+	+	+	+	+	−	−	−	−
Aeromonas sp.	−	−	−	−	−	+	+	+	+	+	+	+	−	−	−	−
Aeromonas sp.	+	+	+	−	+	+	+	+	+	+	+	+	−	+	+	+
Cytophaga hutchinsonii	+	D	D	−	−	D	D	D	D	−	−	−	−	+	−	−
Cytophaga salmonicolor	−	−	−	−	−	−	D	+	+	+	D	+	D	D	D	D
Cytophaga fermentans	+	+	−	−	+	+	+	+	+	+	+	+	−	+	+	+
Flexibacter succinicans	D	+	−	−	−	−	+	−	−	−	−	+	D	−	−	−
Flexibacter aggregans	+	D	−	−	−	+	+	+	+	+	+	+	−	D	D	D

+: >80% positive response; −: <20% positive responses; D: 21–79% positive responses offered by computation. On a cautionary note, it is essential, as far as possible, to mimic the precise testing regime, including composition of media and incubation conditions, used by the original scientist in the formulation of the diagnostic table. The reason centres around the marked differences in results recorded for a given isolate in a range of allegedly similar media or incubation conditions. For example, some media are more suitable for the expression of enzymic activity than others.

4.4 COMPUTER-BASED IDENTIFICATION SCHEMES

It is possible to use diagnostic tables in conjunction with computers. For example, the tables are incorporated into the computer memory, and the results for unknown isolates are compared with each of the possibilities. The highest correlation is taken to infer a correct identification. Such an arrangement has been used with the identification table for coastal marine heterotrophs (see Austin, 1982; Table 4.1). However, such systems are not readily available. Instead, they are centred in a few specialized laboratories.

4.5 SEROLOGY

Although serology has its uses, much confusion has resulted from its adoption in bacterial identification because of problems with the specificity of reagents, namely the antisera. Numerous serological procedures have been described, of which whole cell agglutination, latex agglutination, direct and indirect fluorescent antibody tests and the enzyme linked immunosorbent assay (ELISA) have value in aquatic bacteriology. The main advantage is speed, and, in the case of whole cell agglutination, the result may be recorded in a few minutes. However, antisera for only a few of the organisms found in the aquatic environment have been developed. Therefore, the procedure is restricted in its usefulness for routine identification purposes. Unfortunately, in the case of polyclonal antisera, heterologous reactions may occur, giving rise to misidentification. The problem may be resolved by careful evaluation of polyclonal antisera, or by development of highly specific monoclonal antibodies.

Polyclonal antibodies are inevitably produced in rabbits, of which the long-eared New Zealand white rabbit is a popular breed. Bacterial suspensions are prepared to *ca.* $5 \times 10^8 - 10^9$ cells ml^{-1} in distilled water or saline. These suspensions are fixed, preferably overnight, in formalin (to 0.1% v/v), mixed thoroughly with equal volumes of Freunds complete (or incomplete) adjuvant and 1 ml aliquots injected intramuscularly. Booster doses will need to be applied after 2–3 weeks. Two weeks later, a small

(1–2 ml) sample of blood may be removed by venepuncture, allowed to clot at 37 °C/1 h then overnight at 4 °C, and the serum carefully removed by means of a Pasteur pipette (it may be necessary to centrifuge the clotted blood before removing the serum). The titre of the antiserum may be assessed as doubling dilutions against the homologous antigen (*ca.* 2×10^8 bacterial cells ml^{-1}). If the titre is adequate, i.e. ⩾1:4000, greater volumes of blood may be removed by venepuncture, and the resulting serum stored as separate 0.1 ml volumes at -20 to -70 °C until required. However, all too often, a further booster dose of antigen will need to be administered to achieve an acceptable antibody titre. Once prepared, the antiserum should be carefully evaluated against a wide range of bacterial antigens to confirm specificity. If cross-reactions occur, the unwanted antibodies may be adsorbed to the bacteria in question (i.e. add *ca.* 5×10^7 cross-reacting bacteria ml^{-1} of serum, incubate for 1 h at room temperature, and then remove bacteria by centrifugation) and thence removed.

The serological tests may be carried out in the following ways:

Whole cell agglutination

Place one drop of a bacterial suspension (10^8 cells ml^{-1}) in 0.9% (w/v) saline, onto a clean microscope slide and add one drop of the (diluted) antiserum. Following gentle mixing, for 2–3 min, a positive result is indicated by clumping of the cells.

Latex agglutination

This test involves use of globulins from hyperimmune serum (titre = ⩾1:5000) and sensitized latex particles of 0.81 μm in diameter. Effectively, globulins are precipitated by addition of saturated ammonium sulphate to the antiserum, and the precipitated proteins sedimented by centrifugation. These globulins are dissolved in 0.9% (w/v) saline, dialysed overnight at 4 °C against three changes of saline, and, after centrifugation, the supernatant which contains the globulins is stored frozen at -70 °C until required. Latex particles are sensitized in a globulin solution at 37 °C for 2 h. For the test, 200 μl of the bacterial suspension in glycine buffered saline (0.73% (w/v) glycine; 1% (w/v) NaCl; pH 8.2; supplemented with 1% (w/v) Tween 80) is mixed for 2 min with an equal volume of sensitized latex on a clean glass slide. A positive response is indicated by clumping of the latex.

Fluorescent antibody technique

There are two variations to this test, namely the direct and indirect method. For the direct method, fluorescein isothiocyanate is conjugated with whole or with the IgG fraction of an antiserum. Dilutions (1:5 and 1:8) are next prepared in phosphate buffered saline (PBS; 0.1236% (w/v) Na_2HPO_4, anhydrous; 0.018% (w/v) $NaH_2PO_4 \cdot H_2O$; 0.85% (w/v) NaCl; pH 7.6) and used to standardize the conjugate. For the test, a bacterial suspension (containing *ca.* 10^8 cells ml^{-1} in PBS) is pipetted onto grease-free glass microscope slides, air-dried, and fixed at 60 °C for 2 min. The conjugate is now pipetted onto the bacteria and left in a moist chamber at room temperature for 5 min. Then the excess liquid is drained away, and the slide rinsed thoroughly in PBS. Following air-drying, the smear is covered with a drop of buffered glycerol at pH 9.0, overlayered with a coverslip and quickly examined under a fluorescence microscope. Positive and negative controls are necessary.

For the indirect method, bacterial smears are prepared and fixed, as above. Then (rabbit) antiserum (20 μl volumes) is added to the smear, with incubation for 30 min in a moist chamber at room temperature. A 30 min washing step, using two changes of PBS, is employed, prior to air-drying, and then application of a few drops of fluorescein-labelled sheep anti-rabbit globulin with incubation for 30 min in the moist chamber. The slide is rinsed thoroughly in PBS, air-dried, and mounted in buffered glycerol, with subsequent examination under a fluorescence microscope.

Enzyme linked immunosorbent assay

This is the most sensitive serological technique, which has been described to date. There is a need for specific antisera, which are adsorbed onto plastic surfaces (overnight incubation at 4–15 °C in buffer at pH 7.0), to which bacterial suspensions (up to 10^8 cells ml^{-1}) are reacted for 5–15 min at room temperature. Following thorough washing in tapwater or saline, the next stage involves incubation (for 10–15 min) with an antibody enzyme conjugate (alkaline phosphatase or horse radish peroxidase are commonly used enzymes), then more washing, and finally a reaction stage with an enzyme substrate. A positive result is indicated by a colour change.

Although serological procedures provide useful information for identification/diagnosis, the results should be confirmed by examination of phenotypic traits. Nevertheless, serology is especially useful in epidemiological work, for determining the nature of specific types of pathogens (serotypes or serovars).

4.6 COMMERCIAL KITS

Many manufacturers have developed kits, of which a few types have gained widespread use. However, most of the products have been developed for medical use rather than for the identification of environmental isolates. Often, problems arise when such kits are used for the identification of aquatic bacteria. In this context, it is important to realise that the aquatic environment contains a wider range of taxa than that encountered in medical diagnostic laboratories.

At present, emphasis has been placed on two types of commercial kits, namely those that focus on biochemical responses of the organism, and those based on serology. Of these, the former have taken the larger share of the (mainly medical) market.

Commercially available products, measuring biochemical reactions, centre around rows of microtubes or paper strips impregnated with various freeze-dried test substrates. These are rehydrated upon inoculation with bacterial suspensions, and, after a predetermined incubation period, the results are recorded as colour changes usually following the addition of reagents. Identification can result within 48 h. In this author's experience, there are two products worthy of attention by aquatic bacteriologists. Both of these kits are produced by API (API Laboratory Products, Basingstoke). The API-20E 'rapid' identification system, designed originally for use on medically important Enterobacteriaceae representatives, is of some use with Gram-negative fermentative rods, e.g. *Aeromonas hydrophila* (abundant in fresh water). A few examples of profiles have been included in Table 4.3. The API-20E system comprises 20 biochemical reactions, namely β-galactosidase, arginine dihydrolase, lysine and ornithine decarboxylases, utilization of Simmons citrate, H_2S production, urease, tryptophan deaminase, indole production, the Voges-Proskauer reaction, gelatinase, and acid production from glucose, mannitol, inositol, sorbitol, rhamnose, saccharose, melibiose, amygdalin and arabinose. In addition, the oxidase test needs to be carried out. The microtubes are rehydrated with a few drops of the bacterial suspension, sterile mineral oil added to seal some tubes, i.e. arginine dihydrolase, lysine and ornithine decarboxylase, and H_2S and urease production, incubated at 35–37 °C for 24 or 48 h, and the reactions recorded, sometimes after the addition of reagents, i.e. to the indole, tryptophan deaminase, and Voges Proskauer tests. Thence, the results (including oxidase production) are coded numerically into a 7 digit code for comparison with a data matrix. With aquatic bacteria the incubation temperature (recommended as 37 °C) and duration (should be recorded after 24h and 48h) are seldom appropriate. Moreover, plasmid DNA may wreak havoc with the resulting profile, leading to misidentification.

The API-zym system, which is based on the detection of 19 bacterial enzymes (plus one control tube), is suitable for the identification of Gram-positive organisms (see Table 4.4), such as abound in aquatic sediment samples. Here, the microtubes are rehydrated with two drops

TABLE 4.3 Examples of profiles obtained in the AP1–20E rapid identification system

Taxon	1 β-Galactosidase	2 Arginine dihydrolase	3 Lysine decarboxylase	4 Ornithine decarboxylase	5 Citrate utilization	6 H₂S production	7 Urease production	8 Tryptophan deaminase	9 Indole production	10 Voges-Proskauer reaction	11 Gelatin hydrolysis	12 Acid from glucose	13 Acid from mannitol	14 Acid from inositol	15 Acid from sorbitol	16 Acid from rhamnose	17 Acid from saccharose	18 Acid from melibiose	19 Acid from amygdalin	20 Acid from arabinose	(21) Oxidase
Acinetobacter calcoaceticus	−	−	−	−	D	−	−	−	−	−	−	+	−	−	−	−	−	+	−	+	−
Aeromonas hydrophila	+	+	D	D	D	−	−	−	+	D	+	+	+	−	−	−	+	−	D	D	+
Citrobacter freundii	+	D	−	−	D	D	−	−	−	−	−	+	+	−	+	+	D	D	D	+	−
Enterobacter aerogenes	+	−	+	+	+	−	−	−	−	+	−	+	+	+	+	+	+	+	+	+	−
Escherichia coli	+	−	D	D	−	−	−	−	+	−	−	+	+	−	+	+	D	D	−	+	−
Hafnia alvei	D	−	−	+	−	−	−	−	−	−	−	+	+	−	−	D	−	−	−	+	−
Klebsiella pneumoniae	+	−	D	−	+	−	D	−	−	+	−	+	+	+	+	+	+	+	+	+	−
Plesiomonas shigelloides	+	+	+	+	−	−	−	−	+	−	−	+	−	+	−	−	−	−	−	−	+
Serratia liquefaciens	+	−	+	+	+	−	−	−	−	D	D	+	+	D	+	−	+	D	+	+	−

AP1–20E test no:

S. marcescens	+	+	–	+	+	–	+	D	–	D	+	D	+	–	–
S. rubidaea	+	+	–	D	+	–	D	D	–	D	+	+	+	+	–
Vibrio alginolyticus	–	–	+	+	–	–	D	+	–	–	+	+	–	+	+
V. anguillarum	+	+	+	–	+	–	+	–	+	–	+	+	–	+	+
V. cholerae	+	+	+	D	+	–	+	D	–	–	+	+	–	+	–
V. fluvialis	–	+	–	–	+	–	D	–	–	–	+	+	+	+	+
V. metschnikovii	+	+	–	D	+	–	D	D	+	–	+	+	–	+	–
V. mimicus	–	–	–	–	+	–	+	–	–	–	+	+	–	–	+
V. ordalii	–	–	–	–	–	–	+	–	–	–	+	+	–	–	–
V. parahaemolyticus	–	+	+	D	+	+	+	+	–	+	+	+	–	D	+
V. vulnificus	+	D	+	+	+	–	+	–	+	D	D	+	+	–	–

+, –, and D correspond to ≥80%, ≤20% and 21–79% positive responses, respectively

TABLE 4.4 Distinguishing profiles for some Gram-positive bacteria, found in the aquatic environment, as obtained in the API-zym system after incubation at 15 °C for 18 h.

Taxon	Control 1	Alkaline phosphatase 2	Esterase (butyrate) 3	Esterase (caprylate) 4	Lipase (myristate) 5	Leucine arylamidase 6	Valine arylamidase 7	Cystine arylamidase 8	Trypsine 9	Chemotrypsinase 10	Acid phosphatase 11	Phosphoamidase 12	α-Galactosidase 13	β-Galactosidase 14	β-Glucuronidase 15	α-Glucosidase 16	β-Glucosidase 17	N-acetyl-β-glucosaminidase 18	α-Mannosidase 19	α-Fucosidase 20
Arthrobacter aurescens	−	−	−	+	−	+	−	+	+	−	+	+	−	+	+	+	+	−	+	−
Arth. crystallopoietes	−	−	−	+	−	+	+	−	+	−	−	+	−	−	−	−	−	−	−	−
Bacillus cereus	−	+	+	+	−	+	+	−	+	+	+	+	−	−	−	+	−	−	−	−
Bac. licheniformis	−	+	+	+	−	−	−	−	−	−	+	+	−	−	−	+	−	−	−	−
Bac. megaterium	−	+	+	+	−	+	−	−	−	+	+	+	+	+	−	−	−	−	−	−
Bac. polymyxa	−	+	+	+	−	+	−	−	−	+	−	+	+	+	−	−	−	−	−	−
Bac. sphaericus	−	+	+	+	−	+	−	−	−	+	+	+	−	−	−	−	−	−	−	−
Brevibacterium flavum	−	+	−	+	−	+	+	−	+	−	+	+	−	−	+	−	+	−	−	−
Listeria denitrificans	−	−	−	+	−	+	−	−	−	−	−	+	+	+	+	+	+	−	−	−

Lis. grayi	−	−	(+)	+	−	(+)	(+)	−	−	−	−	−	−	−	−	−
Lis. murrayi	−	−	+	+	+	+	−	−	−	−	+	−	−	−	−	−
Micrococcus luteus	−	+	+	−	+	−	+	−	−	+	+	−	−	−	−	−
Mycobacterium aquae	−	+	+	+	+	+	+	−	−	+	+	+	−	−	−	−
Myc. fortuitum	−	+	+	+	+	+	+	+	−	+	+	−	−	−	−	−
Myc. marinum	−	+	+	+	+	+	+	−	+	+	+	−	−	−	−	−
Nocardia asteroides	−	+	+	−	+	−	+	−	+	−	+	+	−	−	−	−
Staphylococcus epidermidis	−	+	(+)	+	−	−	−	−	−	+	+	+	−	−	+	−

Positive results, in parentheses, may appear to be weak, and, on occasion, could be recorded as negative responses.

(100 μl) of a dense bacterial suspension (*ca.* 5×10^8 cells ml^{-1}) in distilled water or 0.9% (w/v) saline. Following incubation (this author uses 15 °C for 24 h), one drop of each of the two required reagents is added to the microtubes. Then, the results are recorded, as colour changes, after a few minutes incubation in bright light. Unfortunately, extensive data banks are not available for comparative purposes. However, several comparative schemes are available in a few specialized laboratories, of which at least one is computer based. Undoubtedly, this shortcoming will inconvenience many aquatic bacteriologists.

4.7 BACTERIOPHAGE TYPING

A primary interest with bacteriophages is in epidemiological investigations of pathogens. The technique involves use of virulent (lytic) bacteriophages, which are inoculated as drops onto freshly seeded bacterial lawns. With incubation at 15–37 °C for 18–48 h, a positive result is indicated by the presence of zones of clearing (plaques) in the otherwise uniform layer of bacterial growth. The reaction is extremely specific. As regards aquatic bacteria, a comprehensive bacteriophage typing scheme has been proposed for the bacterial fish pathogen, *Aeromonas salmonicida* (Rodgers *et al.*, 1981). However, the system has limited value for general identification of aquatic bacteria.

4.8 CHEMOTAXONOMIC METHODS

Many chemotaxonomic methods appear to be extremely suitable for identification purposes, particularly those techniques that provide quantitative data which could be analysed by computer. Examples include determination of electrophoretic protein and fatty acid profiles, and pyrolysis and mass spectrometry of whole cells. Such techniques could have the advantage of speed and ease of automation over conventional phenotypic tests. However, as yet, little has been done to exploit the technology for the identification of aquatic bacteria. Nevertheless, there is hope for the future.

At present, the approach involves the determination of profiles for reference strains, the data for which are stored as matrices in a computer. Then profiles for fresh isolates are compared to patterns for the reference organisms, and identification is achieved if the patterns show high similarity. The computer analyses are inevitably based on discrimination procedures in which a combination of quantitative characters, that most appropriately differentiate established sets of taxa, are determined.

4.9 HYBRIDIZATION PROBES

With the advent of recombinant DNA technology, the means has been provided for generating large amounts of specific DNA sequences (probes). These may be labelled, either radioactively with ^{32}P or using nucleotides derivatized with biotin so that they can be detected by autoradiography or by linking to an enzyme colorimetric system, respectively. These probes may be used to detect, via hybridization, complementary DNA sequences, which have been spotted onto nitrocellulose filters.

In the 'dot-blot' hybridization procedures, samples are extracted with phenol to purify any nucleic acids which may be present. Then they are immobilized (dotted) onto a nitrocellulose filter in an array using a specialized vacuum (hybridot) manifold. The DNA samples are subsequently hybridized with a labelled probe for a specific organism, and any hybrids are revealed by autoradiography. Positives are readily observed as dark spots.

DNA-based identification is still at an early stage of development. The advantages concern specificity, rapidity and the ability to detect and identify organisms in natural samples without prior isolation. The future seems extremely promising for the technique, and it is hoped that aquatic bacteriology will benefit from the technology.

4.10 CONCLUSIONS

Methods exist for identification of specific groups of organisms within specialized laboratories. Unfortunately, there is a dearth of methods which may be used for the reliable identification of the broad range of bacteria occurring in aquatic habitats. Although there is a promising future for a small number of 'advanced' techniques, it seems unlikely that identification of many bacterial groups will be greatly improved unless there is a concomitant increase in taxonomic studies of bacteria of ecological interest.

4.11 REFERENCES

Allen, D. A., Austin, B. and Colwell, R. R. (1983). Numerical taxonomy of bacterial isolates associated with a freshwater fishery. *Journal of General Microbiology*, **129,** 2043–2062.

Austin, B. (1982). Taxonomy of bacteria isolated from a coastal, marine fish-rearing unit. *Journal of Applied Bacteriology*, **53,** 253–268.

Austin, B. and Priest, F. G. (1986). *Modern Bacterial Taxonomy*, Van Nostrand Reinhold, Workingham.

Cowan, S. T. (1974). *Cowan and Steel's Manual for the Identification of Medical Bacteria*, 2nd edn, Cambridge University Press, Cambridge.

Krieg, N. R. (Ed.). (1984). *Bergey's Manual of Systematic Bacteriology*, vol. 1, Williams & Wilkins, Baltimore.

Rodgers, C. J., Pringle, J, H. McCarthy, D. H. and Austin, B. (1981). Quantitative and qualitative studies of *Aeromonas salmonicida* bacteriophage. *Journal of General Microbiology*, **125**, 335–345.

Sneath, P. H. A. (Ed.). (1986). *Bergey's Manual of Systematic Bacteriology*, vol. 2, Williams & Wilkins, Baltimore.

PART 2

Specialized Environments

Methods in Aquatic Bacteriology
Edited by B. Austin
© 1988 John Wiley & Sons Ltd.

5

Bacteria of Fish

J. E. Sanders and J. L. Fryer*

Department of Microbiology, Oregon State University, Corvallis, OR 97331–3804, USA

5.1 INTRODUCTION

The purpose of this chapter is to describe the bacteria that one might expect to isolate from fin fish, and to comment on those which are pathogenic and those which are known to constitute part of the normal fish microflora. More is understood about the pathogens of fish than the bacteria which compose the normal flora. Infectious diseases caused by bacteria are frequently encountered among populations of cultured fish and many have been isolated and identified. The detection, prevention and control of infectious diseases for fish remains one of the most important considerations for their successful culture. Worldwide, approximately 35 species of bacteria from 21 genera have been isolated from or associated with diseased fish (Table 5.1). Of these, 12 species of bacteria from nine genera are considered major pathogens of fish and five are considered obligate pathogens. Finally, nine of these species from seven genera have been detected in association with bacterial diseases of humans.

The reader should be aware that no one method(s) will detect all the micro-organisms that may be present in a specimen. For example, *Renibacterium salmoninarum* cannot be isolated from infected fish unless it is present in great numbers, the correct medium is used and the requirements of incubation are rigidly followed. Likewise only two species of anaerobic bacteria *Eubacterium tarantellus* and *Clostridium botulinum* have been isolated from diseased fish or their environment. This may represent reduced interest by researchers in the use of specific techniques required to isolate anaerobic micro-organisms, rather than a lack of anaerobic species in the fish microflora. In summary, the known presence of one bacterium

---

* To whom correspondence should be addressed.

TABLE 5.1 Bacterial fish pathogens, the diseases they cause and their mol % guanine plus cytosine

Bacterial species	Disease	Mol % G + C
GRAM-NEGATIVE PATHOGENS		
Vibrio anguillarum[1,2]	Vibriosis	43–45
Vibrio ordalii[1,2,4]	Vibriosis	43–45
Vibrio alginolyticus	Vibriosis	45–47
Vibrio damsela	Vibriosis	43
Vibrio cholerae (non-01)	Vibriosis	—
Vibrio vulnificus (Biogroup 2)	Vibriosis	46–48
Aeromonas salmonicida[1,2]	Furunculosis	58
Aeromonas hydrophila[1,3]	Motile Aeromonad Septicaemia	57–63
Pasteurella piscicida[1]	Pasteurellosis	37–43
Providencia rettgeri[3]	Bacterial Haemorrhagic Septicaemia	39
Edwardsiella tarda[3]	Edwardsiellosis	56–59
Edwardsiella ictaluri[1]	Enteric Septicaemia	53
Yersinia ruckeri[1]	Enteric Redmouth Disease	48
Acinetobacter sp.[5]	Acinetobacterosis	—
Pseudomonas anguilliseptica	Pseudomonas Septicaemia	57
Pseudomonas chlororaphis	Pseudomonas Septicaemia	—
Pseudomonas fluorescens[3]	Pseudomonas Septicaemia	—
Cytophaga psychrophila[1]	Bacterial Coldwater Disease	31
Cytophaga spp.[1]	Fin Rot, Bacterial Gill Disease	—
Flexibacter columnaris[1]	Columnaris	30–36
Flexibacter maritimus	Flexibacterosis	31–33
Sporocytophaga sp.	Salt Water Columnaris	—
Flavobacterium sp.	Bacterial Gill Disease	33–34
GRAM-POSITIVE PATHOGENS		
Renibacterium salmoninarum[1,2]	Bacterial Kidney Disease	53–55
Eubacterium tarantellus	Eubacterial Meningitis	—
Lactobacillus piscicola	Pseudokidney Disease	35
Staphylococcus epidermidis[3]	Staphylococcosis	—
Streptococcus spp.	Streptococcal Septicaemia	—
Clostridium botulinum[6]	Botulism	—
Myxococcus piscicola[7]	White Mouth	—
ACID FAST PATHOGENS		
Mycobacterium marinum[3]	Mycobacteriosis	64
Mycobacterium fortuitum[3]	Mycobacteriosis	64
Mycobacterium chelonei[3]	Mycobacteriosis	63
Nocardia asteriodes[3]	Nocardiosis	—
Nocardia kampachi[1,2]	Nocardiosis	—

[1] Species of bacteria considered major pathogens of fin fish.
[2] Species of bacteria considered obligate pathogens of fin fish.
[3] Species of bacteria also associated with human disease.
[4] Vibrio ordalii is the species name designated for biotype 2 of V. anguillarum (Schiewe et al., 1981).
[5] Schachte, 1974.
[6] This species produces seven antigenically specific neurotoxins.
[7] This is considered an important bacterial pathogen of cultured carp in the Peoples Republic of China (Xu, 1975).

should not preclude a search for other less obvious mic
might be present. Bacteria initially isolated from the s
epizootics are frequently opportunistic or secondary ir
become prevalent after the activity of a primary path

5.2 THE VIBRIONACEAE

Genus I *Vibrio*

Genus II *Photobacterium*

Genus III *Aeromonas*

Genus IV *Plesiomonas*

These bacteria are small, Gram-negative straight to curved rods that are usually motile by means of a single polar flagellum. Species from each of the genera that comprise this family have been isolated from the surfaces and intestinal contents of marine and freshwater fish. These micro-organisms are worldwide in distribution, with species of *Vibrio* and *Aeromonas* considered as normal constituents of the microflora of aquatic animals in salt and freshwater, respectively. Studies have shown that the intestinal microflora of healthy salmonids in freshwater is composed primarily of aeromonads and members of the family Enterobacteriaceae; whereas, migration into saltwater produces a shift, and species of *Vibrio* become predominant (Yoshimizu and Kimura, 1976). *Photobacterium* and *Vibrio* species require Na^+ for growth or the stimulation of growth, whereas *Aeromonas* and *Plesiomonas* species are non-halophilic. Chitin-digesting species of *Vibrio* and *Photobacterium* are believed responsible for many of the lesions seen in the exoskeleton of marine crustaceans. Some photobacteria are symbionts in specialized luminous organs of marine fish.

5.2.1 Genus *Vibrio*

This genus consists of approximately 19 species, of which five have been reported to cause disease in fish and shellfish. *Vibrio anguillarum* has long been known as a major pathogen; *Vibrio ordalii* was recently proposed as a new species designation for the biotype 2 of *V. anguillarum* (Schiewe *et al.*, 1981). Both species cause septicaemic infections although differences in pathology have been described (Ransom *et al.*, 1984). *Vibrio alginolyticus* and *Vibrio damsela* have been isolated from the sea bream (*Sparus aurata*) in the Red Sea and from the damsel fish (*Chromis punctipinnis*) in the United States, respectively. In Japan *Vibrio cholerae* (non-01) has been isolated from the ayu (*Plecoglossus altivelis*) and *Vibrio vulnificus* from the eel (*Anguilla japonica*).

Each of these *Vibrio* species may be isolated from the internal organs, principally the kidney or lesions on diseased fish using standard bacteriological media such as brain heart infusion agar (BHI) or trypticase (tryptone) soy agar (TSA). Supplementation with NaCl (1–3%) or a seawater base is recommended especially for primary isolation. *Vibrio anguillarum* typically grows more rapidly than *V. ordalii* on TSA and after 1–2 days incubation at 22 °C produces circular, raised, yellowish-brown, opaque colonies. *Vibrio ordalii* requires 4–6 days incubation to form circular, white, translucent colonies.

Marine agar and thiosulphate-citrate-bile salt-sucrose agar (TCBS) are also commonly used to isolate and grow vibrios. The latter is a selective medium used for the isolation of enteropathogenic vibrios such as *V. cholerae* and *V. parahaemolyticus. Vibrio damsela* forms green colonies 2–3 mm in diameter on this medium.

Presumptive evidence indicating isolation of a species of *Vibrio* is the presence of Gram-negative straight or curved rods approximately 0.5–1.5 to 3.0 μm and motile by polar flagella. Vibrios are cytochrome oxidase positive, fermentative by the oxidative/fermentative (O/F) glucose test and sensitive to 2,4-diamino 6,7-diisopropyl pteridine phosphate (vibriostat, O/129) and novobiocin. The two most commonly isolated vibrios, *V. anguillarum* and *V. ordalii*, can be separated by a number of tests; the former produces acetylmethylcarbinol (Voges-Proskauer reaction) and arginine dihydrolase, grows on both Simmon's and Christensen's citrate, utilizes starch and sorbitol and shows lipase activity (tributyrin) (Schiewe *et al.*, 1981). Additional biochemical tests give variable results among the species although all utilize D-glucose, D-fructose, maltose and glycerol. The presence of numerous biotypes suggests that additional species will be proposed.

5.2.2 Genus *Aeromonas*

This genus consists of four species; of these *Aeromonas salmonicida* is a major worldwide fish pathogen causing furunculosis primarily in salmon and trout. This bacterium is considered an obligate parasite of fish and its existence outside of the host is very short-lived (Popoff, 1984). *Aeromonas hydrophila*, an opportunistic pathogen of fish, and the remaining motile aeromonads are routinely isolated from the environment and as normal inhabitants of the fish microflora.

Strains of *A. salmonicida* may be isolated from the internal organs or lesions of diseased fish using bacteriological media such as BHI and TSA. After incubation at 25 °C for 2–3 days, colonies of this micro-organism appear circular, raised and translucent. Typical strains produce a brown, water-soluble pigment that diffuses into the medium surrounding the colonies. Isolates of *A. salmonicida* that do not produce pigment have been reported (Paterson *et al.*, 1980a,b).

Aeromonas hydrophila and the other motile aeromonads, *A. caviae* and *A.*

sobria, are isolated from the intestines or external surfaces of fish using a number of bacteriological media such as BHI and TSA or nutrient agar (NA). On NA after 24–48 h of incubation at 25 °C, colonies of these bacteria will be white to buff in colour, circular and convex with an entire margin. On selective media such as McConkey's agar, colonies will appear as lactose negative. Selective media designed specifically for the isolation of *A. hydrophila* have been formulated. Rimler-Shotts (RS) and pril-xylose-ampicillin agar (Rogol *et al.*, 1979) are examples. Both of these contain inhibitors to prevent growth of Gram-positive bacteria but enable the production of characteristically pigmented *A. hydrophila* colonies. On RS medium incubated at 37 °C for 20–24 h, colonies of *A. hydrophila* will be yellow in colour. Species of the Enterobacteriaceae and Pseudomonadaceae are either completely inhibited or produce colonies that are greenish-yellow to green in colour or yellow or green with black centres.

Presumptive evidence for the isolation of a species of *Aeromonas* is indicated by the presence of Gram-negative rods approximately 0.3–1.0 by 1–3.5 μm that are motile by means of a single polar flagellum. If the isolate is non-motile and produces a brown water soluble pigment *A. salmonicida* should be suspected. Both motile and non-motile aeromonads are cytochrome oxidase positive, fermentative by the O/F glucose test and resistant to vibriostat and usually to novobiocin. Additionally the motile aeromonads will grow at 37 °C, whereas strains of *A. salmonicida* will not. *Aeromonas sobria* differs from *A. hydrophila* and *A. caviae* by its inability to hydrolyse aesculin, ferment salicin and utilize L-histidine, L-arabinose and L-arginine as sole carbon sources. Production of hydrogen sulphide from cysteine and gas from the fermentation of glucose by *A. sobria* and *A. hydrophila* are used to distinguish these species from *A. caviae*.

The genus *Plesiomonas* contains one species, *P. shigelloides*, which has been variously identified as a *Pseudomonas* or *Aeromonas* before it was placed in this genus. This bacterium has been isolated from fish and can be grown on TSA or BHI. It has the basic properties of the Vibrionaceae, but differs from both *Vibrio* and *Aeromonas* by having several (2–5) lophotrichous flagella and lacking the enzymes lipase, deoxyribonuclease, gelatinase and caseinase. The fermentation of inositol by *P. shigelloides* is also rare for bacteria in this family, with none of the aeromonads possessing this ability.

5.3 THE PASTEURELLACEAE

Genus I *Pasteurella*   Genus III *Actinobacillus*
Genus II *Haemophilus*

Members of this family are small, Gram-negative, non-motile coccoid to rod-shaped cells. Atypical variants from other families, especially the Enterobacteriaceae and Vibrionaceae, have been confused with species of these genera. For example, the bacterium initially named *Haemophilus piscium* is now recognized as an achromogenic strain of *A. salmonicida* (Paterson *et al.*, 1980a,b). Positive oxidase and alkaline phosphatase reactions and sensitivity to benzylpenicillin are particularly useful tests for differentiating the Pasteurellaceae. Separation of the recognized species of *Pasteurella, Haemophilus* and *Actinobacillus* based on morphological appearance or biochemical tests is not well-defined. DNA relatedness indicate that these bacteria are heterogenous and there is the need for further divisions and generic descriptions.

Pasteurellaceae are parasitic on the mucous membranes and in the digestive tracts of a variety of warm-blooded animals. One species, referred to as *Pasteurella piscicida* (Janssen and Surgalla, 1968), is a pathogen of fish. This bacterium differs from other species of the Pasteurellaceae by its larger genome, lower growth temperature and lack of nitrate reductase; hence, future studies may recommend reclassification.

The disease caused by *P. piscicida* was first described in 1963 from a massive fish kill among white perch (*Roccus americanus*) and striped bass (*R. saxatilis*) in Chesapeake Bay on the east coast of North America. Since then the problem has rarely occurred in the United States; however, a disease of cultured fish in the Orient (Japan and Taiwan) called pseudotuberculosis is apparently caused by this same or a closely related bacterium.

5.3.1 Genus *Pasteurella*

Pasteurella piscicida is routinely isolated from the kidney or lesions of diseased fish on bacteriological media such as BHI or TSA. The bacterium grows over a temperature range from 17–31 °C and requires at least 0.5% NaCl in the medium. Optimum growth is obtained when 1.5% NaCl is added to the medium. Colonies develop within 3 days and are round, convex, entire, and greyish-yellow in colour. Morphologically the cells are encapsulated, about 0.5 by 1.5 μm in size, ovoid and stain giving a bipolar 'safety pin' appearance. Isolates are catalase and oxidase positive and fermentative by the O/F glucose test. *Pasteurella piscicida* differs from the other Pasteurellaceae by its lack of growth at 37 °C, inability to reduce nitrates and the requirement of NaCl.

Pasteurella, Actinobacillus and *Haemophilus* are sensitive to vibriostat, a property which separates them from the aeromonads and *Yersinia*. They differ from the vibrios by their lack of motility. *Actinobacillus* species characteristically are arranged as a mixture of rods and coccal cells giving a 'Morse code' appearance. At primary isolation, they often produce extracellular slime and colonies that are very sticky. Species of *Haemophilus* are generally less than 1 μm in width and pleomorphism marked by threads or filmaments are common.

5.4 THE ENTEROBACTERIACEAE

Genus I *Escherichia*		Genus VIII *Serratia*
Genus II *Shigella*		Genus IX *Hafnia*
Genus III *Salmonella*		Genus X *Edwardsiella*
Genus IV *Citrobacter*		Genus XI *Proteus*
Genus V *Klebsiella*		Genus XII *Providencia*
Genus VI *Enterobacter*		Genus XIII *Morganella*
Genus VII *Erwinia*		Genus XIV *Yersinia*

These micro-organisms are Gram-negative straight rods and if motile possess peritrichous flagella. They are also cytochrome oxidase negative, catalase positive, fermentative in O/F glucose medium and reduce nitrates to nitrites. These bacteria inhabit the intestine of man and other animals and can be isolated from a variety of sources including water, sewage, food and soil. Except for *Erwinia*, which is principally associated with plants, each of these genera may be considered as potential isolates from fish. In one study, species from seven of these genera were found in the alimentary tract of salmonids collected from 16 different freshwater locations in British Columbia, Canada (Trust and Sparrow, 1974). Currently, three species from two genera, *Edwardsiella* and *Yersinia*, of the Enterobacteriaceae are considered important pathogens of fish, although the reader should be continually aware of the disease capabilities and possible isolation of other members from this family.

As a group, the enterics are opportunistic fish pathogens. Two recent examples support this description. An epizootic caused by *Proteus rettgeri* (*Providencia rettgeri* in the 1984 edition of *Bergey's Manual*) in silver carp (*Hypophthalmichthys molitrix*) occurred after the trauma of handling had allowed entrance of the bacterium (Bejerano *et al.*, 1979). *Proteus rettgeri* was present because carp ponds in Israel are regularly fertilized with poultry faeces. In a second example, Amandi *et al.* (1982) indicated that the absence of *Edwardsiella tarda* in freshly killed or spawned chinook salmon (*Oncorhynchus tshawytscha*), contrasted to its presence in dead fish, suggested this micro-organism behaved as an opportunistic pathogen in this fish species.

Colonies produced by the enterics on TSA and BHI incubated at 25 °C for 2–3 days are usually similar. Except for size variation they tend to be smooth, circular, raised or convex and greyish-white in colour. Occasional isolates, usually *Klebsiella*, will develop large mucoid colonies. Swarming, which appears as a bluish-grey confluent growth on the agar surface may occur with certain species of *Proteus*.

Isolates showing the basic characteristics of the Enterobacteriaceae should then be inoculated into triple sugar iron agar and lysine iron agar, and the combination of reactions obtained used for preliminary identification of genera (Martin and Washington, 1980). On triple sugar iron agar, the fish pathogens *Yersinia ruckeri* and *Edwardsiella ictaluri* will give rise to an alkaline or red slant over an acid or yellow butt, and *E. tarda* produces an alkaline slant over an acid butt with gas bubbles and a dark colour indicating hydrogen sulphide production. On lysine iron agar *E. tarda* is the only one of these fish pathogens to produce hydrogen sulphide.

The list of tests to further separate the genera is long and complex and, as might be expected from such a large and intensely studied family of micro-organisms, many variant or atypical strains have been isolated and described. Tests that should be included in any scheme to further identify an isolate are: indole production from tryptone broth, methyl red and Voges-Proskauer tests, gelatin liquefaction at 22 °C, growth on Simmon's and Christensen's citrate, malonate utilization and presence of the enzymes urease, phenylalanine deaminase, lysine decarboxylase, arginine dihydrolase, and ornithine decarboxylase. Comprehensive listings of cultural reactions from these and other tests for the Enterobacteriaceae can be found in the *Manual of Clinical Microbiology* (Martin and Washington, 1980) and the latest edition of *Bergey's Manual of Systematic Bacteriology* (Brenner, 1984).

5.4.1 Genus *Edwardsiella*

This genus consists of three species. Two of these, *E. tarda* and *E. ictaluri*, are pathogenic to fish. Both of these bacterial species have been isolated from species of cultured warm-water fish in the southern United States and southeastern Asia.

Isolation of either species from lesions or internal organs can be accomplished using bacteriological media such as BHI or TSA. On both media, incubation for 2–3 days at 22–26 °C produces small (0.5–2 mm in diameter) transparent, circular and raised to slightly convex colonies. *Edwardsiella ictaluri* tends to be more fastidious, growing slower and producing smaller colonies. Inoculation of kidney tissue into thioglycollate broth and incubation at 22 °C for 4 days prior to subculture on BHI has been reported to improve isolating *E. tarda*.

Initial tests should be selected to determine if the micro-organism isolated is a member of the Enterobacteriaceae. *Edwardsiella tarda* differs from *E. ictaluri* in several respects: it is motile at 37 °C, produces indole from tryptone broth, is methyl red positive, utilizes tartrate and grows on Christensen's citrate. The third species in this genus, *Edwardsiella hoshinae*, differs in its ability to produce acid from trehalose and utilize malonate.

5.4.2 Genus *Yersinia*

This genus consists of seven species. One, *Yersinia ruckeri*, is pathogenic to salmonid fish causing enteric redmouth disease in North and South America, Europe and Australia. On bacteriological media such as BHI and TSA, this bacterium produces small, smooth, circular and slightly raised colonies after 48 h incubation at 20–25 °C. A differential media containing Tween 80, sucrose and bromthymol blue indicator has recently been described (Waltman and Shotts, 1984). Colonies of *Y. ruckeri* are green and surrounded by a zone of hydrolysis, while those of both *Edwardsiella* species are also green but without zones of hydrolysis. Colonies of *A. hydrophila* and *Enterobacter* are yellow and with or without zones of hydrolysis.

Suspected isolates of *Y. ruckeri* should have characteristics of the enterics such as Gram-negative motile rods, oxidase negative, catalase positive, fermentative in O/F glucose medium and reduce nitrates to nitrites. *Yersinia ruckeri* differs from the other species of this genus by having lysine decarboxylase and gelatinase enzymes. Test results that will differentiate this bacterium from both *Edwardsiella* species have been described in detail by Hawke *et al.* (1981). These include growth of *Y. ruckeri* on Simmon's citrate agar, gelatin liquefaction at 22 °C and trehalose utilization. Both *Edwardsiella* species are unable to grow on Simmon's citrate agar, liquefy gelatin and utilize trehalose. Additionally, gas is produced from glucose medium incubated at 25 °C by each of the *Edwardsiella* species but not *Y. ruckeri*.

5.5 THE PSEUDOMONADACEAE

The pseudomonads are straight to slightly curved Gram-negative rods that are motile by means of polar flagella. In contrast to the Enterobacteriaceae which they resemble morphologically, these micro-organisms are strict aerobes that are oxidative in O/F glucose medium. Species in this family are frequently isolated from the surface and intestinal contents of fish and may cause disease.

5.5.1 Genus *Pseudomonas*

This large and complex genus consists of approximately 27 recognized species, two of which have been reported to cause disease in fish. A third proposed species, *Pseudomonas anguilliseptica*, has been isolated from diseased eels in Japan. *Pseudomonas chlororaphis* has been reported to cause high mortality in Amago trout (*Oncorhynchus rhodurus*) in Japan; however, only *Pseudomonas fluorescens* has consistently been shown to cause disease although mortalities are limited and occur only after stress or injury to the fish.

Species of *Pseudomonas* can be isolated from the internal organs of diseased fish using bacteriological media such as BHI, TSA or NA and incubation temperatures of 20–25 °C. *Pseudomonas chlororaphis* grows on NA producing a green pigment which forms needle-like crystals in the colonies. *Pseudomonas fluorescens* colonies are round, glistening and produce yellowish-green diffusible water soluble pigments. Colonies of both species are fluorescent when exposed to ultraviolet light.

Presumptive evidence for the isolation of a species of *Pseudomonas* is the presence of a Gram-negative straight or slightly curved rod approximately 0.5–1.0 by 1.5–5.0 μm and motile by polar flagella. Pseudomonads are strictly aerobic giving an oxidative reaction with the O/F glucose test. Strains are also catalase and cytochrome oxidase positive and produce arginine dihydrolase. The presence of fluorescent pigments is a particularly valuable criteria for identification.

5.6 THE CYTOPHAGACEAE

Genus   I *Cytophaga*   Genus IV *Flexithrix*
Genus II *Flexibacter*   Genus V *Saprospira*
Genus III *Herpetosiphon*   Genus VI *Sporocytophaga*

Members of this family are slender, Gram-negative, weakly refractile, flexible rods or filaments that lack flagella and exhibit gliding motility on solid surfaces. All members of this family form carotenoid pigments and produce colonies that are shades of brown, yellow, red or orange. Individual cells are approximately 1 μm in width and depending on the species vary in length from 5 to over 100 μm. Cell ends may be rounded or tapered.

The detection of gliding motility is the single most important characteristic of these bacteria. Because isolation media used are not selective, individual colonies should be subcultured prior to examination. Gliding motility can be determined by several methods to include the hanging-drop and wet-mount preparations; however, the best procedure is to place a coverslip over the peripheral regions of a colony growing on an agar surface and observe using light or phase contrast microscopy (400–1000 × magnification).

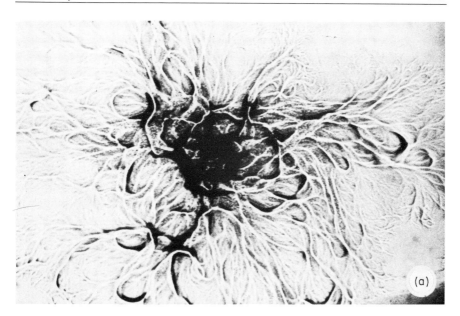

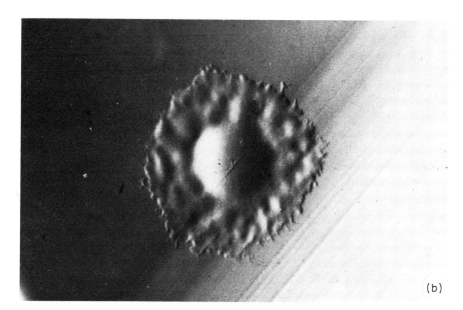

Fig. 5.1 (a). Colony morphology of *Flexibacter columnaris* on cytophaga agar showing the characteristic spreading nature and rhizoid edge (approximately 15×). (b) Photomicrograph showing typical colony morphology of *Cytophaga psychrophila* on cytophaga agar (approximately 25×).

Separation of genera within the family Cytophagaceae is based on the formation of microcysts (*Sporocytophaga*), filaments and sheaths of cells, and the ability to hydrolyse agar, cellulose or chitin. However, growth of pigmented colonies on cytophaga agar with spreading irregular edges and whose cells exhibit gliding motility describes members of the Cytophagaceae.

Members of this family are common inhabitants of soil and both the marine and freshwater environments. Species of *Herpetosiphon*, *Flexithrix* and *Saprospira* have not been isolated from fish. Among the Cytophagaceae, two species are considered significant pathogens of freshwater fish; they are *Flexibacter columnaris* (Fig. 5.1(a)) and *Cytophaga psychrophila* (Fig. 5.1(b)). *Flexibacter maritimus* and a *Sporocytophaga*-like bacterium have been described as pathogens of marine fish. *Cytophaga psychrophila*, as its name suggests, is a pathogen among cold water fish species, particularly the Salmonidae. *Flexibacter columnaris* is ubiquitous in freshwater and causes epizootics in cultured and wild fish especially at temperatures above 12°C (Bullock *et al.*, 1971 and Holt *et al.*, 1975). Bacteria identified as *F. maritimus* and requiring saltwater for growth have been isolated from external lesions on marine fish (Wakabayashi *et al.*, 1986).

5.6.1 Genus *Cytophaga*

Cytophaga psychrophila forms external lesions, and internally the spleen and kidney are the best sources for isolation of the micro-organism. Cytophaga agar is widely used for cultivation of this bacterium. This medium is non-selective and will support the growth of a wide variety of bacteria; hence, knowledge of cell and colony morphology is essential. Colonies of *C. psychrophila* formed after 3–5 days incubation at 15–18°C are bright yellow with a raised convex centre and a thin spreading irregular periphery giving what has been described as a 'fried egg' appearance (Fig. 5.1b). Occasional strains may produce colonies with an entire edge. *Cytophaga psychrophilia* grows at temperatures from 5 to 20°C with only rare strains growing at 25°C.

5.6.2 Genus *Flexibacter*

Flexibacter columnaris causes yellowish external lesions on the gills and body surfaces. The bacterium seen on bits of tissue collected from these lesions is arranged in column-like or 'haystack' masses of cells. Internally the bacterium can be isolated from the kidneys, but this is often only after the disease has reached an advanced stage. On cytophaga agar, colonies of *F. columnaris* that develop after 3–5 days incubation at 25°C are yellow with spreading, convoluted centres, rhizoid edges and adhere to the agar surface (Fig. 5.1a). Atypical colonies that may occasionally be isolated are yellow with rhizoid edges, but differ by having mucoid centres and not adhering to the agar surface. *Flexibacter columnaris* grows at temperatures from 10 to 35 °C with some strains also growing at 37 °C. The isolation of marine gliding bacteria from fish requires the incorporation of either natural or artificial seawater into cytophaga medium at concentrations of 30% or greater. Isolates obtained produce thin, pigmented colonies (usually orange or yellow) with the spreading, uneven edges typical of gliding bacteria (Hikida *et al.*, 1979 and Wakabayashi *et al.*, 1986).

5.6.3 Genus *Flavobacterium*

Members of this genus are slender, Gram-negative, non-motile rods which are strict aerobes (oxidative in O/F glucose medium) and uniformly catalase, oxidase and phosphatase positive. These bacteria, except for their inability to glide or swarm on agar surfaces, are taxonomically similar to members of the family Cytophagaceae, in particular the genera *Cytophaga* and *Flexibacter*. Flavobacteria and *Cytophaga*-like bacteria are commonly isolated from the external gill surfaces of cultured salmonids, especially when fish show signs of the condition referred to as 'bacterial gill disease'; however, their role as the causative pathogens of this condition has not been conclusively demonstrated. Wakabayashi *et al.* (1980) examined 15 strains of flavobacteria isolated from fish with clinical signs of bacterial gill disease in Japan and the United States and concluded they appeared to represent a new species of *Flavobacterium*; however, no species designation was proposed.

These bacteria can be isolated by streaking infected gill tissue on to either cytophaga agar or very dilute (20 fold) TSA. After 5 days incubation at 18 °C, colonies appear as round (0.5–1 mm in diameter), transparent, smooth, and light yellow in colour. Growth occurs over a temperature range of 10–25 °C. The addition of NaCl to the medium is not required. Morphologically, individual bacterial cells are 5–8 μm in length and usually occur in chains of two or three rods. As is characteristic of flavobacteria, they are non-motile with neither gliding motility nor swarming growth on the agar surface. The isolates from fish differ from most other species of *Flavobacterium* by their lack of growth at 37 °C and their ability to hydrolyse starch. *Flavobacterium aquatile* which does not grow at 37 °C and *F. balustinum*, strains of which hydrolyse starch, appear related to the fish flavobacteria. Some strains of *F. balustinum*, like the fish isolates, will grow at 5 °C, a characteristic not found with the six remaining species of flavobacteria.

5.7 THE NON-SPOREFORMING GRAM-POSITIVE RODS

Genus *Renibacterium*   Genus *Eubacterium*
Genus *Lactobacillus*

The term 'coryneform' has been widely applied to any non-sporeforming, Gram-positive, pleomorphic rods. These bacteria are readily isolated and identified when studying the normal microflora of fish and comprise a significant portion of the non-pathogenic bacterial population. However, the fish pathogens that morphologically resemble coryneform bacteria are very fastidious and require specific methods for isolation and culture.

5.7.1 Genus *Renibacterium*

This genus contains a single species, *Renibacterium salmoninarum*, the causative agent of bacterial kidney disease in salmonids (Sanders and Fryer, 1980; and 1986). It is found in wild and hatchery-reared salmonids in Europe, Japan, North America and recently in South America (Chile). This bacterium has only been isolated from salmonids and, like *A. salmonicida*, is thought to be an obligate parasite of fish.

Renibacterium salmoninarum is isolated from the internal organs (primarily the kidney) or lesions using KDM-2 agar. Excellent growth has also been obtained by replacing the fetal bovine serum in KDM-2 with 0.1% activated charcoal. Other media used to grow renibacteria have included Mueller-Hinton supplemented with 0.1% L-cysteine, and KDM-2 modified by the addition of selected antibiotics. Although the serum content of the medium can be altered or eliminated, studies show renibacteria has an absolute requirement for cysteine.

Colonies of renibacteria on KDM-2 agar are of varying sizes, circular, convex and white to creamy yellow. The optimal growth temperature is 15 to 18 °C and several weeks (5–6) of incubation may be required for visible colonies to appear, especially on primary isolation. The asceptic excising of colonies of faster growing contaminants from the agar surface before they overgrow the medium improves isolation.

Morphologically, cells appear as short, strongly Gram-positive, non-motile rods, 0.3–1.0 by 1.5 µm, often occurring in pairs. Pleomorphic forms are most common among bacterial cells obtained at primary isolation from the host tissue. This bacterial species is aerobic, catalase positive, and cytochrome oxidase negative. In KDM-2 broth modified by the addition of glucose and phenol red indicator, the pH becomes basic indicating that the organism utilizes proteins in the medium rather than producing acid from glucose.

Even with the use of media designed specifically for its growth, *R.*

salmoninarum is seldom isolated from fish other than those showing overt pathological signs of bacterial kidney disease. The supplementation of KDM-2 with antibiotics has been used to improve isolation from contaminated tissue samples and from the environment.

Serological procedures that attempt to specifically detect either the bacterium itself or products (antigens) produced by its growth have been extensively tested. These tests when compared to detection by the Gram stain and bacterial isolation have yielded widely variable results (Cipriano *et al.*, 1985). Currently, the fluorescent antibody and the enzyme linked immunosorbent assay (ELISA) tests are probably the most widely used serological methods for early detection of bacterial kidney disease. Both these tests require antibodies produced against an antigen unique to *R. salmoninarum*. Until such an antigen is conclusively demonstrated, perhaps through the use of monoclonal antibody techniques, results from serological tests will be suspect.

5.7.2 Genus *Eubacterium*

This large genus contains one species, *Eubacterium tarantellus*, responsible for deaths among estuarine fish in the southern United States (Henley and Lewis, 1976; Udey *et al.*, 1977). The mortality occurred primarily in the grey (striped) mullet (*Mugil cephalus*) and redfish (*Sciaenops ocellata*); although, subsequent studies found the organism in 10 additional species of estuarine fish.

Strains of *E. tarantellus* are isolated from brain tissue using either thioglycollate medium or BHI incubated anaerobically. Colonies of the bacterium are visible on BHI agar after 48 h incubation at 25 °C. They are 2–5 mm in diameter, flat, translucent and rhizoid with a distinct 'pinwheel' appearance (Udey *et al.*, 1977). On BHI plates containing sheep blood, β haemolysis occurs. Morphologically, primary isolates form long chains or filaments (>100 μm) of pleomorphic rods. Upon subculture these tend to break up into individual rods 10–17 μm in length. The bacteria are Gram-positive, non-motile, and do not produce spores. The inability to form spores is a key characteristic separating this genus from the clostridia.

Differentiation of species within the genus is based initially on the end products produced from carbohydrate or peptone fermentation. *Eubacterium tarantellus* produces acetic and formic acids and traces of lactic or succinic acid in contrast to the majority of eubacteria which produce butyric acid. Additional tests used to separate this species from the other eubacteria include indole production, acid from raffinose, glycogen, inositol and maltose, and isovaleric acid and hydrogen gas production (Moore and Moore, 1986).

5.7.3 Genus *Lactobacillus*

Members of this genus are Gram-positive, non-motile, non-sporeforming rods that occur singly or in chains. Lactobacilli have been identified as part of the normal flora of both marine and freshwater fish. Kvasnikov *et al.* (1976) isolated seven species of *Lactobacillus* from the digestive tracts of the carp (Cyprinidae), pike (Esocidae), and perch (Percidae) collected from reservoirs and ponds in Russia. In marine fish, *Lactobacillus* species have been isolated from the mackerel (Scombridae), hake (*Merluccius gayi*) and Atlantic cod (*Gadus morhua*), and *Lactobacillus plantarum* from the saithe (*Gadus virens*). Several freshwater species of the Salmonidae have been reported to harbour *Lactobacillus* sp. in internal organs such as the heart, liver and kidney.

In this large genus, only the recently named species, *Lactobacillus piscicola*, is known to cause pathology in fish. In adult salmonids, it is often observed shortly after spawning and is thought to be stress related (Cone, 1982; Hiu *et al.*, 1984). The bacterium has been isolated from juvenile salmonids; however, the mortality was negligible.

Strains of *L. piscicola* can be isolated from internal organs, principally the kidney, or lesions on diseased fish using bacteriological media such as BHI and TSA. Incubation at 25 °C for 2 days produces non-pigmented (less than 2 mm in diameter), white, round, entire colonies. Morphologically, individual rods measure approximately 0.5 by 1–1.5 μm. Short chains of two to three cells are often seen. Young cultures are Gram-positive but those more than 24 h old frequently become Gram-variable.

Other characteristics of the *Lactobacillus* are the lack of cytochrome oxidase and catalase, and the inability to reduce nitrates to nitrites and to produce hydrogen sulphide on triple sugar iron agar. Taxonomically this genus has been divided into three main subgenera, the thermobacteria, streptobacteria and betabacteria. The majority of *Lactobacillus* species isolated from fish have been identified with the streptobacteria by their ability to grow at 15 but not 45 °C, ferment ribose and produce carbon dioxide from gluconate. Although *Lactobacillus piscicola* is identified most closely with the streptobacteria, it differs by its lack of carbon dioxide production from gluconate.

5.8 GRAM-POSITIVE COCCI

Family I Micrococcaceae
  Genus   I *Micrococcus*
  Genus II *Stomatococcus*
  Genus III *Planococcus*
  Genus IV *Staphylococcus*

  Genus *Streptococcus*
  Genus *Aerococcus*

With the Gram-positive cocci the plane of cell division causes the appearance of clusters or packets of cells with the Micrococcaceae and tetrads or chains of cells with the genus *Streptococcus*. However, morphological appearance alone cannot always be used and final separation requires tests for catalase and cytochromes. The absence of cytochromes is demonstrated by a negative benzidine reaction and provides the definitive evidence required to identify a member of the streptococci. A positive benzidine test indicates a member of the Micrococcaceae.

Gram-positive cocci from the genera *Micrococcus*, *Staphylococcus* and *Streptococcus* have been detected during numerous examinations of the external surfaces and intestines of fish; however, none are recognized as major pathogens of fish. The genus *Aerococcus* contains the single species, *A. viridan*, which is pathogenic to lobsters (*Homarus americanus* and *Homarus vulgaris*) and causes the disease gaffkemia.

5.8.1 Genus *Staphylococcus*

There has been one report of a bacterium identified as *Staphylococcus epidermidis* from disease outbreaks in 1976 and 1977 in yellowtail (*Seriola quinqueradiata*) and red seabream (*Chrysophrys major*) in Japan (Kusuda and Sugiyama, 1981). These isolates developed colonies on BHI after incubation for 48 h at 25 °C which were circular, convex, entire and white to whitish yellow in colour. Individual cells ranged from 0.6 to 1.8 μm in diameter and replicated to form irregular clusters.

The genus *Staphylococcus* differs from the other genera in the Micrococcaceae on the basis of motility, production of acid anaerobically, and sensitivity to lysostaphin. All of the motile cocci are placed in the genus *Planococcus*. Staphylococci, but not micrococci, can produce acid under anaerobic conditions from media containing either glucose or glycerol plus 0.4 μg of erythromycin ml^{-1} (Schleifer and Kloos, 1975). The ability of micrococci to grow in media containing lysostaphin at a concentration of 200 μg ml^{-1} is another feature separating them from the staphylococci.

Differentiation between *S. epidermidis* and the human pathogen, *Staphylococcus aureus*, is based on the coagulase-positive property of the latter; however, with other staphylococci, a series of tests are required. A detailed discussion of these procedures and expected results can be found in the *Manual of Clinical Microbiology* (Kloos and Smith, 1980).

5.8.2 Genus *Streptococcus*

Streptococcus species have been described from diseased fish in both freshwater and saltwater in Japan and the United States and possibly in rainbow trout (*Salmo gairdneri*) from South Africa; however, no species designations have been proposed. Colonies of streptococci isolated from internal organs, usually the kidney, or lesions and growing on BHI at 30 °C for 2–3 days are small, viscous, raised and white in colour. Supplementation with blood frequently improves isolation and growth and allows the determination of haemolytic activity. Individual cocci range from 0.6–0.9 μm in diameter and form chains of from two to six cells. Subculture in broth such as Todd-Hewitt often enhances chain formation.

The genus *Streptococcus* contains a large number of species; however, determination of the Lancefield group antigen can often be used as a rapid method of identification. The streptococci from disease outbreaks in the southern United States during 1974 were identified as Lancefield group B or *Streptococcus agalactiae* (Plumb *et al.*, 1974). The isolates from yellowtail cultured in salt water in Japan (Kusuda and Komatsu, 1978; Kitao, 1982) could not be classified by this method, necessitating a series of further tests. These isolates grew at 10 and 45 °C, at pH 9.6, in 6.5% NaCl, and in 0.1% methylene blue milk and were classified as biochemically similar to *Streptococcus faecalis* and *Streptococcus faecium*. The first streptococci isolated in Japan in freshwater from rainbow trout by Hoshina *et al.* (1958) were also identified as Lancefield group B; however, more recent isolates have had different biochemical characteristics and have not contained Lancefield group B antigens. These isolates have also differed from the saltwater streptococci in several properties including growth at 45 °C, at pH 9.6, and in 0.1% methylene blue milk (Kitao *et al.*, 1981). Detailed procedures for the extraction of Lancefield group antigens and other tests separating the streptococci are described in the *Manual of Clinical Microbiology* (Facklam, 1980).

5.9 THE ENDOSPORE-FORMING GRAM-POSITIVE RODS AND COCCI

Genus *Bacillus*   Genus *Desulfotomaculum*
Genus *Sporolactobacillus*   Genus *Sporosarcina*
Genus *Clostridium*   Genus *Oscillospira*

Members of this group are either rods or cocci (*Sporosarcina*) and all share the ability to produce endospores. Species of *Sporolactobacillus*, *Desulfotomaculum*, *Sporosarcina* and *Oscillospira* have not been obtained from fish. Species of the genus *Bacillus* are among the non-pathogenic microorganisms isolated from the external surfaces and digestive tract of fish.

Deaths caused by the type E toxin of *Clostridium botulinum* have occurred at salmonid hatcheries in the United States, United Kingdom and Denmark

(Cann and Taylor, 1982; Eklund *et al.*, 1984). Seven specific neurotoxins (A–G) have been detected from different strains of this bacterium. Five of these (B–F) have been detected in tissues of either marine or freshwater fish. The *C. botulinum* bacterium is non-invasive and deaths result from ingestion of the preformed toxin. The type E outbreaks which have been described apparently have occurred after live fish cannibalized dead fish decomposing on the pond bottom under anaerobic conditions.

5.9.1 Genus *Bacillus*

Isolates of these bacteria are grown on laboratory media such as BHI, TSA or NA at 20–25 °C. The appearance of colonies are variable. Endospores can be demonstrated by subculture of selected colonies.

Morphologically, the cells of bacilli are rods approximately 0.5–1.5 by 1–10 µm, motile, and form endospores. Cultures are Gram-positive during early growth and frequently become Gram-variable or even negative after 24 h of incubation. The salient features separating this genus from the clostridia are aerobic growth and presence of catalase activity.

5.9.2 Genus *Clostridium*

Morphologically, the cells of *C. botulinum* and the other clostridia are similar to those of the bacilli. Anaerobic growth, absence of catalase activity, and endospore formation are characteristics that permit identification of *Clostridium*.

Isolates of *C. botulinum* have been obtained from the intestinal contents and tissues of fish by strict anaerobic culture at 25–30 °C on enriched media such as egg yolk and blood agar and trypticase, peptone, glucose liquid medium (Eklund *et al.*, 1967 and Eklund *et al.*, 1984). Types B, E and F of *C. botulinum* grow and produce toxin at temperatures as low as 3.3 °C. Sediments from ponds in which a type E toxin-induced fish mortality has occurred have also been used to culture this bacterium.

Colony characteristics are somewhat variable among the types of *C. botulinum*; in general, they are circular to slightly raised, translucent to semi-opaque, and have a matt to semi-glossy surface. Types B and E have been described as producing a pearly (iridescent) layer covering and surrounding each colony.

The presence of *C. botulinum* type E toxin can be demonstrated by intraperitoneal injection of mice or fish with supernatant fluids from tissue homogenates or culture filtrates. Injected animals develop signs of botulism and death. Further confirmation can be obtained by mixing these preparations with type E antitoxin and again injecting into test animals.

Failure to develop botulism is a further corroboration of the presence of type E toxin in the original preparation. Similar procedures can be of value to detect the other toxins produced by types of C. botulinum.

5.10 THE MYCOBACTERIACEAE

5.10.1 Genus Mycobacterium

Members of this family are acid-fast, slightly curved to straight, non-motile, often pleomorphic, 0.2–0.6 by 1.0–10.0 μm rods. The observation of acid-fastness is the single most important morphological characteristic of these bacteria. Acid-fast bacteria stain Gram-positive; however, this characteristic is of no importance because alcohol used in the Gram stain cannot remove crystal violet after it has entered the mycobacterial cell.

Micro-organisms identified as mycobacteria, often only by the property of acid-fastness, have been observed worldwide in over 150 species of marine and freshwater fishes (Nigrelli and Vogel, 1963). Three of the mycobacteria capable of causing disease in fish have been described sufficiently to be given species status. These are *Mycobacterium marinum*, *Mycobacterium fortuitum*, and *Mycobacterium chelonei*. A fourth species, *Mycobacterium salmoniphilum*, is now believed to be a variant of *M. fortuitum* (Gordon and Mihm, 1959). Each of these species is also capable of causing non-tuberculous skin and pulmonary lesions in man which on occasion have proven fatal.

Growth of mycobacteria requires the addition of selected growth factors to a basal medium often containing coagulated eggs, glycerol and potato flour. Four widely used formulations are those of Lowenstein-Jensen, Petragnani, Ogawa and Sauton. Isolates are usually obtained from the culture of lesions (resembling tubercles) which are frequently found on or in kidney, liver and spleen tissue. Initial isolation of these bacteria may require prolonged incubation periods, of up to 30 days. Mycobacteria may occasionally grow on BHI and TSA forming waxy, raised colonies that may be pigmented. Growth on these media requires a large inoculum probably containing some tissue from the specimen.

Cultivation of *M. chelonei* and *M. fortuitum* on Ogawa's egg medium at 25–30°C produces rapid growth within 7 days with smooth, moist, shiny and creamy to buff coloured colonies (Arakawa and Fryer, 1984). On Sauton's medium, colonies of *M. fortuitum* are rough and creamy to buff in colour while those of *M. chelonei* are initially smooth but tend to become rough after prolonged incubation and those of *M. marinum* are smooth and moist with a bright yellow-orange colour when grown in the light or exposed to light for 24 h after incubation in the dark. Other media may produce further colony variations; however, the important characteristics are (1) pleomorphic rods that maintain the acid-fast property upon subculture, (2) growth on medium within 7 days at either 25 or 37 °C and (3) pigment production. Each of these three species are considered rapid growers (visible growth within 7 days) and only *M. marinum* has been shown to be photochromogenic. Not all strains will grow at 37 °C. Three tests are especially useful to separate *M. chelonei* and *M. fortuitum*. The latter utilizes sucrose, reduces nitrate to nitrite (Tsukamura, 1965) and shows iron uptake. *Mycobacterium chelonei* is negative for each of these tests. Distinctive mycolic acid patterns produced by two-dimensional thin-layer chromatography may also be useful for separating these species (Arakawa and Fryer, 1984).

5.11 THE NOCARDIOFORMS

5.11.1 Genus *Nocardia*

Members of this genus are strictly aerobic and non-motile bacteria with mycelium production consisting of either aerial or substrate hyphae or both. The *Nocardia* do not produce spores on differentiated hyphae and mycelial fragments function as reproductive bodies. Two species of *Nocardia* have been isolated from fish: *Nocardia asteroides* from aquarium fish and the proposed *Nocardia kampachi* from cultured yellowtail in Japan (Kusuda *et al.*, 1974).

These bacteria can be isolated from the internal organs or lesions (often in the gills) of diseased fish on bacteriological media such as BHI, TSA or NA incubated at 20–30 °C. On NA, growth of *N. asteroides* appears in 4–5 days as folded, granular colonies which may vary from white to yellow-orange and with aerial hyphae that are most prevalent along the colony edge.

Mycelial fragments (about 1 μm in diameter), coccoid or spore cells, and bacilli are Gram-positive and may also be acid-fast. Preparations should be examined closely for the presence of branched mycelia. This characteristic of *Nocardia* can be used to separate them from filamentous bacteria. The acid-fast characteristic is frequently very weak and difficult to establish; although, some species of *Nocardia* will become strongly acid-fast when grown in litmus milk for several weeks. Coccoid or spore cells produced by *Nocardia* are not as resistant as spores of the bacilli but will, nevertheless, survive temperatures of 80 °C for several hours. In older cultures, bacilli, coccoid cells, and hyphae fragments are commonly found in the same preparation.

5.12 CULTURE MEDIA

The majority of media and reagents indicated in the text are routinely used for the isolation and growth of many species of bacteria and along with the details of preparation are available commercially. Those described in detail in this section are either used primarily, or formulated specifically, for culture and identification of the aquatic micro-organisms of fin fishes and cannot be obtained in commercial formulation.

Cytophaga agar

Pacha, R. E., and Ordal, E. J. (1967). Histopathology of experimental columnaris disease in young salmon. *Journal of Comparative Pathology*, **77**, 419–423.

Tryptone	0.5 g
Yeast extract	0.5 g
Sodium acetate	0.2 g
Beef extract	0.2 g
Agar	9–11 g
Distilled water	1000 ml

Ingredients in the medium are dissolved by heating and the pH adjusted to 7.2–7.4 prior to autoclaving.

Kidney disease medium-2 (KDM-2)

Evelyn, T. P. T. (1977). An improved growth medium for the kidney disease bacterium and some notes on using the medium. *Bulletin del' Office International der Epizootie*, **87**, 511–513.

Peptone	10 g
Yeast extract	0.5 g
Cysteine-HCl	1 g
Agar	15 g
Distilled water	800 ml

Peptone, yeast extract and cysteine are dissolved in the distilled water and adjusted to pH. 6.5 with NaOH. Agar is added, dissolved by heating and this base medium sterilized. Immediately prior to use the medium is dissolved, cooled to 45 °C and 20% v/v fetal calf serum added. Base medium should not be stored longer than 3 months. The slow growth of *Renibacterium salmoninarum* necessitates that inoculated plates be tightly wrapped to ensure adequate surface moisture during incubation.

Kidney disease medium: charcoal

Daly, J. G. and Stevenson, R. M. W. (1985). Charcoal agar, a new growth medium for the fish disease bacterium *Renibacterium salmoninarum*. *Applied and Environmental Microbiology*, **50**, 868–871.

Peptone	10 g
Yeast extract	0.5 g
L-cysteine HCl	1 g
Agar	15 g
Activated charcoal	1 g
Distilled water	1000 ml

The ingredients are dissolved by heating, adjusted to pH 6.8 with NaOH and autoclaved. To obtain a clear broth medium the charcoal is placed in dialysis tubing and loosely tied to prevent rupture during autoclaving.

Selective kidney disease medium

Austin, B., Embley, T. M. and Goodfellow, M. (1983). Selective isolation of *Renibacterium salmoninarum*. *FEMS Microbiology Letters*, **17**, 111–114.

Tryptone	10 g
Yeast extract	0.5 g
Cycloheximide	0.05 g
Agar	10 g
Distilled water	900 ml

The ingredients are dissolved by heating and the pH adjusted to 6.8. After autoclaving, fetal calf serum (10% v/v) and filter sterilized (0.22 μm) solutions of L-cysteine HCl (0.1% w/v), D-cycloserine (0.00125% w/v), polymyxin B sulphate (0.0025% w/v) and oxolinic acid (0.00025% w/v) are added aseptically to the cooled basal medium.

Ogawa's egg medium

Tsukamura, M. (1967). Identification of mycobacteria. *Tubercle*, **48**, 311–338.

Sodium glutamate	1 g
KH$_2$PO$_4$	1 g
Distilled water	100 ml
Whole eggs	200 ml
Glycerol	6 ml
Malachite green	6 ml of 2% (w/v) solution

Dissolve the sodium glutamate and KH_2PO_4 in distilled water and mix in the remaining ingredients. Iron uptake can be demonstrated by the growth of orange-red colonies on basal medium containing 2% (w/v) ferric ammonium citrate. Adjust the pH to 6.8 and sterilize as slopes in culture tubes at 90 °C for 1 hour.

Rimler–Shotts medium

Shotts, E. B. and Rimler, R. (1973). Medium for the isolation of *Aeromonas hydrophila. Applied Microbiology,* **26,** 550–553.

L-lysine-HCl	5 g
L-ornithine-HCl	6.5 g
Maltose	3.5 g
Sodium thiosulphate	6.8 g
L-cysteine-HCl	0.3 g
Bromothymol blue	0.03 g
Ferric ammonium citrate	0.8 g
Sodium deoxycholate	1 g
Novobiocin	0.005 g
Yeast extract	3 g
Sodium chloride	5 g
Agar	13.5 g
Distilled water	1000 ml

The ingredients are dissolved by stirring and the pH adjusted to 7.0. The preparation is then heated to a boil for 1 min, cooled to 45 °C and dispensed. Prepared plates are refrigerated until needed.

Sauton's medium

Tsukamura, M. (1965). Salicylate degradation test for differentiation of *Mycobacterium fortuitum* from other mycobacteria. *Journal of General Microbiology,* **41,** 309–315.

Glycerol	50 ml
Sodium glutamate	4 g
KH_2PO_4	0.5 g
$MgSO_4 \cdot 7H_2O$	0.5 g
Sodium citrate	2 g
Ferric ammonium citrate	0.05 g
Agar	30 g
Distilled water	950 ml

The ingredients are dissolved in the distilled water and the pH adjusted to

7.0 with 10% (w/v) KOH. Sterilization is by autoclaving at 115 °C for 30 min.

Shotts–Waltman medium

Waltman, W. D. and Shotts, E. B. (1984). A medium for the isolation and differentiation of *Yersinia ruckeri*. *Canadian Journal of Fisheries and Aquatic Sciences*, **41**, 804–806.

Tryptone	2 g
Yeast extract	2 g
Tween 80	10 ml
Sodium chloride	5 g
Calcium chloride (CaCl$_2 \cdot$2H$_2$O)	0.1 g
Brom thymol blue	0.003 g
Agar	15 g
Distilled water	980 ml

The ingredients are dissolved by heating and the pH adjusted to 7.4. After sterilization and cooling to 50 °C, 10 ml of 0.5 g ml^{-1} filter sterilized sucrose is added.

Acknowledgements

Preparation of this chapter was supported by the Oregon Department of Fish and Wildlife under the Anadromous Fish Act PL 89304 and by Oregon Sea Grant through NOAA Office of Sea Grant, Department of Commerce, under grant no. NA85AA-O-SG095 (project No. R/FSD-10). This is Technical Paper No. 7691, Oregon Agricultural Experiment Station.

5.13 REFERENCES

Amandi, A., Hiu, S. F., Rohovec, J. S. and Fryer, J. L. (1982). Isolation and characterization of *Edwardsiella tarda* from fall chinook salmon (*Oncorhynchus tshawytscha*). *Applied and Environmental Microbiology*, **43**, 1380–1384.
Arakawa, C. K. and Fryer, J. L. (1984). Isolation and characterization of a new subspecies of *Mycobacterium chelonei* infectious for salmonid fish. *Helgoländer Meeresuntersuchungen*, **37**, 329–342.
Bejerano, Y., Sarig, S., Horne, M. T. and Roberts, R. J. (1979). Mass mortalities in silver carp *Hypophthalmichthys molitrix* (Valenciennes) associated with bacterial infection following handling. *Journal of Fish Diseases*, **2**, 49–56.
Brenner, D. J. (1984). Enterobacteriaceae. In Bergey's Manual of Systematic Bacteriology, vol. 1, N. R. Krieg (Ed.) Williams and Wilkins, Baltimore, pp. 408–516.
Bullock, G. L., Conroy, D. A. and Snieszko, S. F. (1971). Bacterial diseases of fish.

Diseases of Fishes, book 2A S. F. Snieszko and H. R. Axelrod (Eds), T. F. H. Publications Inc. Ltd, Hong Kong.

Cann, D. C. and Taylor, L. Y. (1982). An outbreak of botulism in rainbow trout, Salmo gairdneri Richardson, farmed in Britain. Journal of Fish Diseases, 5, 393–399.

Cipriano, R. C., Starliper, C. E. and Schachte, J. H. (1985). Comparative sensitivities of diagnostic procedures used to detect bacterial kidney disease in salmonid fishes. Journal of Wildlife Diseases, 21, 144–148.

Cone, D. K. (1982). A Lactobacillus sp. from diseased female rainbow trout, Salmo gairdneri Richardson, in Newfoundland, Canada. Journal of Fish Diseases, 5, 479–485.

Eklund, M. W., Poysky, F. T. and Wieler, D. I. (1967). Characteristics of Clostridium botulinum type F isolated from Pacific Coast of the United States. Applied Microbiology, 15, 1316–1323.

Eklund, M. W., Poysky, F. T., Peterson, M. E., Peck, L. W. and Brunson, W. D. (1984). Type E botulism in salmonids and conditions contributing to outbreaks. Aquaculture, 41, 293–309.

Facklam, R. R. (1980). Streptococci and Aerococci. In Manual of Clinical Microbiology, 3rd edn, E. H. Lennette, A. Balows, W. J. Hausler, and J. P. Truant (Eds), American Society for Microbiology, Washington, DC, pp. 88–110.

Gordon, R. E. and Mihm, J. M. (1959). A comparison of four species of mycobacteria. Journal of General Microbiology, 21, 736–748.

Hawke, J. P., McWhorter, A. C., Steigerwalt, A. G. and Brenner, D. J. (1981). Edwardsiella ictaluri sp. nov., the causative agent of enteric septicemia of catfish. International Journal of Systematic Bacteriology, 31, 396–400.

Henley, M. W. and Lewis, D. H. (1976). Anaerobic bacteria associated with epizootics in grey mullet (Mugil cephalus) and redfish (Sciaenops ocellata) along the Texas gulf coast. Journal of Wildlife Diseases, 12, 448–453.

Hikida, M., Wakabayashi, H., Egusa, S. and Masumura, K. (1979). Flexibacter sp., a gliding bacterium pathogenic to some marine fishes in Japan. Bulletin of The Japanese Society of Scientific Fisheries, 45, 421–428.

Hiu, S. F., Holt, R. A., Sriranganathan, N., Seidler, R. J. and Fryer, J. L. (1984). Lactobacillus piscicola, a new species from salmonid fish. International Journal of Systematic Bacteriology, 34, 393–400.

Holt, R. A., Sanders, J. E., Zinn, J. L., Fryer, J. L. and Pilcher, K. S. (1975). Relation of water temperature to Flexibacter columnaris infection in steelhead trout (Salmo gairdneri), coho (Oncorhynchus kisutch) and chinook (O. tshawytscha) salmon. Journal of the Fisheries Research Board of Canada, 32, 1553–1559.

Hoshina, T., Sano, T. and Morimoto, Y. (1958). A Streptococcus pathogenic to fish. Journal of the Tokyo University of Fisheries, 44, 57–68.

Janssen, W. A. and Surgalla, M. J. (1968). Morphology, physiology, and serology of a Pasteurella species pathogenic for white perch (Roccus americanus), Journal of Bacteriology, 96, 1606–1610.

Kitao, T. (1982). The methods for detection of Streptococcus sp., causative bacteria of streptococcal disease of cultured yellowtail (Seriola quinqueradiata)—especially, their cultural, biochemical and serological properties. Fish Pathology, 17, 17–26.

Kitao, T., Aoki, T. and Sakoh, R. (1981). Epizootic caused by β-haemolytic Streptococcus species in cultured freshwater fish, Fish Pathology, 15, 301–307.

Kloos, W. E. and Smith, P. B. (1980). Staphylococci. In Manual of Clinical Microbiology, 3rd edn, E. H. Lennette, A. Balows, W. J. Hausler, and J. P. Truant (Eds), American Society for Microbiology, Washington, DC, pp.83–87.

Kusuda, R. and Komatsu, I. (1978). A comparative study of fish pathogenic *Streptococcus* isolated from saltwater and freshwater fishes. *Bulletin of the Japanese Society of Scientific Fisheries*, **44**, 1073–1078.

Kusuda, R. and Sugiyama, A. (1981). Studies on the characters of *Staphylococcus epidermidis* isolated from diseased fishes. I. On the morphological, biological and biochemical properties, *Fish Pathology*, **16**, 15–24.

Kusuda, R., Taki, H. and Takeuchi, T. (1974). Studies on norcardia infection of cultured yellowtail—II Characteristics of *Norcardia kampachi* isolated from a gill tuberculosis of yellowtail. *Bulletin of the Japanese Society of Scientific Fisheries*, **40**, 369–373.

Kvasnikov, E. I., Kovalenko, N. K and Materinskaya, L. G. (1976). Lactic acid bacteria of freshwater fish. *Mikrobiologiya*, **46**, 755–760.

Martin, W. J. and Washington II, J. A. (1980). Enterobacteriaceae. In *Manual of Clinical Microbiology*, 3rd edn, E. H. Lennette, A. Balows, W. J. Hausler and J. P. Truant (Eds), American Society for Microbiology, Washington, DC, pp. 195–219.

Moore, W. E. C. and Holdeman Moore, L. V. (1986). *Eubacterium*. In *Bergey's Manual of Systematic Bacteriology*, vol. 2, P. H. A. Sneath (Ed.), Williams and Wilkins, Baltimore, pp. 1353–1373.

Nigrelli, R. F. and Vogel, H. (1963). Spontaneous tuberculosis in fishes and other cold-blooded vertebrates with special reference to *Mycobacterium fortuitum* Cruz from fish and human lesions. *Zoologica*, **48**, 131–144.

Paterson, W. D., Douey, D. and Desautels, D. (1980a). Isolation and identification of an atypical *Aeromonas salmonicida* strain causing epizootic losses among Atlantic salmon (*Salmo salar*) reared in a Nova Scotian hatchery. *Canadian Journal of Fisheries and Aquatic Science*, **37**, 2236–2241.

Paterson, W. D., Douey, D. and Desautels, D. (1980b). Relationships between selected strains of typical and atypical *Aeromonas salmonicida*, *Aeromonas hydrophila* and *Haemophilus piscium*. *Canadian Journal of Microbiology*, **26**, 588–598.

Plumb, J. A., Schachte, J. H., Gaines, J. L., Peltier, W. and Carroll, B. (1974). *Streptococcus* sp. from marine fishes along the Alabama and northwest Florida coast of the Gulf of Mexico. *Transactions of the American Fisheries Society*, **103**, 358–361.

Popoff, M. (1984). *Aeromonas*. In *Bergey's Manual of Systematic Bacteriology*, vol. 1, N. R. Krieg (Ed.), Williams and Wilkins, Baltimore, pp. 545–550.

Ransom, D. P., Lannan, C. N., Rohovec, J. S. and Fryer, J. L. (1984). Comparison of histopathology caused by *Vibrio anguillarum* and *Vibrio ordalii* in three species of Pacific salmon. *Journal of Fish Diseases*, **7**, 107–115.

Rogol, M., Sechter, I., Grinberg, L. and Gerichter, C. B. (1979). Pril-xylose-ampicillin agar, a new selective medium for the isolation of *Aeromonas hydrophila*. *Journal of Medical Microbiology*, **12**, 229–231.

Sanders, J. E. and Fryer, J. L. (1980). *Renibacterium salmoninarum* gen. nov., sp. nov., the causative agent of bacterial kidney disease in salmonid fishes. *International Journal of Systematic Bacteriology*, **30**, 496–502.

Sanders, J. E. and Fryer, J. L. (1986). Renibacterium. In *Bergey's Manual of systematic Bacteriology*, vol 2, P. H. A. Sneath (Ed.), Williams and Wilkins, Baltimore, pp. 1253–1254.

Schachte, J. H. (1974). Bacterial destruction of catfish eggs. *Catfish Farmer*, 4 pp.

Schiewe, M. H., Trust, T. J. and Crosa, J. H. (1981). *Vibrio ordalii* sp. nov.: A causative agent of vibriosis in fish. *Current Microbiology*, **6**, 343–348.

Schleifer, K. H. and Kloos, W. E. (1975). A simple test system for the separation of staphylococci from micrococci. *Journal of Clinical Microbiology*, **1**, 337–338.

Shotts, E. B. and Rimler, R. (1973). Medium for the isolation of *Aeromonas hydrophila*. *Applied Microbiology*, **26**, 550–553.

Trust, T. J. and Sparrow, R. A. H. (1974). The bacterial flora in the alimentary tract of freshwater salmonid fishes. *Canadian Journal of Microbiology*, **20**, 1219–1228.

Tsukamura, M. (1965). Salicylate degradation test for differentiation of *Mycobacterium fortuitum* from other mycobacteria. *Journal of General Microbiology*, **41**, 309–315.

Udey, L. R., Young E. and Sallman, B. (1977). Isolation and characterization of an anaerobic bacterium, *Eubacterium tarantellus* sp. nov., associated with striped mullet (*Mugil cephalus*) mortality in Biscayne Bay, Florida. *Journal of the Fisheries Research Board of Canada*, **34**, 402–409.

Wakabayashi, H., Egusa, S. and Fryer, J. L. (1980). Characteristics of filamentous bacteria isolated from a gill disease of salmonids. *Canadian Journal of Fisheries and Aquatic Science*, **37**, 1499–1504.

Wakabayashi, H., Hikida, M. and Masumura, K. (1986). *Flexibacter maritimus* sp. nov., a pathogen of marine fishes. *International Journal of Systematic Bacteriology*, **36**, 396–398.

Waltman, W. D. and Shotts, E. B. (1984). A medium for the isolation and differentiation of *Yersinia ruckeri*. *Canadian Journal of Fisheries and Aquatic Science*, **41**, 804–806.

Xu, H. (1975). Studies on the gill disease of the grass carp (*Ctenopharyngodon idelluls*). I. Isolation of a myxobacterial pathogen. *Acta Hydrobiologica Sinica*, **5**, 315–329.

Yoshimizu, M. and Kimura, T. (1976). Study on the intestinal microflora of salmonids. *Fish Pathology*, **10**, 243–259.

Methods in Aquatic Bacteriology
Edited by B. Austin
© 1988 Crown copyright
Published by John Wiley & Sons Ltd.

6

Bacteria of Aquatic Invertebrates

P. A. West

Aquatic Environment Protection Division 2, Directorate of Fisheries Research, Ministry of Agriculture, Fisheries and Food, Burnham-on-Crouch, Essex CM0 8HA, England*

6.1 INTRODUCTION

Natural aquatic ecosystems usually support diverse, heterogenous and relatively stable bacterial populations (Rheinheimer, 1985). The bacteria can be considered either indigenous (autochthonous) to the habitat or occur as a result of introduction from other habitats (allochthonous). These bacteria may be free-living in the water, attached to sediment particles, or specifically associated with the internal and external surfaces of organisms at higher trophic levels. The development of highly specific consortia of bacteria in certain ecological microniches, such as the surfaces of invertebrates, is not unusual (Schlegel and Jannasch, 1981).

The bacterial flora of aquatic invertebrates appears to be a significant component in the overall contribution of micro-organisms to macro-scale productivity in aquatic ecosystems such as freshwater lakes or parts of the ocean (Tietjen, 1980). Evidence increasingly suggests that some bacteria are a direct food source for phytoplankton (Bird and Kalff, 1986) as well as for invertebrates at higher trophic levels in aquatic environments (Birkbeck and McHenery, 1982; Carman and Thistle, 1985). Accordingly, the specific interactions between bacteria and aquatic invertebrates of the meiofauna (e.g. microscopic metazoans) and macrofauna (e.g. amphipods) at the base of food chains may substantially influence the flow of energy through subsequent higher trophic levels (Pomeroy, 1980; Cavanaugh, 1983; Jannasch, 1985).

Unfortunately, the information is usually fragmentary for bacterial interactions with those aquatic invertebrates likely to be of little or no

---

*Present address: Ministry of Agriculture, Fisheries and Food, Fish Diseases Laboratory, Weymouth, Dorset, DT4 8UB, England.

economic importance. Despite characterization of several diverse bacteria–invertebrate interactions featuring, for example, luminous bacteria (Nealson and Hastings, 1979; O'Brien and Sizemore, 1979), the large spirochaetes (*Cristispira*) of molluscs (Kuhn, 1981), bacteria with antifungal properties (Fisher, 1983) or the highly pigmented bacteria in cuttlefish (Van Den Branden *et al.*, 1980), their role and significance to the general ecology of aquatic environments remains obscure.

In contrast, the emphasis of most published work is directed towards bacteria associated with those invertebrates, such as shellfish, which are commercially exploited by aquaculture techniques. For example, the settlement and metamorphosis of planktonic larval stages of molluscan shellfish has been shown to be enhanced by the presence of attractant chemicals produced by bacterial films on the surface of structures in aquatic environments (Weiner *et al.*, 1985). Some bacteria–invertebrate interactions can, however, be detrimental and result in severe economic loss. Large mortalities of shellfish can occur when there is disturbance to the delicate association between these animals and their specific adherent bacterial population. Under these circumstances, the commensal saprophytic bacteria, derived from the surrounding water, become opportunist pathogens for the animal which they have colonized (Sindermann, 1984). Additionally, there are some bacteria which are obligate pathogens of shellfish and so the need for their control and eradication has prompted extensive characterization (Sindermann, 1970).

Shellfish can rapidly accumulate human enteric pathogens from harvesting waters which receive effluent from sewage treatment plants and, without adequate treatment before consumption, may cause outbreaks of gastroenteritis (West, 1986). Characterization of the bacterial flora of shellfish is widely employed to assess the sanitary quality of live shellfish (Hood *et al.*, 1983). Interestingly, there are some bacteria which are indigenous to aquatic environments but which occasionally cause infections in humans. These bacteria belong predominantly to the genera *Vibrio* and *Aeromonas* and their ability to cause disease in humans critically depends on attachment and growth on the surfaces of chitinous invertebrates to numbers high enough for initiation of disease (Colwell, 1984).

Bacteria–invertebrate interactions are therefore significant in aquatic environments for a variety of reasons including mediation of energy transfer through trophic levels in food webs, economic prosperity of shellfish production and transmission of infection amongst humans. To illustrate the various techniques available for studying bacteria–invertebrate interactions, examples have been selected and reviewed from these procedures which have been used successfully in previous specific studies. This general representative range of procedures can be modified as appropriate to suit individual requirements. Although emphasis has been placed on the marine and estuarine environment, most procedures are applicable to freshwater studies with an appropriate adjustment of the salinity and osmotic pressure of culture media or test conditions.

Rickettsiae and rickettsia-like organisms are not considered here in detail. Gulka and Chang (1984) have presented a review of several rickettsia–invertebrate associations as well as specific descriptions of methods useful to characterize rickettsia in scallops. A summary of the studies describing the occurrence of rickettsiae in marine bivalve molluscs is given by Lauckner (1983a).

6.2 COLLECTION AND HANDLING OF AQUATIC INVERTEBRATES

Wherever possible, it is advisable to seek advice or consult literature on the basic biological characteristics of the invertebrate species to be studied to ascertain whether the method of collection will influence the composition of the bacterial flora. Useful sources of information on the biology of aquatic invertebrates have been assembled by Hickman (1967) and Purchon (1977).

6.2.1 Copepods and related meiofauna

These invertebrates may be removed from aquatic samples by sieving techniques. Meiofauna often proliferate around plant life and will be recovered from the water column by repeatedly dipping a net of appropriate mesh size into the water. Alternatively, trawl nets can be used. Trawl speed must not be too fast as to cause damage to the invertebrates if they are required alive for subsequent study. Trawling at 2 knots speed is usually adequate. A fine trawl net will have a mesh size in the range around 70 μm and coarse mesh sizes are around 300 μm. Simidu *et al.* (1971) described the simultaneous use of two different mesh sizes to collect different size fractions of plankton from seawater whereby the larger mesh size (320 μm) net was placed inside the smaller mesh size (112 μm) net and then both were trawled as a single unit. Where trawling with plankton nets is impractical due to shallow-water or obstructions, the use of pumps to filter large volumes of water through fixed plankton nets should be considered (Dixon and Robertson, 1986).

Huq *et al.* (1983) collected planktonic copepods by trawling in order to demonstrate that *Vibrio cholerae* would adhere to the surface of live chitinous invertebrates. In order to remove the indigenous bacterial flora before inoculation with *V. cholerae*, freshly caught copepods were washed by gently pouring up to 4 l of sterilized seawater over approximately 3000 copepods retained on a steel gauze filter. Examination of the copepod surface by scanning electron microscopy revealed virtually complete removal of bacteria with no structural damage to the copepod.

For collection of sediment-bound (benthic) meiofauna invertebrates, the use of metal sieves of appropriate grid size is recommended with preliminary sieving through a grid of large mesh size (usually over 1500 μm) to remove debris and macrofauna (McIntyre and Warwick, 1984).

Samples should be kept in inert plastic containers or glass bottles containing water from the site of collection and transported quickly to laboratory facilities. The storage time for live animals will be shortened if the sample is kept at the extremes of their thermal tolerance range. Samples should be analysed as soon after collection as possible to avoid changes in the proportions of the bacterial population which can occur during storage of samples.

6.2.2 Shellfish

The collective term 'shellfish' includes the crustaceans which are mobile animals with a hard articulated exoskeleton, and the molluscs (excluding cephalopods) with hard external shells but which are sedentary or have limited locomotion. From a commercial standpoint, these molluscs can be divided into the bivalves, which have two shells joined by a hinge, and the gastropods, many of which have a whorled snail-like shell.

6.2.2.1 Crustaceans

Lobsters and crabs may be caught by use of submerged and baited pots which collect several animals. Occasionally, diseased or moribund animals are not attracted to baited pots and trawling the seabed or scuba diving should be used to gather samples.

Crustaceans should be handled carefully to avoid damage to their limbs which would lead to contamination of the haeomolymph and result in alteration of the bacterial flora throughout the animal (Tubiash et al., 1975). The use of thick, protective golves for handling crustaceans is recommended to avoid personal injury. The irritable nature of a freshly caught crab or lobster can be reduced by exposing the animal for short periods (ca. 10 min) to the chilled air inside a − 20 °C deep freezer. Provided the animal does not touch the cold surfaces of the freezer, this does not cause dramatic alteration in the bacterial flora and permits safer removal of samples.

Samples should be transported to the laboratory in a moist chamber to prevent desiccation. Samples should not be stored for lengthy periods in seawater unless it is well aerated and chilled. Where this is not possible, samples may be frozen. Faghri et al. (1984) reported no significant differences in bacterial counts between freshly caught crabs and similar material frozen for up to 2 weeks. Nevertheless, changes in the *proportions* of the components of the bacterial population may arise so freezing of samples prior to bacteriological examination is often undesirable.

Other crustaceans, such as shrimp, can be collected by trawling with a 3 m otter trawl net of appropriate mesh size and handling according to the general procedures listed above for crabs and lobsters.

6.2.2.2 Molluscs

The gathering of several species of molluscan shellfish from some areas is protected by law either for conservation purposes or for protection of public health. It is advisable to investigate whether any local restrictions relating to catch size or quantity, areas for harvesting or close seasons exist prior to sample collection.

Whelks can be either fished using a beam trawl (Eleftheriou and Holme, 1984) or taken from the by-catch of nets used to catch bottom-dwelling (demersal) fish. Winkles are easily picked by hand from rocks in the intertidal zone. Bivalve molluscs such as oysters, mussels, cockles and clams can either be dredged from the seabed or dug from sediments with rakes or tongs. After collection, samples should be rinsed or hosed with water to remove mud or loose fauna from the shells. Samples can be transported in a moist chamber at around 8–10 °C for several hours before changes in the composition of the bacterial flora become apparent (Lovelace *et al.*, 1968). Bivalve molluscan shellfish should not be reimmersed in water during transportation because these animals will continue their filter-feeding activity and so cause dramatic alterations in counts of the bacterial flora within a few hours (Son and Fleet, 1980).

Where bivalve molluscan shellfish are examined for public health purposes, several important handling requirements are needed to avoid contamination of animals after harvesting or substantial changes in their bacterial content. These requirements include removal of mud from shell crevices, discarding animals with cracked shells and constraints on storage temperature and time prior to examination (American Public Health Association, 1985; West and Coleman, 1986).

6.2.3 Other macroinvertebrates

Large animals, such as echinoderms, should be collected by scuba diving, trawling with nets or dredged, or remote-controlled rigs, depending on the ecology and habitat of the species. The use of hand-operated nets or dredges is feasible at low water and these have been described in greater detail by Eleftheriou and Holme (1984).

Specialized equipment may be required for study of specific habitats such as deep-environments. The use of trawls or grabs in these circumstances will often cause unacceptable trauma and decompression to animals during retrieval. There are only a few reports of the use of apparatus specifically designed for recovery of compressed samples from deep-sea regions. Traps for deep-sea amphipods placed at the seabed have been described by Yayanos (1978). These incorporated pressure-retaining devices, water aeration facilities and thermal insulation to protect animals during ascent from cold, deep-sea habitats. It was also feasible to maintain animals under conditions which mimic the deep-sea environment in the traps for several days in order to carry out further studies in the laboratory. In contrast, Jannasch *et al.* (1980) developed apparatus and techniques to study the uptake of radiolabelled foods by amphipods on the deep-sea floor. Baited traps were placed on the seabed using the submersible vessel *Alvin* and recovered later to determine the rate of utilization of foods by these invertebrates.

6.3 EXAMINATION PROCEDURES

Where specific organs are removed from invertebrates for bacteriological analysis, it is essential to avoid cross-contamination during dissection. This is usually difficult as fluids often bathe many of the internal organs of invertebrates and have to be washed away prior to dissection.

6.3.1 Copepods and related meiofauna

Homogenization of the whole animal with a small amount of sterile water, taken originally from the site of sample collection, in a teflon-tipped tissue-grinder is recommended for examination of the bacterial flora associated with whole animals (Boyle and Mitchell, 1981). The period of homogenization is halted when the sample appears visually to be uniformly disintegrated (Simidu *et al.*, 1971; Sochard *et al.*, 1979; Huq *et al.*, 1983).

In some species, it is feasible to dissect the rudimentary internal organs. In these circumstances, the bacterial flora of the animal surfaces should be removed by copious, gentle washing in sterile, flowing water to minimize cross-contamination during removal of the internal organs. Alternatively, invertebrate material can be separated from surface-associated bacteria by centrifugation in a sucrose gradient (Berk *et al.*, 1976). Sochard *et al.* (1979) described a method to remove the guts from small (*ca.* 1 mm) copepods whereby their heads were embedded in agar and short (3 mm) and fine (76 μm diameter) tungsten wires attached to glass rods were used to hook out the digestive tract.

6.3.2 Crustaceans

Examination of crustacean haemolymph (blood) is usually undertaken when animals are unhealthy and refusing food but where there is no overt indication of disease (Welsh and Sizemore, 1985). In addition, the detection of the disease of lobsters called Gaffkaemia requires sampling of the haemolymph to detect the bacterial pathogen *Aerococcus* (*Gaffkya*) *viridans* subsp. *homari* (Stewart, 1984).

The haemolymph from crabs is removed aseptically by puncture, using a sterile needle (18- or 23-gauge) and a 5 ml syringe, into the cardiac sinus through the intersegmental membrane. This membrane is located between the posterior of the carapace and the abdomen. Care should be taken to avoid entry into the hepatopancreas which is adjacent to this region. Alternatively, samples of haemolymph can be removed from the ventral blood sinus by puncture of the uncalcified membrane at the base of each swimming leg (Welsh and Sizemore, 1985). In order to avoid cross-contamination from the shell surfaces, the puncture site should be wiped clean and swabbed with iodine tincture or 70% ethanol before insertion of the needle. Crabs can be safely handled by grasping the base of one of the swim fins with the thumb and bracing the same hand against the carapace as illustrated by Tubiash *et al.* (1975). Large animals should alternatively be strapped to an immobilization platform if necessary.

Removal of haemolymph from lobsters is a similar procedure where the animal is bled from the ventral sinus after swabbing the puncture site with 70% ethanol (Brinkley *et al.*, 1976). The two main lobster claws should be banded with tape to protect the handler from injury. Haemolymph can then be either examined directly by gram-staining or cultured in selective media for the pathogen.

Examination of meats from crustaceans is usually only undertaken when assessing the sanitary quality of animals (Faghri *et al.*, 1984). Since raw crab or lobster meat is rarely sold for consumption, samples are usually taken at each stage of commercial cooking, packing and storage operations to detect entry or multiplication of potential pathogens or spoilage organisms (Phillips and Peeler, 1972). Where it is necessary to remove raw meat from crab shell, especially the claws, samples should be wrapped loosely in sterile cloths and struck with a blunt instrument until the shell cracks sufficiently to aseptically remove the meat. Other methods for removal of crab meat and internal organs include drilling holes in the shells or squeezing meat from dismembered limbs (Lee and Pfeifer, 1975).

Removal of internal organs from crustaceans can only be practically undertaken once the animal is dead and the shell can be prised open or broken with ease. Egidius (1972) reported the deep-freezing of animals before dissection of the various internal organs. It is recommended that the effect of such procedures on the invertebrate bacterial populations is ascertained since freezing and thawing can also eradicate many micro-organisms. If necessary, it is possible to remove some small and loosely attached external organs from live animals using sterile scalpels or forceps (Harper and Talbot, 1984).

It is common to find erosive or necrotic lesions on the exoskeleton of crustaceans which may be due to a variety of chitinolytic bacteria (Sindermann, 1984). Isolation of these bacteria from surface lesions on large crustaceans is accomplished by taking swabs directly from lesions and culturing on suitable media. In order to restrict overgrowth of the lesion-associated bacteria by other surface micro-organisms, it is some-times necessary to dip the infected area briefly in absolute alcohol prior to swabbing (El-Gamal et al., 1986).

Swabbing of small crustacean larvae or animals is not practical so removal of loose attached surface-associated bacteria is achieved by shaking samples in sterile water and culturing aliquots after settling out of the material under examination. Alternatively, small portions of diseased shell should be removed with a sterile instrument and streaked over the surface of a suitable nutrient agar medium (Cook and Lofton, 1973).

The bacterial flora associated with the internal organs of small crustacean larvae or animals can be cultured by first removing external surface-associated bacteria by washing and sterilizing with 0.002% (w/v) iodine before homogenizing the whole sample in a tissue grinder (Austin and Allen, 1982). When there is no requirement to distinguish between external or internal surface-associated bacteria, whole larve should be simply homogenized (Colorni, 1985).

6.3.3 Molluscs

Raw meats from gastropod molluscs, such as whelks and periwinkles, are removed from shells by aseptically breaking the shell and dissecting out various internal organs. Cooked meats of these species shrink in the shell during processing and so can be removed with a sterile needle without the need to break open the shell.

There is no such requirement to break open shells of bivalve molluscan shellfish to remove raw meats since the two shells can be prised open to expose the internal organs once the adductor muscle has been severed. Removal of meats from bivalve molluscan shellfish in this manner is commonly known as 'shucking'. It is necessary to use a shucking knife with a strengthened blade and sharp tip to enter the hinge area between the shells, prise the shells open and sever the adductor muscle attachments (American Public Health Association, 1985; West and Coleman, 1986). If shells are fragile and crack during handling, a sterile object can be wedged between the shells while open when the animal is still filter-feeding in water. Upon removal from water, the shells will not close fully thus permitting easier insertion of the shucking knife blade.

Care is required when severing adductor muscle attachments to prevent trauma to the internal organs and possible cross-contamination by bacteria from one organ to another. In general, the mollusc should be opened with the flat or shallow shell uppermost; its muscle attachments can then be severed, and this shell removed, to leave the internal organs in the deeper shell.

In large bivalve molluscan shellfish, stomach contents can be removed with a syringe and wide-bore needle. Other areas of the disgestive system can be detected visually by a brown ribbon of food towards the stomach if animals have fed recently. Where it is necessary to remove different internal organs for bacteriological examination, the whole animal viscera should be flushed with sterile water to remove surface bacteria and liquor (mantle fluid) associated with the organs (Kueh and Chan, 1985). If required, samples of mantle fluid can be withdrawn using a sterile pipette before removal of organs (Lovelace *et al.*, 1968). Elston and Lockwood (1983) removed surface-associated bacteria from abalone tissue by swabbing with ethanol prior to making an incision into muscle tissue to obtain samples.

Removal of haemocytes or haemolymph samples from viscera can be made using a hypodermic needle and syringe from the anterior adductor muscle sinus (Feng *et al.*, 1971). If it is necessary for animals to remain alive for subsequent re-examination, samples of haemolymph can be removed via holes drilled into shells. When the hole is subsequently plugged with cotton wool and melted paraffin wax, animals may be returned to water to continue filter-feeding activity (Hartland and Timoney, 1979).

Samples of molluscan shellfish meat can be smeared onto the surface of culture media for isolation of the bacterial flora. In general, however, it is necessary to homogenize and dilute samples to achieve a consistency which permits ease and accuracy of inoculation into culture media. This is particularly desirable if bacteria are to be enumerated. Since disintegration of samples is essential for release of the bacterial flora from tissues, homogenization using blender jar units is preferred. A suitable period of treatment is blending at $12000\,\mathrm{rev\,min}^{-1}$ for 15s, with 15-s intervals to prevent heating of samples, for up to 2 min. Processing of samples with 'Stomacher' units should take up to 10 min or until the sample has visually disintegrated completely.

A suitable diluent must be used with these techniques to permit blending and to ensure survival of bacteria until the sample is transferred to an appropriate culture media. The diluent is often filtered and autoclaved water taken from the site of sample collection. The use of diluents containing peptone is advised for recovery of human pathogens from bivalve molluscan shelfish to assist recovery of sublethally-injured bacteria (Mossel and Van Netten, 1984; West and Coleman, 1986). Homogenization of shellfish meats, in particular viscera from bivalve molluscan shellfish, can release bactericidal compounds (Oliver, 1981) which may dramatically influence recovery of specific bacterial populations.

Studies on bacteria associated with cephalopods are limited. Van Den Branden *et al.* (1980) described procedures to culture pigmented bacteria from the nidamental glands of cuttlefish whereby internal organs were swabbed with 98% ethanol to ensure removal of surface-associated bacteria before sampling from the inside of the gland.

6.3.4 Other macroinvertebrates

The procedures described by Unkles (1977) for sea urchins is generally applicable to large invertebrates. The tough urchin shell was opened with sterile scissors to reveal the internal organs. Samples of gut, coelomic fluid and various internal membrane surfaces were aseptically removed by dissection, aspiration using syringes and needles and repeated washing and swabbing of surfaces, respectively.

The use of antibiotics such as penicillin, streptomycin and amphotericin B is recommended when it is desirable to eliminate indigenous bacteria associated with freshly caught animals prior to studies on animals challenged with specific bacterial species or on the bacterial-free behaviour of these invertebrates. Once the antibiotic sensitivity of representative bacteria has been determined, water can be dosed with antibiotics to permit removal of bacteria during storage of animals prior to study (Tietjen, 1980; Guerinot and Patriquin, 1981). Alternatively, the bacterial flora associated with some macroinvertebrates, such as oligochaetes, can be removed by careful washing of samples (Felbeck *et al.*, 1983).

Procedures for extraction of bacteria from the surface of polychaetes using ultrasonication have been briefly described by Alongi (1985). Such methods should be used cautiously to avoid death and destruction of bacterial cells during the period of treatment. Alternatively, procedures to remove surface-associated bacteria from macroinvertebrates by mechanical shaking should be evaluated (Baker and Bradnam, 1976).

6.4 BACTERIOLOGICAL MEDIA AND METHODS

Once suitable homogenization and dilution of the sample has been made, conventional bacteriological procedures are used to inoculate a variety of culture media. In general, most-probable-number (MPN) and spread plate methods are preferred for isolating and enumerating bacteria associated with aquatic invertebrates. The basic principles of these procedures have been described in detail by Koch (1981). A membrane filteration method has been reported by Kelly and Dinuzzo (1985) which involved coarsely mincing shellfish tissue in sterile seawater and culturing the supernatant. This procedure is useful if bactericidal compounds are released from samples during homogenization. Pour plate methods which involve exposure of bacteria to molten agar media are not advised since this causes considerable death of heat-sensitive bacteria and hence an underestimation of counts (Väätänen, 1977).

A variety of media have been devised for isolation of specific types of bacteria from aquatic environments (See Chapter 3). Some media for groups of bacteria such as the faecal coliforms or the pathogenic *Vibrio* species are commercially available. Where it is necessary to assess the batch-to-batch quality of media when studying human or shellfish pathogens, a simple statistical procedure for this purpose has been described by West *et al.* (1982).

It is well established that a large number of bacteria in aquatic environments cannot be cultured on conventional media, usually developed originally for use in clinical laboratories, which are merely supplemented with salts to increase the salinity as appropriate. A useful general growth medium for marine bacteria is seawater agar (Austin and Allen, 1982), comprising (gl^{-1}):Lab-Lemco powder (Oxoid) 8.0 g; Bacteriological Peptone (Oxoid) 10.0 g; Agar no. 3 (Oxoid) 12.0 g and prepared in 750 ml filtered and aged seawater plus 250 ml tapwater.

The commercial brands of 'marine agar' developed from the observations of Oppenheimer and Zobell (1952) are now considered too high in nutrient despite their continued widespread use for isolation of bacteria from marine and estuarine invertebrates. 'Marine agar' will select for fast-growing proteolytic bacteria while inhibiting growth of low-nutrient-preferring (oligotrophic) bacteria. A variety of low-nutrient media, mostly not available commercially at present, have been developed for use in estuarine or coastal marine environments (Goulder, 1976; Väätänen, 1977; Weiner *et al.*, 1980). The estuarine agar devised by Weiner *et al.* (1980) is recommended for general use with aquatic invertebrates from these environments. The medium is easily prepared by addition of 11.22 g marine broth (Difco) and 15 g bacteriological agar to 1000 ml of deionized water. The mixture should be boiled for 1 min then autoclaved for 15 min at 121 °C. By obtaining the major constituent of estuarine agar from a commercial source, it is possible to permit some degree of comparison between bacterial counts obtained using different batches of medium.

The use of Oppenheimer-Zobell Reduced (OZR) medium is recommended for growth of marine bacteria in relatively low levels of organic nutrient compared with commercially available media. One variety of OZR medium comprises $(g\,l^{-1})$:yeast extract (Difco) 1.0 g; trypticase (BBL) 1.0 g; bacteriological agar (Difco) 10 g prepared in 800 ml aged and filtered seawater plus 200 ml distilled water (Tettelbach *et al.*, 1984).

These media using seawater can also be used for samples from slightly brackish waters by increasing the amount of distilled water to adjust the seawater to the appropriate lower salinity. MacLeod *et al.* (1954) have described the composition of artificial seawater mixtures for use with distilled water where plentiful supplies of seawater are not available to prepare media.

For freshwater samples, seawater should be omitted and replaced by water collected at the site of sample collection. The development and composition of a casein–peptone–starch medium suitable for enumeration of bacteria associated with freshwater invertebrates has been described by Jones (1970).

6.4.1 Chitin agar

Chitinolytic bacteria are often associated with diseases of invertebrates, and chitinase activity of individual isolates can be detected using the medium described by West and Colwell (1984).

Crude, unbleached chitin powder (Sigma Chemical Co.) should be cleaned by alternatively soaking for 24 h in 1 M NaOH and 1 M HCl five times; the material is then washed four times with 95% ethanol and allowed to dry completely. Thence, carefully dissolve 20 g of cleaned chitin in 600 ml 50% H_2SO_4 at room temperature with constant stirring, and filter the solution through glass wool. Chitin is precipitated by adding the acidified solution slowly to 10 l of ice-cold deionized water followed by neutralization of the mixture with 10 M NaOH. The suspension should stand for 24 h at 4 °C before centrifuging and washing the precipitate several times with deionized water. The colloidal chitin paste should be suspended to a final concentration of 10% (w/v) and sterilized by autoclaving. For use, 1 ml of colloidal chitin suspension is added to 9 ml of molten bacteriological grade agar and allowed to set as an overlay on a plate of the appropriate nutrient agar medium. Organisms are spot-inoculated onto the medium and incubated to observe zones of clearing around bacterial growth which indicates chitin breakdown.

6.4.2 'Gaffkya' media

For the isolation of *Aerococcus* (*Gaffkya*) *viridans* subsp. *homari* from crustaceans, the phenylethyl-glucose broth described by Stewart *et al.* (1966) should be employed. Phenylethyl-glucose broth contains $(g\,l^{-1})$:

glucose 6.5 g; bacto-yeast extract (Difco) 4.5 g; tryptone 15.0 g; NaCl 6.4 g; phenylethyl alcohol 2.5 g; bromocresol purple 0.008 g and prepared in 1000 ml distilled water. The medium is adjusted to pH 7.4 and autoclaved at 15 psi for 15 min.

For use, 0.5 ml of haemolymph is used to inoculate 4.5 ml broth and incubated for 5 d at 28 °C. Tubes which show acid production should be subcultured to phenylethyl agar (BBL) or a suitable non-selective nutrient medium. Colonies of *Aerococcus* (*Gaffkya*) *viridans* subsp. *homari* characteristically contain Gram-positive cocci in tetrads, and are catalase-negative.

6.4.3 Methods for investigation of diseases in shellfish

Apart from the disease gaffkaemia of lobsters, many of the bacterial infections of shellfish are still poorly characterized (Lightner, 1983). Often a variety of bacteria can be isolated from dead bivalve molluscan shellfish larvae or chitinolytic lesions of crustacean exoskeletons (Brock, 1983; Stewart, 1984). Accordingly, it is usually necessary to screen healthy animals or larvae against a wide variety of bacterial isolates to demonstrate pathogenicity by a specific type before remedial action can be implemented (Garland *et al.*, 1983; Leong, 1983). Chitinolytic lesions of crustaceans are not currently considered a serious economic problem unless widespread on the animal. Where it is desirable to test the chitinolytic potential of bacterial isolates from crustacean shell lesions, the procedure described by Cook and Lofton (1973) can be used. This involved mechanical damage to the shell of a live crustacean with a rasping file, then application of a pure culture of the test bacterium. The method was reported to yield occasionally inconclusive reactions.

In contrast, diseases of bivalve molluscan shellfish and their larvae are common and cause severe economic loss in mariculture operations (Sindermann, 1984). Many bacterial species have been incriminated in diseases of bivalve molluscs and Lauckner (1983b) has summarized the major diseases considered to have a bacterial aetiology.

Tubiash (1974) described procedures for continuous and single exposure of mature bivalve molluscan shellfish to a range of marine vibrios isolated from diseased material. Continuous exposure was relatively simple and involved dosing of tanks containing seawater and shellfish with broth cultures of the organism under test. When testing single dose exposures, the trauma to molluscan shellfish usually associated with injection of the bacteria under test was overcome ingeniously by removing live, healthy oysters from a storage tank, and placing individual animals in a widemouthed vessel with the shellfish hinge towards the base. Each vessel was submerged in seawater until the oyster resumed filtering seawater and opened the shells. At this stage, metal syringe needles (18-gauge) were quickly inserted between the gaping shells to prevent complete closure of the shells on subsequent handling but not such that any internal organs were punctured. The test commenced with the removal of the shellfish and widemouthed vessel from the tank. Aliquots of the bacterial suspension to be tested for pathogenicity were injected into the mantle cavity of each animal and the syringe needle withdrawn. Oysters remained out of water for a few hours to allow the bacteria to permeate the animal, then were returned to flowing seawater in tanks and observed for mortality over a period of weeks.

Pathogenicity testing of bacteria for mollusc larvae has been reported by Di Salvo et al. (1978), Jefferies (1982), Garland et al. (1983) and Tettlebach et al. (1984). In each case, a procedure was described which involved adding varying dilutions of a broth culture of live bacteria to glass beakers or petri dishes containing seawater and live larvae at a concentration of 1 or 2 larvae per ml. Bacterial cultures were rapidly accumulated by the filter-feeding activity of larvae. Oraganisms were considered pathogenic if their presence resulted in larval necrosis and death or severe disturbance of normal functions such as swimming. Significantly, several fractions from cell-free culture supernatants of mollusc bacterial pathogens have been shown to cause mortality. These possible 'virulence factors' were assayed using oyster spat as well as gill tissue excised from mussels. These procedures and their interpretation in virulence testing have been described by Nottage and Birkbeck (1986).

A testing procedure to detect bacterial pathogens of prawn larvae was described by Colorni (1985). Prawn larvae were fed with flakes of fish which had been soaked in broth cultures in order to ensure adequate uptake of the bacteria under test. This method of inoculation may not always be successful with adult shrimps and prawns although Leong (1983) described its successful use with juvenile shrimp. Lightner and Lewis (1975) and Leong (1983) described an alternative testing procedure involving inoculation of shrimp intramuscularly with broth cultures between the 5th and 6th abdominal segments.

6.5 MICROSCOPY TECHNIQUES

Incident-light microscopy procedures which do not use stains to assist visualizing bacteria are generally ineffective for studying bacteria– invertebrate interactions because they cannot distinguish the subcellular structure of living cells from bacteria. However, microscopic examination of tissues is an important procedure for examining bacterial associations, in particular for diagnosis of invertebrate diseases. Accordingly, a variety of histopathologic stains and methods have been developed to selectively stain bacteria and invertebrate tissue (Elston *et al.*, 1981; Elston and Lockwood, 1983; De Burgh and Singla, 1984; Peters, 1984). The development of staining procedures for use with epifluorescence microscopy (Murray and Robinow, 1981) has also permitted detection of bacteria associated with a variety of invertebrate tissues. For example, Baker and Bradnam (1976) described epifluorescence microscopy and acridine orange staining techniques to enumerate bacteria released from tissues of freshwater invertebrates. Deming (1986) presented several illustrations to describe the detection of bacteria in deep-sea samples using epifluorescence microscopy.

A double-staining procedure for detection of living bacteria in protozoa has been described by Sherr and Sherr (1983). Samples were sequentially stained with the DNA-specific fluorochrome, 4',6-diamino-2-phenylindole (DAPI) and a non-specific protein stain fluorescein isothiocyanate (FITC) to demonstrate DNA-rich, bacterium-sized particles in the cytoplasm of protozoa. Cavanaugh (1983) demonstrated the presence of intracellular bacteria in bivalve mollusc tissue using epifluorescence microscopy. These procedures require specialized microscopes equipped with a beam splitter, barrier filter and excitation filters specific for the fluorochrome used. Some staining procedures require the use of microscopes equipped with expensive objectives which permit transmission of fluorescence. Their application for use with many aquatic invertebrates continues to be restricted due to problems with bright background non-specific fluorescence of tissues and cell debris in sample preparations.

One procedure that is useful in alleviating problems of non-specific staining is the utilization of fluorescent-antibody preparations which have a high specificity for particular bacteria. However, these procedures to detect bacteria in invertebrate tissue have been extensively applied only for investigation of routes of infection in shellfish of commercial importance (Elston *et al.*, 1981). The methods described by Elston and Leibovitz (1980) for the detection of pathogenic *Vibrio* infection in oyster larvae by fluorescent-antibody microscopy illustrate one of the procedures available for preparation of the staining reagents.

A fluorescent-antibody procedure to detect the presence of *V. cholerae* in crustacean tissue was described by Huq *et al.* (1986). This technique had originally been developed for use with water samples but was modified for use with crab gut tissue.

Electron microscopy techniques have, in contrast, been used extensively to study a diverse range of bacteria–invertebrate interactions. A variety of preparation methods for aquatic invertebrate samples have been described, all of which required fixation of material in glutaraldehyde- or formaldehyde-based solutions before dehydration through increasing concentrations of acetone or ethanol, and critical point drying. Specific

TABLE 6.1 Some uses of electron microscopy in studies of bacteria–invertebrate associations

Reference	Bacteria–Invertebrate association	Comments or conclusions
Holland and Nealson (1978)	Saprophytic bacteria in echinoderm cuticle layers	• Bacteria only detected by modification of the sample preparation procedure to keep cuticular structures intact • Bacteria appeared to have a Gram-negative cell wall and be symbiotic in the animal
Sleeter et al. (1978)	Bacteria in the digestive tract of wood-boring isopods	• Digestive tract of animals fed on sterile wood was devoid of bacterial flora
Zachary and Colwell (1979)	Bacteria in the digestive tract of wood-boring isopods	• Diverse bacterial flora detected in animals taken from creosote-treated wood pilings considered necessary for utilization of this material
Boyle and Mitchell (1981)	External microflora of wood-boring isopods	• Diverse bacterial flora associated with external surfaces
Austin and Allen (1982)	Bacteria colonizing brine shrimp eggs	• Dry eggs in vacuum-sealed cans are virtually sterile
Deming and Colwell (1982)	Deep-sea bacteria in digestive tracts of holothurians	• Rod-shaped, Gram-negative bacteria intimately associated with epithelial tissue of the holo thurian hindgut
Garland et al. (1982)	Bacteria of internal organs in oysters	• Absence of bacterial flora on ciliated epithelial surfaces of internal organs
Cavanaugh (1983)	Bacteria in bivalve molluscs	• Intracellular sulphur-oxidizing chemoautotrophic bacteria wide spread in bivalves of sulphide-rich habitats

TABLE 6.1 (*Contd.*)

Reference	Bacteria–Invertebrate association	Comments or conclusions
Huq *et al.* (1983)	External microflora of copepods	• Gentle rinsing of copepods removed surface-associated bacteria • Cells of *Vibrio cholerae* can extensively colonize copopod surfaces
Peters *et al.* (1983)	Bacterial flora of tropical corals	• Bacteria living within coral tissues may cause disease and abnormalities
De Burgh and Singla (1984)	Bacterial colonization of the gills of a a limpet mollusc from a hydrothermal vent	• Filamentous bacteria embedded in the epithelial cell surfaces or located intracellularly
Faghri *et al.* (1984)	Bacteria associated with muscle tissue of freshly caught crabs	• Absence of significant bacterial flora in crab muscle considered to indicate an extensive cellular defence system to restrict contamination of muscle tissue
Harper and Talbot (1984)	Bacterial flora of lobster eggs in relation to loss of eggs from the pleopods	• Extensive and diverse bacterial flora of lobster eggs was ubiquitously and unrelated to loss of eggs
Maes and Jangoux (1984)	Micro-organisms causing bald-sea-urchin disease	• Preliminary data on necrotic tissue indicates a bacterial etiology
Nagasawa and Nemoto (1984)	Mortality of chaetognath planktonic species	• Morphological abnormalities and death of plankton were associated with infection by bacteria
Rodrick and Ulrich (1984)	Defence mechanisms in haemolymph of molluscs challenged with bacteria	• Subcellular changes in haemocytes were described during phagocytosis of marine bacteria

(*continued overleaf*)

TABLE 6.1 (Contd.)

Reference	Bacteria–Invertebrate association	Comments or conclusions
Nagasawa et al. (1985)	External microflora of copepods	• Particular sites on copepods were selectively colonized by bacteria
El-Gamal et al. (1986)	Micro-organisms causing black burn spot exoskeleton lesions in prawns	• Lesions were heavily colonized by bacteria which could be removed by antibiotic treatment
Huq et al. (1986)	Attachment of Vibrio cholerae to crab tissue	• Only hindguts colonized indicating requirement for chitinous surfaces for attachment of this bacterium

staining procedures and preparation of material for scanning or transmission electron microscopy are described in each published study listed in Table 6.1, and Maugel et al. (1980) have presented a useful review of electron microscopy staining methods.

6.6 CHARACTERIZATION OF BACTERIA-SPECIFIC ENZYMES AND LIPIDS

In some circumstances, the microscopic observation of bacteria-like structures associated with invertebrate tissues is not sufficiently unequivocal to permit speculation that a significant interaction may exist. Several studies have therefore additionally demonstrated the substantial presence of enzymes highly characteristic for bacteria. In this context, ribulose-1,5-diphosphate carboxylase which features in CO_2 fixation, and sulphite-oxidase, have in particular been utilized for chemoautrotrophic bacteria. The use of these enzyme detection procedures to augment the visual siting of bacteria in tissues of invertebrates has been elegantly demonstrated by Felbeck et al. (1983) for marine oligochaetes and by Cavanaugh (1983) for a bivalve mollusc from environments rich in sulphide. Such studies are likely to be used increasingly when it is not feasible to culture cells or to demonstrate that the bacteria have a symbiotic relationship with the invertebrate rather than being derived from contamination of the sample. However, specialist equipment is required to detect the products of reactions catalysed by specific enzymes, in particular when radiolabelled compounds are used (Tuttle, 1985).

An additional feature which has been used to demonstrate the significant bacterial associations with invertebrates is the presence of lipopolysaccharides characteristic of outer cell walls of bacteria in tissues (Cavanaugh, 1983). In general, these procedures can only be undertaken in laboratories which possess equipment for detection and characterization of lipids by a variety of chromatography procedures adapted for use with aquatic environment samples (Morrison and White, 1980).

6.7 RADIOACTIVITY UPTAKE TO STUDY BACTERIA– INVERTEBRATE INTERACTIONS

The uptake of nitrogen or carbon mediated by bacteria associated with invertebrates has been demonstrated using ^{15}N-labelled compounds (Guerinot and Patriquin, 1981) and [14]CO_2 (Felbeck *et al.*, 1983). Such studies indicate the extent to which invertebrates utilize bacteria to assist meeting their nutritional requirements. Specialized laboratory facilities are not generally required but equipment such as scintillation counters, mass spectrometers and glassware with sealed stoppers is essential (Wood, 1981; Tabor *et al.*, 1982).

The clearance of bacteria labelled with ^{3}H-thymidine from seawater by a bivalve mollusc was described by Birkbeck and McHenery (1982) who postulated that certain bacterial species could contribute significantly to carbon uptake by bivalves. This procedure was also later used to demonstrate inhibition of mollusc filtration activity by some marine vibrios during studies to elucidate the mechanisms by which bacteria are pathogenic for these invertebrates (McHenery and Birkbeck, 1986).

Radiolabelling of mercury compounds was used by Berk and Colwell (1981) to demonstrate the potential for heavy metals to be mobilized by two species of bacteria at the base of a simple food web and transferred to higher trophic levels via ciliated protozoa and planktonic copepods. Suspensions of bacteria which had accumulated labelled mercury were fed to the ciliate and uptake of mercury monitored by liquid scintillation counting procedures.

6.8 METHODS FOR DEEP-SEA INVERTEBRATES

The development of invertebrate populations, which are heavily colonized with bacteria, around deep-sea hydrothermal vents has attracted much interest in recent years (Corliss *et al.*, 1979). Reduced inorganic compounds are emitted from hydrothermal vents and used for chemoautotrophic oxidation by bacteria (Baross and Deming, 1985). This production of microbial biomass serves as the base of a food web which permits development of specific invertebrate communities around the vents (Jannasch, 1985). Many of these newly described marine invertebrates are either primary consumers or hosts of chemoautotrophic bacteria (Cavanaugh, 1985). The use of several techniques described in this chapter (e.g. epifluorescence microscopy, enzyme characterization to detect chemoautotrophic bacteria, and electron microscopy) has enabled some aspects of the ecology of this unusual environment to be elucidated (Karl *et al.*, 1980; De Burgh and Singla, 1984; Deming, 1986).

The association of bacteria with benthic invertebrates retrieved from deep-sea sites other than hydrothermal vents was investigated by Tabor *et al.* (1982) who demonstrated by ^{14}C-glutamate utilization and epifluorescence microscopy that these invertebrates appear to carry a resident bacterial population adapted to deep-sea conditions of temperature and pressure. Subsequent studies of holothurian digestive tracts using epifluorescence microscopy, ^{14}C-glutamate utilization and transmission electron microscopy revealed that most microbial activity was located in the holothurian hindgut (Deming and Colwell, 1982). The contribution of these resident gut bacteria of deep-sea animals was evaluated by Wirsen and Jannasch (1983) who trapped amphipods at the seabed in traps baited with ^{14}C-radiolabelled food. After recovery of the traps from the seabed several days later, amphipods were analysed for the presence of ^{14}C in various fractions of tissue. Comparison of radioactivity counts in each fraction enabled confirmation of previous observations that the gut of these deep-sea animals carries a resident bacterial flora of ecological importance.

6.9 MISCELLANEOUS METHODS FOR STUDY OF BACTERIA–INVERTEBRATE INTERACTIONS

Levy *et al.* (1984) reported a general procedure for demonstrating the potential for survival of bacteria in chlorinated potable waters which was mediated through association with macroinvertebrates. These studies involved attaching faecal bacteria to amphipods before exposure to phosphate-buffer solutions containing varying quantities of free residual chlorine. Bacteria were subsequently enumerated by homogenization of amphipods and culture on selective media. Interestingly, in comparison with other studies, faecal bacteria were difficult to remove from amphipods by repeated washing.

The settlement and metamorphosis of macroinvertebrate larvae on surfaces coated with a variety of microorganisms, in particular bacteria, is of considerable economic importance (Bonar *et al.*, 1986). Biodeterioration of many man-made structures is an adverse aspect of invertebrate–bacteria symbiosis and is initiated by the formation of bacterial films on solid surfaces which subsequently attract macroinvertebrate larvae and boring worms to settle and grow. Several bacteria–invertebrate models have been described for elucidation of this phenomenon and the theory involving lectin attractants has been reviewed by Kirchman and Mitchell (1983). In contrast, the settlement and development of economically valuable molluscan shellfish larvae appears to be significantly dependent on the coating of appropriate surfaces in marine environments with highly specific bacterial populations (Weiner, 1985). Bacteria within these coatings appear to synthesize attractants which enhance settlement of bivalve mollusc larvae. Weiner *et al.* (1985) described the quantitative assessment of oyster larvae settlement on glass slides coated with bacteria which synthesize the attractant compound, or cell-free extracts containing the attractant. Slides were coated and then immersed in tanks of filtered seawater containing 'eyed' oyster larvae. The degree of successful attraction was determined by the number of larvae cementing themselves to the coated surfaces.

6.10 POSSIBLE FUTURE TRENDS

This review has outlined descriptions and provided references for some basic methods for studying bacteria–invertebrate interactions. Where such interactions are of public health or economic importance in aquaculture, many detailed studies have been reported. There has been noticeably less interest in those interactions between bacteria and aquatic invertebrates which remain obscure in function. An exception is the development of hydrothermal vent invertebrate communities which is based almost exclusively on primary production by chemoautotrophic bacteria. In these circumstances, the specific association between bacteria and invertebrates has considerable ecological significance. It is therefore hypothesized that if bacteria–invertebrate interactions are important at the base of food webs in general, then the extent to which factors, such as pollution, impede or enhance these interactions can affect or limit the flow of energy upwards to higher trophic levels requires further investigation.

Improved understanding of the importance of invertebrate–bacteria interactions, in particular those of economic potential such as biofouling or bivalve mollusc larvae settlement, may also permit the application of molecular genetic engineering and biotechnology principles to control such interactions to provide both environmental and economic benefit.

There appears, therefore, to be considerable stimulus for increasing our basic understanding of interactions between aquatic invertebrates and their bacterial flora.

6.11 REFERENCES

Alongi, D. M. (1985). Microbes, meiofauna, and bacterial productivity on tubes constructed by the polychaete *Capitella capitata*. *Marine Ecology Progress Series*, **23**, 207–208.

American Public Health Association (1985). *Laboratory Procedures for the Examination of Seawater and Shellfish*, 5th edn, American Public Health Association, Washington DC.

Austin, B. and Allen, D. A. (1982). Microbiology of laboratory-hatched brine shrimp (*Artemia*). *Aquaculture*, **26**, 369–383.

Baker, J. H. and Bradnam, L. A. (1976). The role of bacteria in the nutrition of aquatic detritivores. *Oceologia*, **24**, 95–104.

Baross, J. A. and Deming, J. W. (1985). The role of bacteria in the ecology of black-smoker environments. *Bulletin of the Biological Society Washington*, **6**, 355–371.

Berk, S. G., Guerry, P. and Colwell, R. R. (1976). Separation of small ciliate protozoa from bacteria by sucrose gradient centrifugation. *Applied and Environmental Microbiology*, **31**, 450–452.

Berk, S. G. and Colwell, R. R. (1981). Transfer of mercury through a marine microbial food web. *Journal of Experimental Marine Biology and Ecology*, **52**, 157–172.

Bird, D. F. and Kalff, J. (1986). Bacterial grazing by planktonic lake algae. *Science, New York*, **231**, 493–495.

Birkbeck, T. H. and McHenery, J. G. (1982). Degradation of bacteria by *Mytilus edulis*. *Marine Biology*, **72**, 7–16.

Bonar, D. B., Weiner, R. M. and Colwell, R. R. (1986). Microbial–invertebrate interactions and potential for biotechnology. *Microbial Ecology*, **12**, 101–110.

Boyle, P. J. and Mitchell, R. (1981). External microflora of a marine wood-boring isopod. *Applied and Environmental Microbiology*, **42**, 720–729.

Brinkley, A. W., Rommel, F. A. and Huber, T. W. (1976). The isolation of *Vibrio parahaemolyticus* and related vibrios from moribund aquarium lobsters. *Canadian Journal of Microbiology*, **22**, 315–377.

Brock, J. A. (1983). Diseases (infectious and noninfectious), metazoan parasites, predators, and public health considerations in *Macrobrachium* culture fisheries. In *CRC Handbook of Mariculture* vol. 1, *Crustacean Aquaculture*, J. P. McVey (Ed.), CRC Press, Boca Raton, pp. 329–377.

Carman, K. R. and Thistle, D. (1985). Microbial food partitioning by three species of benthic copepods. *Marine Biology*, **88**, 143–148.

Cavanaugh, C. M. (1983). Symbiotic chemoautotrophic bacteria in marine invertebrates from sulphide-rich habitats. *Nature, London*, **302**, 58–61.

Cavanaugh, C. M. (1985). Symbioses of chemoautotrophic bacteria and marine invertebrates from hydrothermal vents and reducing sediments. *Bulletin of the Biological Society Washington*, **6**, 373–388.

Colorni, A. (1985). A study on the bacterial flora of giant prawn, *Macrobrachium rosenbergii*, larvae fed with *Artemia salina* nauplii. *Aquaculture*, **49**, 1–10.

Colwell, R. R. (1984). *Vibrios in the Environment*, Wiley, New York.

Cook, D. W. and Lofton, S. R. (1973). Chitinoclastic bacteria associated with shell disease in *Penaeus* shrimp and the blue crab. *Journal of Wildlife Diseases*, **9**, 145–159.

Corliss, J. B., Dymond, J., Gordon, L. I., Edmond, J. M., von Herzen, R. P., Ballard, R. D., Green, K., Williams, D., Bainbridge, A., Crane, K. and van Andel, T. H. (1979). Submarine thermal springs on the Galapagos rift. *Science, New York*, **203**, 1073–1083.

De Burgh, M. E. and Singla, C. L. (1984). Bacterial colonization and endocytosis on the gill of a new limpet species from a hydrothermal vent. *Marine Biology*, **84**, 1–6.

Deming, J. W. (1986). Ecological strategies of barophilic bacteria in the deep ocean. *Microbiological Science*, **3**, 205–211.

Deming, J. W. and Colwell, R. R. (1982). Barophilic bacteria associated with digestive tracts of abyssal holothurians. *Applied and Environmental Microbiology*, **44**, 1222–1230.

DiSalvo, L. H., Blecka, J. and Zebal, R. (1978). *Vibrio anguillarum* and larval mortality in a California coastal shellfish hatchery. *Applied and Environmental Microbiology*, **35**, 219–221.

Dixon, P. and Robertson, A. I. (1986). A compact, self-contained zooplankton pump for use in shallow coastal habitats: design and performance compared to net sampling. *Marine Ecology Progress Series*, **32**, 97–100.

Egidius, E. (1972). On the internal bacterial flora of the European lobster (*Homarus vulgaris* L.) and its susceptibility to gaffkaemia. *Aquaculture*, **1**, 193–197.

Eleftheriou, A. and Holme, N. A. (1984). Macrofauna techniques. In *Methods for the Study of Marine Benthos*, International Biological Programme Handbook No. 16, N. A. Holme and A. D. McIntyre (Eds), Blackwell, Oxford, pp. 140–216.

El-Gamal, A. A., Alderman, D. J., Rodgers, C. J., Polglase, J. L. and MacIntosh, D. (1986). A scanning electron microscope study of oxolinic acid treatment of burn spot lesions of *Macrobrachium rosenbergii*. *Aquaculture*, **52**, 157–171.

Elston, R. and Leibovitz, L. (1980). Pathogenesis of experimental vibriosis in larval American oysters, *Crassostrea virginica*. *Canadian Journal of Fisheries and Aquatic Science*, **37**, 964–978.

Elston, R., Leibovitz, L., Relyea, D. and Zatila, J. (1981). Diagnosis of vibriosis in a commercial oyster hatchery epizootic: diagnosis tools and management features. *Aquaculture*, **24**, 53–62.

Elston, R. and Lockwood, G. S. (1983). Pathogenesis of vibriosis in cultured juvenile red abalone, *Haliotis rufescens* Swainson. *Journal of Fish Diseases*, **6**, 111–128.

Faghri, M. A., Pennington, C. L., Cronholm, L. S. and Atlas, R. M. (1984). Bacteria associated with crabs from cold waters with emphasis on the occurrence of potential human pathogens. *Applied and Environmental Microbiology*, **47**, 1054–1061.

Felbeck, M., Liebezeit. G., Dawson, R. and Giere, O. (1983). CO_2 fixation in tissues of marine oligochaetes (*Phallodrilus leukodermatus* and *P. planus*) containing symbiotic chemoautotrophic bacteria. *Marine Biology*, **75**, 187–191.

Feng, S. Y., Feng, J. S., Burke, C. N. and Khairallah, L. H. (1971). Light and electron microscopy of the leucocytes of *Crassostrea virginica* (Mollusca: Pelecypoda). *Zeitschrift für Zellforschung und mikroskopische Anatomie*, **120**, 222–245.

Fisher, W. S. (1983). Eggs of *Palaemon macrodactylus*: III. Infection by the fungus, *Lagenidium callinectes*. *Biological Bulletin*, **164**, 214–226.

Garland, C. D., Nash, G. V. and McMeekin, T. A. (1982). Absence of surface-associated microorganisms in adult oysters (*Crassostrea gigas*). *Applied and Environmental Microbiology*, **44**, 1205–1211.

Garland, C. D., Nash, G. V., Sumner, C. E., and McMeekin, T. A. (1983). 'Bacterial pathogens of oyster larvae (*Crassostrea gigas*) in a Tasmanian hatchery', *Australian Journal of Marine Freshwater Research*, **34**, 483–487.

Goulder, R. (1976). Evaluation of media for counting viable bacteria in estuaries. *Journal of Applied Bacteriology*, **41**, 351–355.

Guerinot, M. L. and Patriquin, D. G. (1981). The association of N_2-fixing bacteria with sea-urchins. *Marine Biology*, **62**, 197–207.

Gulka, G. and Chang, P. W. (1984). Pathogenicity and infectivity of a rickettsia-like organism in the sea scallop, *Placopecten magellanicus*. *Journal of Fish Diseases*, **8**, 309–318.

Harper, R. E. and Talbot, P. (1984). Analysis of the epibiotic bacteria of lobster (*Homarus*) eggs and their influence on the loss of eggs from the pleopods. *Aquaculture*, **36**, 9–26.

Hartland, B. J. and Timoney, J. F. (1979). *In vivo* clearance of enteric bacteria from the hemolymph of the hard clam and the American oyster. *Applied and Environmental Microbiology*, **37**, 517–520.

Hickman, C. P. (1967). *Biology of the Invertebrates*, C. V. Mosley Co., St. Louis.

Holland, N. D. and Nealson, K. H. (1978). The fine structure of the echinoderm cuticle and the subcuticular associated bacteria of echinoderms. *Acta Zoologica Stockholm*, **59**, 169–185.

Hood, M. A., Ness, G. E., Rodrick, G. E. and Blake, N. J. (1983). Effects of storage on microbial loads of two commercially important shellfish species, *Crassostrea virginica* and *Mercenaria campechiensis*. *Applied and Environmental Microbiology*, **45**, 1221–1228.

Huq, A., Small, E. B., West, P. A., Huq, M. I., Rahman, R. and Colwell, R. R. (1983). Ecological relationships between *Vibrio cholerae* and planktonic crustacean copepods. *Applied and Environmental Microbiology*, **45**, 275–283.

Huq, A., Huq, S. A., Grimes, D. J., O'Brien, M., Chu, K. H., Capuzzo, J. M. and Colwell, R. R. (1986). Colonization of the gut of the blue crab (*Callinectes sapidus*) by *Vibrio cholerae*. *Applied and Environmental Microbiology*, **52**, 586–588.

Jannasch, H. W. (1985). Deep sea life on the basis of chemical synthesis. *Naturwissenshaften*, **72**, 85–290.

Jannasch, H. W., Cuhel, R. L., Wirsen, C. O. and Taylor, C. D. (1980). An approach for *in situ* studies of deep-sea amphipods and their microbial gut flora. *Deep-Sea Research*, **27**, 867–872.

Jefferies, V. E. (1982). Three *Vibrio* strains pathogenic to larvae of *Crassostrea gigas* and *Ostrea edulis*. *Aquaculture*, **29**, 201–226.

Jones, J. G. (1970). Studies on freshwater bacteria: effect of medium composition and method on estimates of bacterial populations. *Journal of Applied Bacteriology*, **33**, 679–686.

Karl, D. M., Wirsen, C. O. and Jannasch, H. W. (1980). Deep-sea primary production at the Galapagos hydrothermal vents. *Science, New York*, **207**, 1345–1347.

Kelly, M. T. and Dinuzzo, A. (1985). Uptake and clearance of *Vibrio vulnificus* from Gulf coast oysters (*Crassostrea virginica*). *Applied and Environmental Microbiology*, **50**, 1548–1549.

Kirchman, D. and Mitchell, R. (1983). Biochemical interactions between microorganisms and marine fouling invertebrates. In *Biodeterioration 5*, T. A. Oxley and S. Barry (Eds), Wiley, New York, pp. 281–290.

Koch, A. L. (1981). Growth measurement. In *Manual of Methods for General Microbiology*, P. Gerhardt (Ed.), American Society for Microbiology, Washington, DC, pp. 179–207.

Kueh, C. W. S. and Chan, K. Y. (1985). Bacteria in bivalve shellfish with special reference to the oyster. *Journal of Applied Bacteriology*, **59**, 41–47.

Kuhn, D. A. (1981). The genus *Cristispira*. In *The Prokaryotes—A Handbook on Habitats, Isolation and Identification of Bacteria*. M. P. Starr, H. Stolp, H. G. Trüper, A. Balows and H. G. Schlegel (Eds), Springer-Verlag, New York, pp. 555–563.

Lauckner, G. (1983a). Diseases of mollusca: bivalvia—agents: Rickettsiae, Chlamydiae and Mycoplasmas. In *Diseases of Marine Animals*, vol. II, O. Kinne (Ed.) Biologische Anstalt Helgoland, Hamburg, pp. 503–510.

Lauckner, G. (1983b). Diseases of mollusca: bivalvia—agents: bacteria. In *Diseases of Marine Animals*, vol. II, O. Kinne (Ed.), Biologische Anstalt Helgoland, Hamburg, pp. 489–503.

Lee, J. S. and Pfeifer, D. K. (1975). Microbiological characteristics of Dungeness crab (*Cancer magister*). *Applied Microbiology*, **30**, 72–78.

Leong, J. K. (1983). A bioassay system for studying diseases in juvenile penaeid shrimp. In *CRC Handbook of Mariculture*, vol. 1, *Crustacean Aquaculture*, J. P. McVey (Ed.), CRC Press, Boca Raton, pp. 321–327.

Levy, R. V., Cheetham, R. D., Davis, J., Winer, G. and Hart, F. L. (1984). Novel method for studying the public health significance of macroinvertebrates occurring in potable water. *Applied and Environmental Microbiology*, **47**, 889–894.

Lightner, D. V. (1983). Diseases of cultured penaeid shrimp. In *CRC Handbook of Mariculture*, vol. 1, *Crustacean Aquaculture*, J. P. McVey (Ed.), CRC Press, Boca Raton, pp. 289–320.

Lightner, D. V. and Lewis, D. H. (1975). A septicemic bacterial disease syndrome of penaeid shrimp. *Marine Fisheries Review*, **37**, 25–28.

Lovelace, T. E., Tubiash, H. and Colwell, R. R. (1968). 'Quantitative and qualitative commensal bacterial flora of *Crassostrea virginica* in Chesapeake Bay', *Proceedings of the National Shellfish Association*, **58**, 82–87.

MacLeod, R. A., Onofrey, E. and Norris, M. E. (1954). Nutrition and metabolism of marine bacteria. I. Survey of nutritional requirements. *Journal of Bacteriology*, **68**, 680–686.

Maes, P. and Jangoux, M. (1984). The bald-sea-urchin disease: a biopathological approach. *Helgolander Meeresuntersuchungen*, **37**, 217–224.

McHenery, J. G. and Birkbeck, T. H. (1986). Inhibition of filtration in *Mytilus edulis* L. by marine vibrios. *Journal of Fish Diseases*, **9**, 257–261.

McIntyre, A. D. and Warwick, R. M. (1984). Meiofauna techniques. In *Methods for the Study of Marine Benthos*, International Biological Programme, Handbook No. 16, N. A. Holme and A. D. McIntyre (Eds), Blackwell, Oxford, pp. 217–244.

Maugel, T. K., Bonar, D. B., Creegan, W. T. and Small, E. B. (1980). Specimen preparation techniques for aquatic organisms. *Scanning Electron Microscopy*, **2**, 57–77.

Morrison, S. J. and White, D. C. (1980). Effects of grazing by estuarine gammaridean amphipods on the microbiota of allochthonous detritus. *Applied and Environmental Microbiology*, **40**, 659–671.

Mossel, D. A. A. and Van Netten, P. (1984). Harmful effects of selective media on stressed microorganisms: nature and remedies. In *The Revival of Injured Microbes*, M. H. E. Andrew and A. D. Russel (Eds), Academic Press, London, pp. 329–369.

Murray, R. G. E. and Robinow, C. F. (1981). Light microscopy. In *Manual of Methods for General Bacteriology*, P. Gerhardt (Ed.), American Society for Microbiology, Washington DC, pp. 6–16.

Nagasawa, S. and Nemoto, T. (1984). X-diseases in the chaetognath *Sagitta crassa*. *Helgolander Meeresuntersuchungen*, **37**, 139–148.

Nagasawa, S., Simidu, U. and Nemoto, T. (1985). Scanning electron microscopy investigation of bacterial colonization of the marine copepod *Acartia clausi*. *Marine Biology*, **87**, 61–66.

Nealson, K. H. and Hastings, J. W. (1979). Bacterial bioluminescence: its control and ecological significance. *Microbiological Reviews*, **43**, 496–518.

Nottage, A. S. and Birkbeck, T. H. (1986). Toxicity to marine bivalves of culture supernatant fluids of the bivalve-pathogenic *Vibrio* strain NCMB 1338 and other marine vibrios. *Journal of Fish Diseases*, **9**, 249–256.

O'Brien, C. H. and Sizemore, R. K. (1979). Distribution of the luminous bacterium *Beneckea harveyi* in a semitropical estuarine environment. *Applied and Environmental Microbiology*, **38**, 928–933.

Oliver, J. D. (1981). Lethal cold stress of *Vibrio vulnificus* in oysters. *Applied and Environmental Microbiology*, **41**, 710–717.

Oppenheimer, C. and Zobell, C. E. (1952). The growth and viability of sixty-three species of marine bacteria as influenced by hydrostatic pressure. *Journal of Marine Research*, **11**, 10–18.

Peters, E. C. (1984). A survey of cellular reactions to environmental stress and disease in Caribbean scleractinian corals. *Helgolander Meeresuntersuchungen*, **37**, 113–137.

Peters, E. C., Oprandy, J. J. and Yevich, P. P. (1983). Possible causal agent of 'white band disease' in Caribbean acroporid corals. *Journal of Invertebrate Pathology*, **41**, 394–396.

Phillips, F. A. and Peeler, J. T. (1972). Bacteriological survey of the blue crab industry. *Applied Microbiology*, **24**, 958–966.

Pomeroy, L. R. (1980). Microbial effects of aquatic food webs. In *Microbiology—1980*, D. Schlessinger (Ed.), American Society for Microbiology, Washington DC, pp. 325–327.

Purchon, R. D. (1977). *The Biology of the Mollusca*, Pergamon, Oxford.

Rheinheimer, G. (1985). *Aquatic Microbiology*, Wiley, New York.

Rodrick, G. E. and Ulrich, S. A. (1984). Microscopical studies on the hemocytes of bivalves and their phagocytic interaction with selected bacteria. *Helgolander Meeresuntersuchungen*, **37**, 167–176.

Schlegel, H. G. and Jannasch, H. W. (1981). Prokaryotes and their habitats. In *The Prokaryotes—A Handbook on Habitats, Isolation and Identification of Bacteria*, M. P. Starr, H. Stolp, H. G. Trüper, A. Balows and H. G. Schlegel (Eds), Springer-Verlag, New York, pp. 43–83.

Sherr, E. B. and Sherr, B. F. (1983). Double-staining epifluorescence technique to assess frequency of dividing cells and bacteriovory in natural populations of heterotrophic microprotozoa. *Applied and Environmental Microbiology*, **46**, 1388–1393.

Simidu, U., Ashino, K. and Kaneko, E. (1971). Bacterial flora of phyto- and zoo-plankton in the inshore water of Japan. *Canadian Journal of Microbiology*, **17**, 1157–1160.

Sindermann, C. J. (1970). *Principal Diseases of Marine Fish and Shellfish*, Academic Press, New York.

Sindermann, C. J. (1984). Disease in marine aquaculture. *Helgolander Meeresuntersuchungen*, **37**, 505–532.

Sleeter, T. D., Boyle, P. J., Cundell, A. M. and Mitchell, R. (1978). Relationships between marine microorganisms and the wood-boring isopod *Limnoria tripunctata*. *Marine Biology*, **45**, 329–336.

Sochard, M. R., Wilson, D. F., Austin, B. and Colwell, R. R. (1979). Bacteria associated with the surface and gut of marine copepods. *Applied and Environmental Microbiology*, **37**, 750–759.

Son, N. T. and Fleet, G. H. (1980). Behaviour of pathogenic bacteria in the oyster, *Crassostrea commercialis*, during depuration, re-laying, and storage. *Applied and Environmental Microbiology*, **40**, 994–1002.

Stewart, J. E. (1984). Lobster diseases. *Helgolander Meeresunters uchungen*, **37**, 243–254.

Stewart, J. E., Cornick, J. W. and Spears, D. I. (1966). Incidence of *Gaffkya homari* in natural lobster (*Homarus americanus*) populations of the Atlantic region of Canada. *Journal of the Fisheries Research Board of Canada*, **23**, 1325–1330.

Tabor, P. S., Deming, J. W., Ohwada, K. and Colwell, R. R. (1982). Activity and growth of microbial populations in pressurized deep-sea sediment and animal gut samples, *Applied and Environmental Microbiology*, **44**, 413–422.

Tettlebach, S. T., Petti, L. M. and Blogoslawski, W. J. (1984). Survey of *Vibrio* associated with a New Haven harbor shellfish bed, emphasizing recovery of larval oyster pathogens. In *Vibrios in the Environment*, R. R. Colwell (Ed.), Wiley, New York, pp. 495–509.

Tietjen, J. H. (1980). Microbial–meiofaunal interrelationships: a review. In *Microbiology—1980*, D. Schlessinger (Ed.), American Society for Microbiology, Washington DC, pp. 335–338.

Tubiash, H. S. (1974). Single and continuous exposure of the adult American oyster, *Crassostrea virginica*, to marine vibrios. *Canadian Journal of Microbiology*, **20**, 513–517.

Tubiash, H. S., Sizemore, R. K. and Colwell, R. R. (1975). Bacterial flora of the hemolymph of the blue crab, *Callinectes sapidus:* most probable number. *Applied Microbiology*, **29**, 388–392.

Tuttle, J. H. (1985). The role of sulphur-oxidising bacteria at deep-sea hydrothermal vents. *Bulletin of the Biological Society Washington*, **6**, 335–343.

Unkles, S. E. (1977). Bacterial flora of the sea urchin *Echinus esculentus*. *Applied and Environmental Microbiology*, **34**, 347–350.

Väätänen, P. (1977). Effects of composition of substrate and inoculation technique on plate counts of bacteria in the Northern Baltic Sea. *Journal of Applied Bacteriology*, **42**, 437–443.

Van Den Branden, C., Gillis, M. and Richard, A. (1980). Carotenoid producing bacteria in the accessory nidamental glands of *Sepia officinalis* L. *Comparative Biochemisty and Physiology*, **66B**, 331–334.

Weiner, R. M. (1985). Microbial films and invertebrate settlement and metamorphosis. In *Biotechnology of Marine Polysaccharides*, R. R. Colwell, E. R. Pariser and A. J. Sinskey (Eds), McGraw-Hill, New York, pp. 115–130.

Weiner, R. M., Hussong, D. and Colwell, R. R. (1980). An estuarine agar medium for enumeration of aerobic heterotrophic bacteria associated with water, sediment and shellfish. *Canadian Journal of Microbiology*, **26**, 1366–1369.

Weiner, R. M., Segall, A. M. and Colwell, R. R. (1985). Characterization of a marine bacterium associated with *Crassostrea virginica* (the Eastern oyster). *Applied and Environmental Microbiology*, **49**, 83–90.

Welsh, P. C. and Sizemore, R. K. (1985). Incidence of bacteremia in stressed and unstressed populations of the blue crab, *Callinectes sapidus*. *Applied and Environmental Microbiology*, **50**, 420–425.

West, P. A. (1986). Hazard Analysis Critical Control Point (HACCP) concept: application to bivalve shellfish purification systems. *Journal of the Royal Society of Health*, **106**, 133–140.

West, P. A. and Coleman, M. R. (1986). A tentative national reference procedure for isolation and enumeration of *Escherichia coli* from bivalve molluscan shellfish by most-probable-number method. *Journal of Applied Bacteriology*, **61**, 505–516.

West, P. A. and Colwell, R. R. (1984). Identification and classification of

Vibrionaceae—an overview. In *Vibrios in the Environment*, R. R. Colwell (Ed.), Wiley, New York, pp. 285–363.

West, P. A., Russek, E., Brayton, P. R. and Colwell, R. R. (1982). Statistical evaluation of a quality control method for isolation of pathogenic *Vibrio* species on selected Thiosulfate-Citrate-Bilesalts-Sucrose agars. *Journal of Clinical Microbiology*, **16**, 1110–1116.

Wirsen, C. O. and Jannasch, H. W. (1983). *In-situ* studies on deep-sea amphipods and their intestinal microflora. *Marine Biology*, **78**, 69–73.

Wood, W. A. (1981). Physical methods. In *Manual of Methods for General Bacteriology*, P. Gerhardt (Ed.), American Society for Microbiology, Washington, DC, pp. 309–318.

Yayanos, A. A. (1978). Recovery and maintenance of live amphipods at a pressure of 580 bars from an ocean depth of 5700 meters. *Science, New York*, **200**, 1056–1059.

Zachary, A. and Colwell, R. R. (1979). Gut-associated microflora of *Limnoria tripunctata* in marine creosote-treated wood pilings. *Nature, London*, **282**, 716–717.

Methods in Aquatic Bacteriology
Edited by B. Austin
© 1988 John Wiley & Sons Ltd.

7

Epiphytic Bacteria

J. H. Baker*

Freshwater Biological Association, River Laboratory, East Stoke, Wareham, Dorset BH20 6BB, England

7.1 INTRODUCTION

Epiphytes may be defined as organisms growing on the surface of living plants. Epiphytes include bacteria, algae, fungi, protozoa and small metazoans and, in nature, are not confined to the aquatic environment. However, this review is only concerned with bacterial epiphytes of vascular plants and macrophytic algae in marine and freshwater systems. Stems, leaves and, to a lesser extent, roots are all discussed, but bacteria associated with moribund or decaying plant material have been omitted. It has also been necessary to specifically exclude consideration of the many bacteria often seen on the surface of unicellular algae.

The chapter is divided into sections, each dealing with methods associated with a particular aspect of the study of epiphytic bacteria. Methods for observing the bacteria are dealt with first, followed by methods of enumeration including relevant statistical considerations. Next, there is a discussion on how to determine some important chemical parameters at the plant surface followed by a few cautionary comments on methods for the identification of epiphytic bacteria. There follows a section on the determination of bacterial activity on plant surfaces, and lastly problems associated with the role of epiphytic bacteria as a food resource are described.

Emphasis is placed on recent contributions to the field, particularly those published since the last review on this subject by Fry and Humphrey (1978). Although it has occasionally been necessary to repeat some of the material found in that paper, in general the present work should be regarded as complementary to that material rather than superseding it.

---

* Present address: Natural Environment Research Council, Polaris House, North Star Avenue, Swindon, Wiltshire, SN2 1EU, England.

7.2 DIRECT OBSERVATION USING THE LIGHT MICROSCOPE

Direct observation of bacteria on the plant surface through the light microscope has been reported rather infrequently. It is more common for epiphytic bacteria to be viewed either by electron microscopy or indirectly after removal from the plant material. This is unfortunate because much information on the distribution of the bacteria, both with respect to parts of the plant and to other epiphytic organisms, is lost if a direct method is not used. Moreover, light microscopy of aquatic epiphytes is relatively simple unless the epiphytic community is particularly thick.

The recommended stain for epiphytic bacteria under bright field illumination is phenolic aniline blue (phenol 3.75 g, water-soluble aniline blue 0.05 g, acetic acid 20% v/v 100 ml), and no fixing of the material is necessary. Note that water-soluble aniline blue is synonymous with cotton blue. The preparation should be passed through a Whatman No. 1 paper filter before use. It is stable at room temperature and may be kept for at least 3 months. Phenolic aniline blue was devised 50 years ago (Maneval, 1936) for staining algae and fungi, and popularized by Jones and Mollison (1948) for staining soil preparations. It appears to have been first used for bacteria on aquatic macrophytes by Hossell and Baker (1979a), following its use for epiphytic fungi on aerial leaves by Preece (1962).

For many aquatic macrophytes, such as *Zostera marina* (eel grass), *Veronica beccabunga* (brooklime) leaves and *Apium nodiflorum* (fools cress) leaves, staining with phenolic aniline blue for 2–3 min is the only treatment necessary for light microscopy. The plant material may be mounted in the stain, a coverslip applied and the bacteria observed with an oil-immersion objective. A photograph of *Lemna minor* stained in this way is shown in Fig. 7.1. The staining time is not critical and if the preparation needs to be kept it can be sealed with a proprietary sealing agent such as Glyceel (BDH) or nail varnish. Samples which cannot be examined within a few hours of collection can be preserved before staining in an aqueous solution of thiomersal (BDH, $0.2 \, gl^{-1}$).

Certain plants, such as *Glyceria maxima*, are so densely pigmented that they transmit insufficient light for the observation of epiphytes even when the microscope illumination is turned to maximum. Such material needs to be decolorized, sometimes erroneously called clearing (Baker, 1981), before being stained with phenolic aniline blue. Methanol is a good decolourizing agent for aquatic macrophytes, but chloroform may also be used. The plant material is simply soaked in the methanol in the dark. The time taken for complete or sufficient decolorization varies with the nature of the plant material, but leaving overnight is usually adequate and convenient. The process may be accelerated, if required, by gentle warming.

The disadvantage of using methanol for decolorization is that some

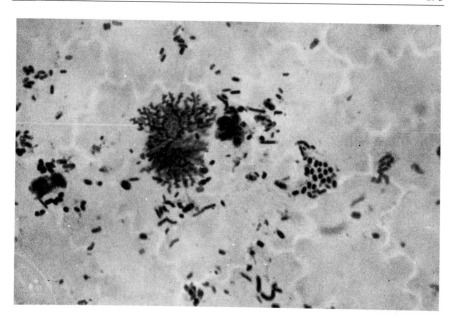

Fig. 7.1 Surface of *Lemna minor* under the light microscope showing the outline of the epidermal cells and bacterial epiphytes with an actinomycete colony, stained with phenolic aniline blue (× 1200).

epiphytic bacteria are washed off the plant surface and hence generally not taken into account. This source of error can be eliminated by using a gaseous decolorizing agent such as chlorine gas. The chlorine is best produced *in situ* from a solution of a bleaching agent. Of several bleaching agents tested by Daft and Leben (1966) for terrestrial leaves dichloro-s-triazine-2, 4, 6- (1H, 3H, 5H)-trione sodium salt dihydrate (Koch-Light) was the most effective. A few grams of this bleach are placed in a small (approximately 200 ml) screw-capped bottle. About 20 ml of distilled water are added and a bridge of metal or plastic gauze placed into the bottle so that the plant material can be held clear of the solution. The specimen is then placed on the bridge and the bottle tightly capped. The period required for adequate decolorization varies between 2 and 24 hours. However, on a few aquatic plants such as *Ranunculus penicillatus* var. *calcareus* (water crowfoot), epiphytic bacteria are less easily seen when bleached in this way than when decolorized with methanol. For such plants, the latter method is preferred.

Another procedure, which is occasionally helpful is prestaining. Pre-staining is necessary when the plant takes up the phenolic aniline blue as readily as the epiphytic microbes so that differentiation is impossible. It is a relatively rare phenomenon, but is easily overcome by prestaining the plant with fresh eosin yellowish (BDH $0.2\,g\,l^{-1}$) solution for an hour. After staining the bacteria with phenolic aniline blue, they appear dark red against a light orange background.

A totally different approach to the problem of densely pigmented material is to illuminate the specimen from above (incident or epi-illumination) rather than trying to force the specimen to transmit sufficient light from below. Stains used in bright-field work, such as phenolic aniline blue, are usually invisible under incident illumination, so it is essential to use fluorescent stains which emit light in the direction of the microscope eyepiece when illuminated with a suitable wavelength light. This technique is commonly called epifluorescence. The wavelength of incident light required is generally very short (400 to 500 nm) and although tungsten halogen lamps do produce some light in this region it is generally better to use a high pressure mercury vapour lamp which is approximately ten times more intense.

Unfortunately cholorophyll and lignin are also fluorescent under certain conditions without the addition of any stain. This phenomenon is known as autofluorescence and may swamp the observation of fluorescing bacteria. Epifluorescence systems have a series of adjustable filters built into the microscope to reduce the interference from autofluorescing components. Nevertheless, it remains a problem and for that reason the technique might be more suitable for the observation of bacteria on root surfaces where chlorophyll is naturally absent. Although there are many reports of epifluorescence being applied to roots to terrestrial plants (e.g. Van Vuurde and Elenbaas, 1978) it has not apparently been used on roots of aquatic plants.

Bulky objects such as stems of *Nasturtium officinale* (watercress) are often impossible to mount whole under a coverslip so that they cannot be satisfactorily observed by any of the above techniques. However, such material usually can be rendered amenable by simply taking tangential sections with a sharp scalpel or razor. A more sophisticated method for removing just the cuticle and epidermis of apple leaves has been used by Preece (1962). He forced pectinase into leaf discs by the ingenious use of a vacuum desiccator and enzymically removed the surface layers. It should be remembered that many aquatic macrophytes have extremely thin cuticles (Sculthorpe, 1967) so that this method may not be directly applicable.

7.3 OBSERVATION BY ELECTRON MICROSCOPY

The general characteristics of the scanning electron microscope (SEM) as used in aquatic microbiology have been described by Todd *et al.* (1973). The

SEM has a practical resolution of about 25 nm compared with approximately 300 nm for the conventional light microscope, and a depth-of-field at least 300 times greater. Thus, scanning electron microscopy has great theoretical advantages. Moreover, significant improvements in the preparation of specimens for the SEM during the early 1970s (Sieburth *et al.*, 1974) has meant that present techniques are reliable, relatively simple and reveal a wealth of detail (Sieburth, 1979)

A typical procedure for the observation of epiphytic bacteria on a freshwater plant through the S.E.M is as follows. Immediately after collection the material is cut into small (<1 cm) pieces and preserved in 5% (v/v) phosphate buffered glutaraldehyde (pH 7.2) at approximately 4 °C (i.e. refrigerated). Before use, the preservative is removed by washing for short time (~10 min) in 0.05 M sodium cacodylate buffer (pH 7.2) and the material is then dehydrated through a graded ethanol series (30%, 50%, 70%, 90%, 100%). After passing through acetone, the specimens are critical point dried in liquid carbon dioxide (Cross *et al.*, 1977). The pieces of dried plant are then stuck on aluminium specimen holders (usually called stubs) with colloidal silver paint and coated with a very thin layer of gold (or gold/palladium alloy) in a sputter coater. The gold is electron dense and renders everything coated with it visible in the SEM. A photograph of bacteria on the surface of *L. minor* (duckweed) prepared in this way is shown in Fig. 7.2.

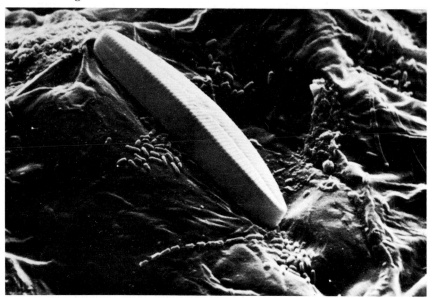

Fig. 7.2 Scanning electron micrograph of bacteria and a diatom (*Navicula*) on *Lemna minor* (duckweed) (× 1800).

Bacterial epiphytes can affect the internal structure of aquatic plant tissue even when that tissue is still fairly young and apparently healthy. Scanning electron microscopy is only useful for morphological studies and the examination of anatomical changes requires the use of transmission electron microscopy (TEM). Rogers and Breen (1981) used both SEM and TEM methods in an investigation of the effects of epiphytic bacteria on *Potamogeton crispus* (curled pondweed). Their method for TEM was as follows. The material was preserved as for SEM (above). The preserved material was washed for 30 min in three changes of 0.05 M cacodylate buffer (pH 7.2). The specimens were then fixed in 2% osmium tetroxide and dehydrated through a graded acetone series (10, 20, 30, 60 and 100%). Next, the material was soaked in propylene oxide before being embedded in araldite resin over 48 h at 70 °C. Transverse sections were cut (approximately 70 μm thick) with a diamond knife microtome and the sections stained with uranyl acetate and lead citrate. Viewing through the T.E.M. showed extensive inward swelling and disorganization of the epidermal cell wall. Zuberer (1984) described the use of light microscopy, SEM and TEM on duckweed epiphytes.

7.4 ENUMERATION INCLUDING SOME STATISTICAL CONSIDERATIONS

In order to count bacteria on aquatic plants, they can be treated by one of the methods detailed in the previous two sections and observed directly on the plant surface by either light or electron microscopy. However, it is also possible to wash the bacteria off the plant surface and either encourage them to grow on agar plates or filter the washings and stain them on the filter. Counting colonies grown on agar is sometimes called a viable count, which is misleading because it may suggest to some readers that the bacteria which do not grow are (or were) not viable. This is not necessarily true simply because the agar supplied may not be a suitable growth medium for them all. It is therefore better to call this method a plate count.

The plate count method is common to many branches of bacteriology and its advantages and disadvantages have been adequately reviewed several times (e.g. Postgate, 1969). Hence, it is unnecessary to describe it at length here. Nevertheless it may be helpful to point out that it is particularly useful for counting subpopulations with a specific physiological attribute. For example, alginate-utilizing bacteria could be enumerated on a selective medium containing alginate as the sole carbon source. This might be a very important part of the epiphytic flora which could not be counted by a direct method (Mow-Robinson and Rheinheimer, 1985). The choice of medium is naturally vital, and for maximum plate counts from freshwater plants, casein-peptone-starch medium is recommended (Fry and Humphrey, 1978).

Dislodging bacteria attached to plants prior to filtration or plate counting is not easy. Traditionally, homogenization has been used (Chan and McManus, 1967), as in soil microbiological studies. However, care must be taken to ensure that the temperature is prevented from rising excessively high and that the shear stress is not sufficient to damage the bacteria (Fry and Humphrey, 1978). When Laycock (1974) counted bacteria on *Laminaria* (kelp), she determined that the optimum conditions were homogenization for 5 min at 3000 rev min^{-1}. In order to avoid the problems associated with homogenization, Fry and Humphrey (1978) recommended using a Stomacher. A Stomacher (A. J. Seward & Co. Ltd) is a machine with two reciprocating paddles, which alternately pummel the specimen contained in a sterile plastic bag with an appropriate diluent. They showed that in seven out of ten trials stomaching for 5 min released significantly more bacteria from plant material than did homogenization.

Velji and Albright (1986) used another method for the removal of epiphytic bacteria. They suggested that the bacteria should be first fixed in 3.7% (v/v) formaldehyde, then the plant samples are incubated in 0.01 M sodium pyrophosphate for 30 min followed by sonication for 45 s at 100 W. They stated that the formaldehyde strengthens the bacterial cells, presumably so that they are not disrupted by the sonication, and that the pyrophosphate promoted dispersion. After removing the bacteria from the plant, they are generally collected on a 0.2 μm black membrane filter and stained with a fluorescent dye. Many stains could be used, but the commonest are acridine orange and 4',6-diamidino-2-phenylindole (DAPI). The bacteria are then counted using the epifluorescence method described earlier.

Regardless of whether homogenization, stomaching or sonication is used, it is necessary to know how many bacteria have been left on the plant surface, i.e. the efficiency of the removal process. This problem is frequently ignored, which is unfortunate because Fry, *et al.* (1985) found that only about 40% of the epiphytic bacteria were removed by stomaching. They determined the efficiency of bacterial removal from three species of riverine plants by estimating bacterial population density using two methods. On one subsample they stained the bacteria with phenolic aniline blue and counted the bacteria directly on the plant surface. A duplicate subsample was washed by stomaching and the filtered bacteria counted after staining with acridine orange. Their results are given in Table 7.1. Note that the variation in their results was large, so that they do not recommend that their factors be applied to other studies without verification.

Once the bacteria have been removed from the plant, they can be treated

TABLE 7.1   Population density of epiphytic bacteria on three freshwater plant species determined by the phenolic aniline blue method, and the percentage removal after 5 min agitation in a stomacher (from Fry *et al.*, 1985, by permission of Blackwell Scientific Publications Ltd

Plant species	*n*	Bacterial numbers ($\times 10^7 \text{ cm}^{-2}$)		Percentage removed	
		Mean	CV[1]	Mean	CV[1]
Nasturtium officinale	6	2.7	39	43	49
Apium nodiflorum	3	3.4	56	44	64
Callitriche sp.	4	0.58	37	36	65

[1] CV: coefficient of variation as a percentage.

like any other bacterial suspension, but care still has to be exercised in the experimental design. The number of microscopic fields counted per membrane filter, the number of filtrations performed per subsample of the original suspension and the number of subsamples used, must all be chosen. Frequently, these choices are arbitrary or pragmatic, but Kirchman *et al.* (1982) have shown that these decisions can greatly affect the precision of the result obtained. Moreover, if the optimum scheme is adopted there can be a considerable saving in time. Unfortunately, the optimum scheme varies with the sample, but in the example given by Kirchman *et al.* (1982) it was two subsamples: one filtration per subsample and seven fields per filter.

When the average number of epiphytic bacteria is calculated, it is generally desirable to give some indication of precision and one of the best ways of doing this is to provide 95% confidence limits. The calculation of confidence limits requires a knowledge of the underlying statistical distribution. The likely distribution can be inferred from the ratio of the variance to the mean. When this ratio is significantly greater than unity, a contagious (over-dispersed) distribution is indicated. This means that the bacteria are aggregated together in clumps. When the variance is approximately equal to the mean, the distribution is random and can probably be described by a Poisson distribution. It is unlikely that bacterial counts will result in the variance being much less than the mean (uniform or regular distribution). The goodness-of-fit to any distribution can be further tested using a chi-squared method or G-test (Sokal and Rohlf, 1981).

Frequently bacteria on membranes follow a Poisson distribution (Jones, 1979) so that the confidence limits are a function of the total count. Hence when totals of 100 or 400 bacteria are counted, the 95% confidence limits are ±20% and ±10% of the counts, respectively. However, counts of bacteria by the phenolic aniline blue method generally follow a clumped distribution and the means are usually correlated with the variance. For such data, the advice in statistical texts is to transform the data in order to normalize it, or words to that effect. While this advice is sound and should be followed, it does not always work. In the author's experience, counts of bacteria directly on the plant surface often produce data which cannot be adequately transformed by any of the common transformations (Baker and Orr, 1986). An example is given in Table 7.2 in which the Kolmogorov-Smirnov goodness of fit test has been applied to bacterial counts obtained using the phenolic aniline blue method. The Kolmogorov-Smirnov test is commonly more powerful than the chi-square test (Sokal and Rohlf, 1981), and is designated by the d_{max} statistic; d_{max} is significant on the untransformed data (Table 7.2) as would be expected, but is also significant ($p < 0.01$) when square root, logarithmic and inverse hyperbolic sine transformations are applied. The inverse hyperbolic sine transformation ($\sinh^{-1}(\sqrt{x})$) is sometimes more effective than the logarithmic transformation (Davies, 1971). Table 7.2 also shows that the failure to normalize the data is due both to skewness of the frequency distribution and to Kurtosis. Skewness means that one tail is much larger than the other, whereas kurtosis means that the peak is either excessively tall and thin, or abnormally flattened. The results are essentially similar for both young and mature leaves although bacterial population density is much greater on the older leaves (Hossell and Baker, 1979b). It is often possible to avoid these problems of non-normality by using non-parametric statistics which are

TABLE 7.2 Effects of three transformations on bacterial population data from young and mature *Ranunculus penicillatus* leaves. All the d_{max} values are significant ($p < 0.01$), i.e. none of the transformations was entirely satisfactory

Leaf stage	n^1	Transformation	d_{max}	Significance level for Skewness	Kurtosis
Young	823	Untransformed	0.44	1%	n.s.
		square root	0.34	1%	1%
		logarithmic	0.38	1%	1%
		$\sinh^{-1}(\sqrt{x})$	0.44	1%	1%
Mature	500	Untransformed	0.26	1%	1%
		square root	0.16	1%	n.s.
		logarithmic	0.17	n.s.	1%
		$\sinh^{-1}(\sqrt{x})$	0.19	1%	1%

[1] n: number of microscopic fields counted.

distribution-free (i.e. do not depend on any underlying distribution). However, some useful parametric analyses, e.g. regression, do not have a readily available parametric equivalent. When such tests are used, the best common transformation (often logarithmic) should still be applied with the explicit proviso that normality of the data has nevertheless not been achieved. Such a statement is neither an admission of failure nor a reason for non-publication, rather it highlights problems associated with the quantitative analysis of highly variable natural populations.

Numbers of epiphytic bacteria can be reported in units of dry weight, wet weight or surface area. Of these three, the only unit which is directly comparable between plant species is surface area, which is therefore preferred. Units of wet (fresh) weight should be avoided due to the errors associated with the quantity of water involved even when all determinations are made by the same operator. When phenolic aniline blue or SEM methods are used, the results are automatically collected per unit area and need only be converted to numbers cm^{-2}. When the epiphytic bacteria are washed off the plant and counted on a filter it is necessary to determine the relationship between surface area and dry weight of the plant, which varies with plant species (Fry and Humphrey, 1978).

What then is the best method for counting epiphytic bacteria associated with aquatic plants? The plate count method, in common with other plate count applications, severely underestimates the total number of bacteria. It should, therefore, be restricted to studies where further information on the bacteria is required, such as physiological characteristics or specific identification. Counting bacteria on filters after a washing technique requires that at least some control samples are examined by direct microscopy to reveal the efficiency of the washing technique (Fry *et al.* 1985). Moreover, in any washing technique, some information on the spatial distribution of the bacteria is always lost. For example, bacteria may be concentrated on one side of the leaf only or may be aggregated into the grooves frequently found on leaf surfaces above the adjoining walls of epidermal cells. Such details as these may be crucial to understanding the ecology of the epiphytes, but cannot be deduced after the bacteria have been removed from the surface. Conversely, observing the bacteria directly on the plant surface has been described as too laborious for quantitative assessment (Last and Warren, 1972). However, Rovira *et al.* (1974) stated that for a given time period direct microscopy gives a more precise estimate of population size than traditional methods. On balance, direct microscopy seems to have more advantages than disadvantages compared with the other methods.

There still remains the choice between SEM and the phenolic aniline blue method. To some extent the choice depends on the ease of access to a scanning electron microscope, but that is not the only difference. As Baker

(1981) emphasized, the minimum number of epiphytic bacteria detected by SEM is very much larger than the limit detected by the phenolic aniline blue method, other things being equal because of the higher magnification used in the electron microscope. Hence many more fields of view need to the be counted under SEM to obtain the same precision as that obtained with light microscopy, assuming that bacteria are seem with equal clarity by both methods. This assumption is sometimes invalid as S.E.M. often provides a clearer image. In Novak's (1984) careful S.E.M. study of the epiphytes on *Posidonia oceanica* (a Mediterranean seagrass), he found it necessary to count 3000 fields on a single shoot. Comparisons between shoots were therefore impractical. To obtain the best of both worlds it would be possible to use S.E.M. at approximately the same magnification as light microscopy (x1000), but this compromise does not yet seem to have been adopted.

7.5 WATER CHEMISTRY AT A SUBMERGED LEAF SURFACE

It would not be appropriate to describe here the many available methods for determining the major chemical variables in the bulk water phase. However, aquatic plants profoundly affect the chemistry of the water, particularly close to the plant surface. These changes in turn affect the epiphytic microflora. The chemical variables principally concerned are pH, oxygen concentration and dissolved organic carbon (DOC) concentration. Methods for determining these variables close to aquatic plant surfaces will therefore be discussed.

Green plants are net producers of oxygen in the light, but consumers of oxygen in the dark. Even in oxygen saturated water, the oxygen concentration can be reduced to zero at the surface of submerged leaves (Sand-Jensen *et al.*, 1985). Thus very steep oxygen gradients occur within a few millimetres of the leaf surface and conventional oxygen probes are much too big to measure differences over such small distances. An oxygen electrode with a tip diameter of only 2.5 to 10 μm, which can be used for this purpose, has been described in detail by Revsbech and Ward (1983). It comprises a modified Clark electrode with a gold cathode protected by a soda glass micropipette sealed with a silicone rubber membrane. The gold is deposited on a platinum wire by a simple electroplating technique. Gold is preferred to platinum as a cathode because it is less easily poisoned. The current is measured with a picoammeter. In use, the oxygen electrode is attached to a micro-manipulator so that measurements can be made at 10 μm intervals if required. Only readings taken as the probe is moved towards the plant surface are reliable. Measurements taken as the probe is removed reflect the disturbance caused by the manipulation.

Sand-Jansen *et al.* (1985) have used this technique on plants supporting both thick and thin layers of epiphytes associated with marine and freshwater plants. They described the significant changes in oxygen concentration observed within a millimetre or two of the plant surface at various light intensities even when epiphytes were apparently absent. These workers have also measured pH close to macrophyte surfaces. Very small pH probes are made commercially (Microelectrodes Inc. available from V.A. Howe), but also require positioning with a micro-manipulator or similar device. Significant pH changes can be found around freshwater plants particularly those growing in hard water. In such locations bicarbonate is often used by the plants as an alternative carbon source to dissolved carbon dioxide. When bicarbonate is taken up hydroxyl ions are released. In certain aquatic plants, e.g. *Potamogeton lucens*, bicarbonate is taken in through the abaxial leaf surface (underside) and hydroxyl ions are released at the adaxial (upper) leaf surfaces (Prins *et al.*, 1982). This results in a much higher pH at the adaxial leaf surface, and Baker and Orr (1986) suggested that this difference in pH might help to explain the different sized epiphytic bacterial populations found on the two leaf surfaces of certain plants.

The majority of aquatic bacteria are heterotrophic, relying on dissolved organic carbon (DOC) compounds for both energy and growth. Aquatic plants are known to release DOC into the water (Filbin and Hough, 1985; Baker and Farr, 1987). The epiphytic bacteria are in an ideal position to utilize this plant-derived DOC. Hence, it is important to know how much DOC is produced by the plants and how much is present in their absence. There are two equally good methods for determining DOC, namely high temperature combustion and photochemical oxidation. The persulphate oxidation method, which has also been used, is no longer recommended (Sharp, 1973; Gershey *et al.*, 1979).

In the high temperature combustion procedure, the acidified water sample is evaporated to dryness and the deposit oxidized in a furnace. The carbon dioxide produced is then measured in a non-dispersive infrared gas analyser. In the photochemical method, the water sample is passed through a long coil (approximately 30 m) of silica glass tubing surrounding an ultraviolet lamp. The UV oxidizes dissolved organic carbon to carbon dioxide, which is quantified, as in the combustion method. Machines utilizing both methods are made commercially. However, as a few millilitres of sample are required for each determination it is not easy to obtain a profile of DOC values as the plant surface is approached. Nevertheless, the proportion of carbon dioxide fixed by the plant which is then released into the water can be determined on axenic plant cultures using either absolute DOC measurements as above or radiolabelling (Baker and Farr, 1987).

7.6 SPATIAL AND TEMPORAL CHANGES

In order to determine spatial and temporal changes in epiphytic bacterial populations, the methods already described for enumeration should be used. However, considerable care has to be taken to ensure that the experimental design is adequate to answer the questions being asked. For example, there is now considerable evidence that older leaves have larger bacterial populations than young leaves (Hossell and Baker, 1979b, c; Novak, 1984). Hence, if it is desired to compare epiphytic populations on plants from different places (or at different times), leaves of the same age must be used, otherwise significant differences might be due to leaf age rather than a difference of position or time. This implies a knowledge of the growth rate of the plants (which may change with time of year), but sometimes it is possible to use the position of a leaf on a plant as a measure of its age relative to the other leaves (Baker and Orr, 1986). However, if one was determining bacterial populations as a food source for protozoans, for example, it might be sensible to compare adjacent leaves regardless of age.

Another complication is produced by different parts of leaves. On *Nasturtium officinale* (watercress) there are significantly more bacteria on the abaxial (lower) surface compared with the adaxial surface (Baker and Orr, 1986). Whereas on *Posidonia oceanica*, the opposite is true (Novak, 1984). Moreover, there can be significant differences between the distal and proximal parts of long narrow leaves characteristic of monocotylednous plants (Novak, 1984) and on different parts of kelp fronds (Mazure and Field, 1980). Thus, when comparing leaves, the same parts of different leaves have to be considered. In order to compare the variance between leaves with the variance within leaves the intraclass correlation coefficient is recommended (Snedecor and Cochran, 1967). Differences in population densities between leaves and stems sampled at the same time have also been reported for some lotic plants (Rimes and Goulder, 1987).

7.7 IDENTIFICATION

Methods for the identification of epiphytic bacteria are similar to those used for other aquatic bacteria, and both traditional schemes (e.g. Zeltner *et al.*, 1978) and modern computer assisted methods (see Chapter 4) are applicable. Therefore little more needs to be said here apart from a few cautionary comments. Microscope observations as well as physiological tests should be used if the maximum number of epiphytic bacteria are to be identified. If direct observation of the plant surface is omitted genera which do not grow well in common media, but may nevertheless be a numerically important part of the bacterial popoulation, may be missed. For example,

Hossell and Baker (1979b) observed large numbers of *Siderocapsa* on *Ranunculus penicillatus* (water crowfoot) and Howard-Williams *et al.* (1979) observed that *Caulobacter* were dominant on *Potamogeton pectinatus* in brackish water. Neither genus would normally be isolated unless exceptional measures were taken. It should also be noted that the bacterial species composition has been shown to vary as leaves age (Howard-Williams *et al.*, 1979) and that different plant species can support quite different bacterial species (Wahbeh and Mahasneh, 1984). Moreover, although Rogers and Breen (1981) assert that the diversity of epiphytes increases with leaf age the evidence is not conclusive and requires further investigation. The species composition of epiphytes has been shown to be similar to that of the suspended bacterial population (Hossell and Baker, 1979b), and this is to be expected if the suspended population is largely composed of detached epiphytes as has been shown to be possible in a vegetated stream (Rimes and Goulder, 1986).

7.8 ATTACHMENT AND DETACHMENT

Methods for determining rates of attachment to inanimate surfaces in water are given in Chapter 15 and while a determined effort in recent years has resulted in considerable progress in this field, attachment of bacteria to plant surfaces has some pecularities worthy of specific comment. The cuticle of many submerged angiosperms is greatly reduced, but still present (Sculthorpe, 1967). The cuticle consists of polyesters of long-chain hydroxy-fatty acids associated with a mixture of non-polar lipids (Gould and Northcote, 1985). It is this layer to which the aquatic bacterium attaches itself. Bacterial attachment is strongly affected by surface charge density and hydrophobicity of the surface, and these properties are quite different for cuticle compared with glass or polystyrene for example. Attachment rates to plant surfaces are, therefore, unlikely to be similar to attachment rates onto inert surfaces of markedly different chemical composition. Another factor of probable importance is the surface roughness which has been shown to substantially affect bacterial attachment rates (Baker, 1984).

Recent studies on the attachment of bacteria to terrestrial plant surfaces have concentrated on genera of economic importance, i.e. *Rhizobium* and *Agrobacterium* (Matthtysse, 1985; Pueppke and Hawes, 1985). Curiously leguminous plants are absent from temperate waters so *Rhizobium* has no direct role in aquatic bacteriology. Moreover, its association with plants is so specialized that no general conclusions can be drawn. *Agrobacterium* enters plants through wounds and does not seem to be found very frequently in water so that information, too, is of little value in aquatic studies. No work has apparently been published on the rates of attachment of specific bacteria to aquatic plants, but there is some information on the attachment rates of natural mixed populations to two lotic plant species (Hossell and Baker, 1979c; Rimes and Goulder, 1985).

In any determination of attachment rates a clear distinction must be drawn between attachment and colonization. Colonization is the sum of attachment and growth, minus losses due to detachment and predation. Colonization rates are relatively easily determined in natural systems, but it is much more difficult to accurately assess the component parts. Rimes and Goulder (1985) avoided some of the problems by determining the increase in bacterial population size on the surface of *Apium nodiflorum* over short periods only (maximum 10 h). They found that attachment rates were correlated with the suspended bacterial concentration and equal to 1.7×10^4 bacteria cm^{-2} h^{-1} for each 10^5 bacteria ml^{-1} in suspension.

Detachment rates have received even less attention than attachment rates. Detachment can take place during the initial reversible phase which precedes permanent attachment, and it can also occur when a permanently attached cell divides, the daughter cell being released into the water column. Hossell and Baker (1979c) determined detachment rates from *Lemna minor* by a simple laboratory technique. They transferred a number of plants to fresh sterile water every hour and counted the bacteria released into the water during that time. The method assumed that no detectable growth occurs which is reasonable for a few hours. Indeed, Hossell and Baker (1979c) found that the detachment rate was constant for the first 10 h, after which it rose exponentially, just like the lag phase and logarithmic growth of a batch culture. The constant detachment rate, which was thought to be the most similar to the *in situ* detachment rate, was 60 000 h^{-1} from 20 plants, i.e. approximately 85 000 h^{-1} cm^{-2} plant surface area. The exponential detachment rate was used in an attempt to apportion the four components of a colonization curve listed above. This detachment rate was subtracted from the overall colonization rate and the result was divided into an attachment rate and a growth rate of the epiphytic bacteria by assuming that attachment was an arithmetic process while growth was exponential (Hossell and Baker, 1979c). Predation losses, due in particular to protozoa feeding, could not be assessed and had to be ignored. The attachment rate derived from this analysis was 5.7×10^5 bacteria cm^{-2} day^{-1}.

7.9 ACTIVITY ESTIMATES

Methods for determining the overall activity of non-epiphytic aquatic bacteria, in particular the heterotrophic uptake method, are discussed in

Chapter 13. Measurements of specific activities, such as nitrogen fixation, have been reported for bacteria on the surface of aquatic plants (e.g. Zuberer, 1982), but the methods are described elsewhere (see Chapters 10 and 11) and apply to epiphytes with little modification. It is, therefore, only necessary to describe here briefly a few other methods which have been specifically applied to epiphytic bacteria. Three methods have been used to determine general metabolic activity of epiphytes. They are: (a) the demonstration on an active electron transport system (Zimmerman *et al.*, 1978); (b) the ability to cleave fluorescein diacetate (Lungren, 1981); and (c) production of elongated cells due to the inhibition of cell division (Kogure *et al.*, 1979; see chapter 13). The electron transport system only operates in actively respiring cells, but it can also be used to reduce 2-(4-iodophenyl)-3-(4-nitrophenyl)-5-phenyltetrazolium chloride (INT) intracellularly to red formazan, which is visible by light microscopy. Quinn (1984) used homogenized *Elodea* (Canadian pondweed) as his source of epiphytic bacteria. He incubated the homogenate with INT (0.02%) for 20 min in the dark and then added acridine orange (final concentration 0.001%). After filtering onto a membrane and examining each field under alternate bright-field and epifluorescent illumination, the number of respiring cells (containing a red deposit) could be calculated as a percentage of the total bacteria (cells fluorescing under UV).

Kidd-Haack *et al.* (1985) also used the INT method on epiphytic bacteria, but instead of obtaining total counts by epifluorescence these workers used malachite green. In this method, entire leaves were incubated in INT and then fixed in formaldehyde. The fixed leaf was then placed in molten gelatin (40 °C) on a microscope slide and held in place with a coverslip. After the gelatin had solidified the slide was dried in a desiccator for 1–2 days. The coverslip and leaf were then removed leaving a gelatin film trapping the epiphytic bacteria which were stained with malachite green (1%w/v) for 30 s. The difficulty with this method, as with all impression techniques, is that the number of bacteria left on the leaf and not transferred to the gelatin is unknown.

The fluorescein diacetate method depends on the fact that this compound is only fluorescent after the diacetate has been removed by hydrolysis. This process is assumed to take place in all active bacteria. The method consists simply of incubating a suspension of cells with fluorescein diacetate (10 μg ml^{-1}) for 3 min before filtering on a black membrane and observing with epifluorescent illumination.

Nalidixic acid is a specific inhibitor of DNA polymerase and therefore of cell division. Thus when a cell suspension is incubated with yeast extract (50 μg ml^{-1}) and nalidixic acid (20 μg ml^{-1}) active cells grow unusually long and swollen without dividing. All the bacteria are stained with acridine orange in the usual way and the proportion of elongated cells is easily determined. Quinn (1984) has compared all three methods on *Elodea* homogenate. He found that the fluorescein diacetate method indicated a substantially lower proportion of active cells compared with the other two methods, but the INT and nalidixic acid methods produced essentially similar results.

7.10 EPIPHYTIC BACTERIA AS A FOOD SOURCE

The importance of epiphytic bacteria as a food resource is a sadly neglected field. Aquatic food webs cannot be properly understood until this problem has been adequately tackled and yet the paucity of information currently available means that this has to be a plea for specific methods to be developed, rather than a description of them. Epiphytic bacteria are consumed principally by protozoa, snails, mayfly larvae and chironomid larvae in freshwater. In the sea the insects' place is taken by crustaceans. The invertebrates seem to be non-selective feeders, ingesting the entire epiphytic community. Thus, it is necessary to differentiate between the relative importance of bacteria in that community and, for example, the diatoms (Baker and Bradman, 1976). It has been suggested that, despite their large numbers, epiphytic bacteria have such a small collective biomass that they are relatively unimportant as a food source (Baker and Bradman, 1976). While this is probably true in terms of carbon, it may be that epiphytic bacteria supply some other essential nutrient such as nitrogen, as appears to be the case for some detritivores (Findlay and Tenore, 1982). It has also been suggested that bacterial products are more important than the epiphytic bacteria themselves in the nutrition of freshwater snails (Thomas *et al.*, 1985).

Although invertebrates may be the most obvious consumers on aquatic plant surfaces, protozoa are probably much more significant. Some protozoa feed solely on bacteria, and a simple calculation based on their relative sizes indicates that several thousand bacteria must be consumed for each protozoan division. Numbers of epiphytic flagellates, ciliates and amoebae are known to be large in southern English rivers (Baldock *et al.* 1983) and yet doubling times for these organisms *in situ* are quite unknown. The lack of methods to tackle these questions is a serious problem which warrants resolute action.

7.11 CONCLUSIONS

Good methods exist for the observation and enumeration of bacteria directly on the surface of freshwater and marine plants. However, great care needs to be taken with the experimental design so that suitable statistical tests can be employed to analyse the very variable data usually encountered. The chemistry of the water adjacent to aquatic plants is often very different to the bulk phase, but probes have been developed capable of determining differences on a submillimetre scale. It is within this specialized chemical environment that the activity of bacterial epiphytes takes place. Several methods are available for determining the activity of these bacteria, but no single method can be recommended unequivocally. Wherever possible several activity methods should be applied to the same or replicate samples. Methods used for determining the attachment of bacteria to model substrates such as glass slides can be applied to plant surfaces, but due consideration must be given to the physicochemical characteristics of the cuticle which are very different to those of artificial surfaces. Moreover, care must be taken to distinguish between attachment rates and colonization rates.

7.12 REFERENCES

Baker, C. D., Bartlett, P. D., Farr, I. S. and Williams, G. I. (1974). Improved methods for the measurement of dissolved and particulate organic carbon in fresh water and their application to chalk streams. *Freshwater Biology*, **4**, 467–481.
Baker, J. H. (1981). Direct observation and enumeration of microbes on plant surfaces by light microscopy. In *Microbial Ecology of the Phylloplane*, J. P. Blakeman (Ed.), Academic Press, London, pp. 3–14.
Baker, J. H. (1984). Factors affecting the bacterial colonization of various surfaces in a river. *Canadian Journal of Microbiology*, **30**, 511–515.
Baker, J. H. and Bradman, L. A. (1976). The role of bacteria in the nutrition of aquatic detritivores. *Oecologia*, **24**, 95–104.
Baker, J. H. and Farr, I. S. (1987). Importance of dissolved organic matter produced by duckweed (*Lemna minor*) in a southern English river. *Freshwater Biology*, **17**, 325–330.
Baker, J. H. and Orr, D. R. (1986). Distribution of epiphytic bacteria on freshwater plants. *Journal of Ecology*, **74**, 155–165.
Baldock, B. M., Baker, J. H. and Sleigh, M. A. (1983). Abundance and productivity of protozoa in chalk streams. *Holarctic Ecology*, **6**, 238–246.
Chan, E. C. S. and McManus, E. A. (1967). Development of a method for the total count of marine bacteria on algae. *Canadian Journal of Microbiology*, **13**, 295–301.
Cross, R. H., Allanson, B. R., Davies, B. R. and Howard-Williams, C. (1977). Critical point drying as a preparative technique for scanning electron microscopy and its application in limnology. *Journal of the Limnological Society of South Africa*, **3**, 59–62.
Daft, G. C. and Leben, C. (1966). A method for bleaching leaves for microscopic investigation of the microflora on the leaf surface. *Plant Disease Reporter*, **50**, 493.
Davies, R. G. (1971). *Computer Programming in Quantitative Biology*, Academic Press, London.
Filbin, G. J. and Hough, R. A. (1985). Photosynthesis, photorespiration, and productivity in *Lemna minor*. *Limnology and Oceanography*, **30**, 322–334.

Findlay, S. and Tenore, K. (1982). Nitrogen source for a detrivore: detritus substrate versus associated microbes. *Science*, **218**, 371–373.

Fry, J. C., Goulder, R. and Rimes, C. A. (1985). A note on the efficiency of stomaching for the quantitative removal of epiphytic bacteria from submerged plants. *Journal of Applied Bacteriology*, **58**, 113–115.

Fry, J. C. and Humphrey, N. C. B. (1978). Techniques for the study of bacteria epiphytic on aquatic macrophytes. In *Techniques for the Study of Mixed Populations*, D. W. Lovelook and R Davies (Ed.), Academic Press, London, pp. 1–29.

Gershey, R. M., McKinnon, M. D., Williams, P. J. le B. and Moore, R. M. (1979). Comparison of three oxidation methods used for the analysis of dissolved organic carbon in seawater. *Marine Chemistry*, **7**, 289–306.

Gould, J. and Northcote, D. H. (1985). Characteristics of plant surfaces. In *Bacterial Adhesion*, D. C. Savage and M. Fletcher (Ed.), Plenum Press, New York, pp. 89–110.

Hossell, J. C. and Baker, J. H. (1979a). A note on the enumeration of epiphytic bacteria by microscopic methods with particular reference to two freshwater plants. *Journal of Applied Bacteriology*, **46**, 87–92.

Hossell, J. C. and Baker, J. H. (1979b). Epiphytic bacteria of the freshwater plant *Ranunculus penicillatus:* enumeration, distribution and identification. *Archiv für Hydrobiologie*, **86**, 322–337.

Hossell, J. C. and Baker, J. H. (1979c). Estimation of the growth rates of epiphytic bacteria and *Lemna minor* in a river. *Freshwater Biology*, **9**, 319–327.

Howard-Williams, C., Davies, B. R. and Cross, R. H. M. (1979). The influence of periphyton on the surface structure of a *Potamogeton pectinatus* leaf. *Aquatic Botany*, **5**, 87–91.

Jones, J. G. (1979). *A Guide to Methods for Estimating Microbial Numbers and Biomass in Fresh Water*, Freshwater Biological Association, Ambleside.

Jones, P. C. and Mollison, J. E. (1948). A technique for the quantitative estimation of soil micro-organisms. *Journal of General Microbiology*, **2**, 54–69.

Kidd-Haack, S., Bitton, G. and Laabes, D. (1985). Epiphytic bacteria: development of a method for determining respiring bacteria on leaves. *Journal of Applied Bacteriology*, **59**, 545–548.

Kirchman, D., Sigda, J., Kapuscinski, R. and Mitchell, R. (1982). Statistical analysis of the direct count method for enumerating bacteria. *Applied and Environmental Microbiology*, **44**, 376–382.

Kogure, K., Simidu, U. and Taga, N. (1979). A tentative direct microscopic method for counting living marine bacteria. *Canadian Journal of Microbiology*, **25**, 415–420.

Last, F. T. and Warren, R. C. (1972). Non-parasitic microbes colonizing green leaves: their form and functions. *Endeavour*, **31**, 143–150.

Laycock, R. A. (1974). The detrital food chain based on seaweeds. Bacteria associated with the surface of *Laminaria* fronds. *Marine Biology*, **25**, 223–231.

Lungren, B. (1981). Fluorescein diacetate as a stain of metabolically active bacteria in soil. *Oikos*, **36**, 17–22.

Maneval, W. E. (1936). Lactophenol preparations. *Stain Technology*, **11**, 9–11.

Matthysse, A. G. (1985). Mechanisms of bacterial adhesion to plant surfaces. In *Bacterial Adhesion*, D. C. Savage and M. Fletcher, Plenum Press, New York, pp. 255–278.

Mazure, H. G. F. and Field, J. G. (1980). Density and ecological importance of bacteria on kelp fronds in an upwelling region. *Journal of Experimental Marine Biology and Ecology*, **48**, 173–182.

Mow-Robinson, J. M. and Rheinheimer, G. (1985). Comparison of bacterial populations from the Kiel Fjord in relation to the presence and absence of benthic vegetation. *Botanica Marina*, **28**, 29–39.

Novak, R. (1984). A study in ultra-ecology: microorganisms on the seagrass *Posidonia oceanica*. *Marine Ecology*, **5**, 143–190.

Postgate, J. R. (1969). Viable counts and viability. In *Methods in Microbiology*, J. R. Norris and D. W. Ribbons. Academic Press, London, pp. 611–628.

Preece, T. F. (1962). Removal of apple leaf cuticle by pectinase to reveal the mycelium of *Venturia inaequalis* (Cooke) Wint. *Nature*, **193**, 902–903.

Prins, H. B. A., O'Brien, J. and Zanstra, P. E. (1982). Bicarbonate utilization in aquatic angiospores, pH and CO_2 concentrations at the leaf surface. In *Studies on Aquatic Vascular Plants*, J. J. Symoens, S. S. Hooper and P. Compère (Eds), Royal Botanical Society of Belgium, Brussels, pp. 112–119.

Pueppke, S. G. and Hawes, M. C. (1985). Understanding the binding of bacteria to plant surfaces. *Trends in Biochemistry*, **3**, 310–313.

Quinn, J. P. (1984). The modification and evaluation of some cytochemical techniques for the enumeration of metabolically active heterotrophic bacteria in the aquatic environment. *Journal of Applied Bacteriology*, **57**, 51–57.

Revsbech, N. P. and Ward D. M. (1983). Oxygen microelectrode that is insensitive to medium chemical composition: use in an acid microbial mat. *Applied and Environmental Microbiology*, **45**, 755–759.

Rimes, C. A. and Goulder R. (1985). A note on the attachment rate of suspended bacteria to submerged aquatic plants in a calcareous stream. *Journal of Applied Bacteriology*, **59**, 389–392.

Rimes, C. A. and Goulder R. (1986). Quantitative observations on the ability of epiphytic bacteria to contribute to the populations of suspended bacteria in two dissimilar headstreams. *Freshwater Biology*, **16**, 301–311.

Rimes, C. A. and Goulder, R. (1986). Temporal variation in density of epiphytic bacteria on submerged vegetation in a calcareous stream. *Letters in Applied Microbiology*, **3**, 17–21.

Rogers, K. H. and Breen, C. M. (1981). Effects of epiphyton on *Potamogeton crispus* leaves. *Microbial Ecology*, **7**, 351–363.

Rovira, A. D., Newman, E. J., Bowen, K. J. and Campbell, R. (1974). Quantitative assessment of the rhizoplane microflora by direct microscopy. *Soil Biology and Biochemistry*, **6**, 211–216.

Sand-Jensen, K., Revsbech, N. P. and Jorgensen, B. B. (1985). Microprofiles of oxygen in epiphyte communities on submerged macrophytes. *Marine Biology*, **89**, 55–62.

Sculthorpe, C. D. (1967). *The Biology of Aquatic Vascular Plants*, Edward Arnold, London.

Sharp, J. (1973). Total organic carbon in seawater—comparison of measurements using persulphate oxidation and high temperature combustion. *Marine Chemistry*, **1**, 211–229.

Sieburth, J. McN. (1979). *Sea Microbes*, Oxford University Press, New York.

Sieburth, J. McN., Brooks, R. D., Gessner, R. V., Thomas, C. D. and Tootle, J. L. (1974). Microbial colonization of marine plant surfaces as observed by scanning electron microscopy. In *Effect of the Ocean Environment on Microbial Activities*, R. R. Colwell and R. Y. Morita (Eds), University Park Press, Baltimore, pp. 418–432.

Snedecor, G. W. and Cochran, W. G. (1967). *Statistical Methods*, State University Press, Iowa.

Sokal, R. R. and Rohlf, F. J. (1981). *Biometry*, Freeman, San Francisco.

Thomas, J. D. Nwanko, D. J. and Sterry, P. R. (1985). The feeding strategies of juvenile and adult *Biomphalaria glabrata* (Say) under simulated natural conditions and their relevance to ecological theory and snail control. *Proceedings of the Royal Society of London, Series B,* **226,** 177–209.

Todd, T. L., Humphreys, W. J. and Odum, E. P. (1973). The application of scanning electron microscopy to estuarine microbial research. In *Estuarine Microbial Ecology,* L. H. Stevenson and R. R. Colwell (Eds), University Press, South Carolina, pp. 115–125.

Van Vuurde, J. W. L. and Elenbaas, P. F. M. (1978). Use of fluorochromes for direct observation of micro-organisms associated with wheat roots. *Canadian Journal of Microbiology,* **24,** 1275–75.

Velji, M. I. and Albright, L. J. (1986). Microscopic enumeration of attached marine bacteria of seawater, marine sediment, fecal matter, and kelp blade samples following pyrophosphate and ultrasound treatments. *Canadian Journal of Microbiology,* **32,** 121–126.

Wahbeh, M. I. and Mahasneh, A. M. (1984). Heterotrophic bacteria attached to leaves, rhizomes and roots of three seagrass species from Jordan. *Aquatic Botany,* **20,** 87–96.

Zeltner, G.-K., Ottow, J. and Kohler, A. (1978). Der Bakterienaufwuchs auf submersen Makrophyten in Abhangigkeit von der Nährstoffbelastung. *Verhandlungen der Gesellschaft für Ökologie, Kiel,* 257–260.

Zimmerman, R., Iturriaga, R. and Becker-Birck, J. (1978). Simultaneous determination of the total number of aquatic bacteria and the number thereof involved in respiration. *Applied and Environmental Microbiology,* **36,** 926–935.

Zuberer, D. A. (1982). Nitrogen fixation (acetylene reduction) associated with duckweed (Lemnaceae) mats. *Applied and Environmental Microbiology,* **43,** 823–828.

Zuberer, D. A. (1984). Microbial colonization of some duckweeds: examination by scanning and transmission electron and light microscopy. *Aquatic Botany,* **18,** 275–285.

Methods in Aquatic Bacteriology
Edited by B. Austin
© 1988 John Wiley & Sons Ltd.

8

Deep-Sea Bacteria

B. Austin

*Department of Brewing and Biological Sciences, Heriot-Watt University,
Chambers Street, Edinburgh EH1 1HX, Scotland*

8.1 INTRODUCTION

Seas comprise the largest continuous expanse of water on this planet. In brief, it has been estimated that seas cover $361 \times 10^6 \, km^2$, and have an average depth of 3800 m. There are three interconnected oceans, namely the Atlantic, Indian and Pacific, and many seas, namely the Arctic, Bering, Black, Caspian, Japan, Mediterranean, Okhotsk, Ross and Weddell. The Pacific is the largest ocean, being approximately equal to the combined area of the Atlantic and Indian Oceans. Moreover, the Pacific Ocean has the greatest average depths, i.e. 3940 m, and the deepest trenches of approximately 11 000 m (Table 8.1).

From the coastal regions and extending outwards, the seas gradually increase in depth before abruptly sloping downwards. This shallow coastal zone is referred to as the *Continental Shelf*, and reaches a maximum depth of approximately 200 m. The continental shelf extends for a width of approximately 65 km from shore, before the downward slope (this is referred to as the *Continental Slope*) leads into the deep water of the ocean basins. These basins, with an average depth of water of 4000–6000 m, occupy almost one-third of the earth's surface. Moreover, it should be emphasized that hydrostatic pressure increases by one atmosphere for evey 10 m of depth.

At the bottom of the sea, there is inevitably a layer of sediment, comprising rock fragments and animal debris, namely pieces of shell and skeleton. The depth of sediment varies considerably, being approximately 600 m thick in the Pacific Ocean (this is regarded as the average thickness of deep sea sediment) increasing to >9 km in thickness within the Puerto Rican Trench. The sediments may be described as comprising *sand*, if

TABLE 8.1   Maximum depth of deep-sea trenches (data from Grant Gross, 1976)

Ocean	Trench	Depth (m)
Atlantic	Puerto Rican	9200
	South Sandwich	8400
Indian	Javan	7460
Pacific	Aleutian	8100
	Japan	9800
	Kermadec-Tongan	10800
	Kuril-Kamchatkan	10500
	Marianas	11000
	Peru-Chile	8050
	Philippine	10000

individual particles exceed 62 μm in diameter, or *mud*, in which case the particles are <62 μm in diameter. In deep-sea sediments, the dominant component is mud, with sand usually comprising <10% of the total constituents. Furthermore, the sediment particles may be characterized by origin, i.e. biogenous (from living organisms), hydrogenous (from inorganic chemical reactions in water), or lithogenous (from erosion of rocks). Briefly, biogenous sediments are derived from the remains of organisms, e.g. shell debris. Thus, they are rich in either calcium carbonate ('calcareous'), silicate ('siliceous') or phosphate ('phosphatic'). To be described as biogenous, a sediment should contain >30% of biogenous particles. Such sediments are common, with calcareous muds covering half of the bottom of the deep sea.

Reference has already been made to the deep sea, and this begs the question about what is meant by the deep sea. Seemingly, deep in this context appertains to depths exceeding 1000 m. Thus, 88% of the area of the sea may be considered as 'deep'. Expressed another way, 75% of the total volume of the oceans comprise deep seas. To date in microbiological studies of the deep sea, a wide range of depths have been sampled, from 1450 m (Jannasch *et al.*, 1971) to 10480 m (Zobell and Morita, 1959). Historically, Certes (1884) was the first investigator to recover bacteria from the deep sea. This feat was achieved during the Travaillier and Talisman Expeditions of 1882–1883. With the discovery of agar as a gelling agent for microbiological media, its use was rapidly adopted in studies of the deep sea. Thus during a transatlantic crossing by passenger ship in 1886, a few colonies resulted from samples collected at depths of >1100 m (Fischer, 1894). Thereafter improvements in the knowledge of deep-sea biology progressed slowly, and it was only after the end of the Second World War that rapid developments ensued, due largely to the stirling efforts of C. E. Zobell and co-workers. This team initiated work on the effect of hydrostatic pressures on bacterial activity (Zobell and Johnson, 1949). Thereafter, participation in the Danish Galathea Expedition of 1950–1952 started Zobell on his pioneering work on deep-sea microbiology.

8.2 THE DEEP-SEA ENVIRONMENT

Although it may be a popular misconception that the deep sea is a dark inhospitable environment, devoid of food and life, evidence shows that both food and unique life-forms exist even at the bottom of the deepest trenches. Apart from the sporadic arrival of large chunks of food, such as dead fish, there is a more continuous settling to the sea floor of minute organic particles, comprising the so-called 'marine snow' (Silver *et al.*, 1978). These particles include highly recalcitrant molecules, faecal matter and components of phyto- and zoo-plankton. However, there is a low input of organic carbon, with only approximately 1% of the photosynthetic organic carbon reaching the bottom of the deep sea.

Much of the deep sea is cold, with temperatures of $3° \pm 1°C$, and is under great pressure. Thus, it may be assumed that psychrophiles and barophiles predominate. The latter term, coined by Zobell and Johnson (1949), refers to organisms which only grow at pressures exceeding atmospheric pressure. The waters and sediments are also well oxygenated, a phenomenon which is attributed to the extremely low rates of oxygen consumption by deep-sea organisms.

8.2.1 Hydrothermal vents

On the ocean bottom, active volcanoes add to the topography of the seafloor. Many of these volcanoes are active, emitting hot lava into the marine environment. The freshly emitted lava contracts during cooling, and allows seawater to percolate downward by a distance of several kilometres into the newly formed crust (Jannasch and Taylor, 1984). Here in the presence of high pressures and temperatures ($>350°C$), the seawater reacts with the underlying basalt rock to form a highly reduced and acidic liquid, known as 'hydrothermal fluid'. This is enriched with hydrogen, hydrogen sulphide and metals, notably carbon, copper, iron, calcium, magnesium and manganese. At this stage, one of two possibilities occurs. Either, the hydrothermal fluid rises and makes contact with oxygen when mixing with the cooler seawater in the porous subsurface lava. With this situation, the result is the emergence of oxygenated (or non-oxygenated) *warm* vents at temperatures of 8–23°C, with discharge rates of 1–2 cm s^{-1}. Alternatively, the hydrothermal fluid may discharge without any prior mixing, into the bottom waters as oxic *hot* vents at temperatures of _ca_. 350 °C, and with discharge rates of 1–2 m s^{-1} (Jannasch and Mottl, 1985). This hot hydrothermal fluid, which contains particulates and hydrogen sulphide concentrations of 0.5–10 mM, mixes with cold seawater upon discharge from the vent. In this situation, there is concomitant rapid and dense precipitation of grey-black polymetasulphide particles, which generate a smoke-like plume. These are the so-called '_black-smokers_'.

8.3 COLLECTION OF SAMPLES

Deep-sea microbiology has surely satisfied the inventive ingenuity of its practitioners. Experimental designs have ranged from the very simple to the extremely complex. Either attempts are made to collect and maintain samples at _in situ_ pressures (and perhaps temperatures) or water and/or sediment may be returned and depressurized with ascent to the surface. In this latter case, it is questionable whether or not the micro-organisms could suffer a fate analogous with the 'bends' in human beings. Some workers have even repressurized decompressed samples. The samples may be obtained in bags, grabs and corers from lines descending from surface ships. The usefulness of such samples for explaining deep sea microbiology is highly suspect. Sinking particulates may be collected in sediment traps (Rowe and Gardner, 1979; Deming, 1985). Sophisticated pressure chambers have also been developed for the collection of deep-sea samples (Fig. 8.1). Tabor and Colwell (1976) described a sampler of two interlocking stainless steel chambers, in which a sequenced cam-driven trigger and a spring driven pump were incorporated into one of the two subunits. The second unit consisted of a 400 ml capacity teflon-lined sample chamber. In operation, the sample chamber was filled with distilled water to which a harmless dye was added (later, this served to see whether or not the sampler filled with seawater). Thence, the sampler was autoclaved, cooled to 3–5 °C and, using a sterilized hand pump, the hydrostatic pressure in the chamber was increased to that of slightly in excess of the anticipated pressure at the desired sampling depth. In operation, the deep-sea sampler, which was inevitably attached to a ship-board line, was triggered by a messenger weight, whereupon seawater entered the sterilized chamber at a fixed rate resulting from the pressure differential and shear force taken up by the sterile distilled water passing through a small orifice. Following collection of the sample, radiolabelled nutrients could be added in order to determine microbial activity at _in situ_ conditions. However, for some, the risk of losing such expensive and valuable equipment is considered to be unacceptable. Moreover, an obvious drawback is that only a few samples may be collected, and the turn-around time before use may be measured in weeks. With increasing financial resources, deep-sea

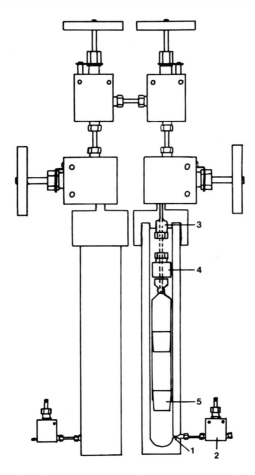

Fig. 8.1. A pressurized deep-sea sampler. To remove a subsample without depressurization, the transfer system is connected to the pressurized sample and pressurized through the reactor port (1) and miniature valve (2). The receiving and incubation chamber is a modified 50 ml syringe, secured with pressure fittings (3, 4) and plugged with a rubber stopper (5). Sterile seawater is added to the pressurized sample to compensate for pressure changes arising from a change in volume (from Tabor *et al.*, 1981; reproduced by permission of Springer-Verlag).

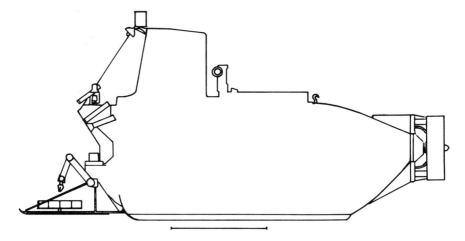

Fig. 8.2 Schematic representation of the deep-sea submersible (Alvin).

submersibles of which the Alvin (Fig. 8.2) is an example, may be hired for the collection of deep-sea samples. However, if these are returned to the surface for examination at atmospheric pressure, the extra cost of the operation is difficult to justify. Alternatively, such samples may be maintained at *in situ* temperature and pressure, either by use of pressure chambers or by carrying out the subsequent experiments in the deep sea, itself. For example, independent vehicles, known as 'tripods' or 'bottom landers' have been assembled. These descend freely to the seafloor for a variety of automated roles, returning to the surface after the timed release of an anchor (Isaacs, 1969).

8.4 RECOVERY OF DEEP-SEA BACTERIA

8.4.1 Recovery in the absence of pressure

For depressurized samples, the subsequent examinations proceed as for any other marine samples. Thus, a dilution step may occur in autoclaved seawater or artificial seawater mixes. Such mixes include salt solutions, e.g. comprising $MgSO_4 \cdot 7H_2O$, 0.698% (w/v); KCl, 0.075% (w/v); NaCl, 2.38% (w/v); pH 7.2 (after Schwarz *et al.*, 1974) and artificial seawater, which comprises $CaCl_2$, 0.2% (w/v); KBr, 0.015% (w/v); KCl, 0.075% (w/v); $MgCl_2 \cdot 6H_2O$, 1.22% (w/v); $MgSO_4 \cdot 7H_2O$, 0.58% (w/v); NaCl, 2.38% (w/v); $NaHCO_3$, 0.03% (w/v); Na_2SO_4, 0.36% (w/v) and $SrCl_2 \cdot 6H_2O$, 0.005% (w/v) (after Schwarz and Colwell, 1975). The diluted samples will then be inoculated onto nutrient-limited (seawater gelled with agar) or

TABLE 8.2   Recipe for modified seawater yeast extract agar (after Schwarz *et al.*, 1974)

Ingredient	Composition (w/v)
Proteose peptone	0.1
Yeast extract	0.1
KCl	0.075
MgSO$_4 \cdot$7H$_2$O	0.698
NaCl	2.38
Agar	2.0
pH 7.2	

nutrient-rich media (e.g. marine 2216E agar or modified seawater yeast extract agar, Table 8.2), with incubation at 4–25 °C for periods of up to 28 days.

8.4.2 Recovery of barophiles

Dietz and Yayanos (1978) pioneered a method for the isolation of barophilic bacteria. In those experiments, attention focused on a nutrient-rich silica gel shake tube medium (Table 8.3), which was used successfully to recover a spirillum from a decomposing deep-sea amphipod. Aliquots (0.1 ml) of diluted samples were inoculated into the medium, which was subsequently sealed, e.g. with parafilm, and incubated under pressure at 3–4 °C for 21 days (Deming, 1985). Pure cultures may then be obtained by serial dilution in broth (Deming and Colwell, 1981). Alas with this technique, it is necessary to decompress the container in order to carry out microbiological manipulations with the cultures. Nevertheless, it was later realized that agar plating techniques could be used in conjunction with pressurized vessels containing oxygen–helium mixtures to recover bacteria in pure culture (Taylor, 1979; Jannasch *et al.*, 1982). Here the chamber was first pressurized before introduction of the sample via specially constructed transfer devices into reception vials. Then, an electrically sterilized inoculation loop was manipulated (from outside the chamber) to receive inoculum for streaking onto the agar plate (such as marine 2216E agar). This procedure was aided by a viewing window (Jannasch *et al.*, 1982). Following incubation, well-isolated colonies were inoculated into vials containing liquid media for subsequent growth experiments. It may be concluded that the technique requires considerable dexterity. For inexperienced personnel, it may well be prone to mishap. It should be emphasized that, as a general rule, the optimum pressure for growth of barophiles corresponds closely to the *in situ* pressure from which the samples were obtained (Jannasch and Taylor, 1984).

TABLE 8.3   Silica-gel medium for the recovery of barophilic bacteria (after Dietz and Yayanos, 1978)

Preparation of silica gel column

1. A strong cation exchange resin (Amberlite 120-H-C-P, medium porosity) was packed to form a 4.4 cm diameter × 43 cm high column.
2. The column was regenerated with 250 ml 4 N HCl, rinsed with 1 l of distilled water, drained of water, and filled with 500 ml of 0.5 M Na₂SiO₃·9H₂O (ion exchange was effective for up to 100 ml/min).
3. The first 200 ml of effluent was discarded, and the remaining was collected in a 1 l cylinder.
4. With silicate in the column, distilled water was added until a total of 800 ml of effluent obtained.
5. Effluent volume was topped up to 1 l with distilled water; the pH adjusted to 1.5 with 5 ml of 4 N HCl, and the column sterilized (121 °C/20 min. This was stable for up to one month at temperatures of 3 °C to room temperature).

Preparation of silica gel medium

1. The silica gel column was cooled to 4 °C, and the pH adjusted to 8.7 with 5 ml of sterile 1 N NaOH.
2. To 100 ml volumes of the silica, 10 ml of nutrients were added. The nutrients comprised (in distilled water)
   tryptone 5.0% (w/v)
   yeast extract 2.5% (w/v)
   glucose 1.0% (w/v)
3. Then, 10 ml of artificial seawater was added (24% w/v NaCl, 0.7% w/v KCl, 5.3% w/v MgCl₂·6H₂O, 7% w/v MgSO₄·7H₂O).

Preparation of silica gel medium:

4. The completed medium was dispensed into test tubes, which had been seeded with 0.1 ml aliquots of inoculum. The tubes were covered with parafilm, and incubated.

For hot hydrothermal vents, the use of a novel gelling agent, i.e. Gelrite (Lin and Casida, 1984), has been advocated for incorporation in a nutrient-deficient medium. This was designed for the cultivation of thermophilic, black-smoker bacteria at 120 °C (Table 8.4; Deming and Baross, 1986). To prepare this medium, the salts and trace elements need to be dissolved in distilled water, and the pH adjusted to 5.5 before addition of the gelling agent, and thence sterilization at 121 °C/30 min. In the initial publication, mention was made that the medium was prepared in 100 ml aliquots in 250 ml volume Erlenmeyer flasks. Immediately after sterilization, the medium was transferred to a bath at 100 °C. Thence 10 ml volumes were transferred, via pre-heated (80 °C) pipettes to 30 ml capacity pre heated glass tubes or 10 ml capacity glass syringes, which had been previously seeded with small quantities (0.03–0.5 ml) of the inoculum. At

TABLE 8.4 A medium for the cultivation of 'black smoker' bacteria (from Deming and Baross, 1986)

Ingredient	Composition (w/v)
$FeSO_4$	0.001
KCl	0.08
$MgSO_4 \cdot 7H_2O$	0.2
$(NH_4)_2SO_4$	0.1
NaCl	2.5
Na_2HPO_4	0.05
$Na_2S_2O_3$	0.3
Sodium acetate	0.5
Trace elements (Rippka *et al.*, 1979)	0.3
Gelrite (Difco)	0.8
pH 5.5; 121 °C/30 min	

this point, it was imperative to ensure rapid mixing before the setting process ensued (this occurred rapidly upon cooling below 80 °C). The inoculated tubes of medium were subsequently covered before incubation at 70–120 °C with or without hydrostatic pressure. Unfortunately, there is a need to decompress any pressurized medium prior to examination of the colonies.

Recovery of bacteria from warm vents is more straightforward, insofar as enrichment culture, e.g. with 1 mM thiosulphate for sulphur bacteria, has sufficed (Jannasch *et al.*, 1976).

8.5 EXPERIMENTS WITH DEEP-SEA BACTERIA

In situ experiments on the seafloor have overcome some of the problems of recovery of samples. Such experiments include the degradation of organic materials, such as chitin and wood, and the assessment of heterotrophic activity using ^{14}C-labelled substrates (e.g. Schwartz and Colwell, 1975). If available, deep-sea submersibles may be used to position material on the seafloor, with subsequent inspection at predetermined intervals. Comparative experiments may also be carried out at atmospheric pressure (Jannasch and Wirsen, 1973).

Microbial activity upon dissolved substrates may be determined by means of elaborate devices, such as the so-called 'syringe array'. Here, 6×200 ml syringes, containing radiolabelled compounds, may be filled with seawater *in situ* through a common inlet nozzle by action of a mechanical arm on the submersible. Thence, incubation is allowed to ensue *in situ*, as well as control experiments at atmospheric pressure in a surface laboratory. This technique has been used successfully to measure carbon dioxide fixation in the deep sea (Jannasch, 1984).

Other *in situ* experiments have utilized independent vehicles, such as 'tripods' or 'bottom landers'. These descend freely to the seafloor for a variety of automated roles, returning to the surface after the eventual timed release of an anchor (Isaacs, 1969). Such equipment has been used to measure microbial growth (Seki *et al.*, 1974), nitrate metabolism (Wada *et al.*, 1975), transformation of radiolabelled organic compounds into the upper layers of sediment (Jannasch and Wirsen, 1980), decomposition of compounds in the sediments, and metabolism of deep sea amphipods, and their intestinal microflora (Jannasch *et al.*, 1980; Wirsen and Jannasch, 1983).

Microscopy (phase contrast or epifluorescence) of deep sea samples, whether passed through bacterial retaining filters or unfiltered, provides useful information about the size of microbial populations. For example, Turner (1979) observed bacteria on the surface of faecal pellets, and concluded that many of the microbes in the deep sea were recent arrivals from surface waters.

Within the laboratory, experiments may be carried out with samples contained within pressurized chambers, for example, starvation studies (to determine the effect upon cell size and shape) and the utilization of substrates. Biomass may be studied by means of epifluorescence microscopy and ATP determinations.

8.6 CONCLUSIONS

Apart from sophisticated sampling and isolation procedures, studies on deep-sea bacteria involve the host of other techniques available for use in aquatic microbiology. However, the study of deep-sea microbes requires sizeable financial commitments, including funds for ship time and access to submersibles. Consequently, the opportunities are available to only a few dedicated laboratories.

8.7 REFERENCES

Certes, A. (1884). Sur la culture, à l'abri des germes atmosphériques, des eaux et des sédiments rapportés par les expéditions du Travailleur et du Talisman. *Compte Rendus hebdomadaire des Séances de l' Académie des Sciences*, **98**, 690–693.

Deming, J. W. (1985). Bacterial growth in deep-sea sediment trap and boxcore samples. *Marine Ecology—Progress Series*, **25**, 305–312.

Deming, J. W. and Baross, J. A. (1986). Solid medium for culturing black smoker bacteria at temperatures to 120 °C. *Applied and Environmental Microbiology*, **51**, 238–243.

Deming, J. W. and Colwell, R. R. (1981). Barophilic bacteria associated with deep-sea animals. *BioScience*, **31**, 507–511.

Dietz, A. S. and Yayanos, A. A. (1978). Silica gel media for isolating and studying bacteria under hydrostatic pressure. *Applied and Environmental Microbiology*, **36**, 966–968.

Fischer, B. (1894). Die Bakterien des Meeres nach den Untersuchungen der Plankton. Expedition unter gleichzeitiger Berucksichtigung einiger alterer and neuerer Untersuchungen. *Zentralblatt für Bakteriologie*, **15**, 657–666.

Grant Gross, M. (1976). *Oceanography*, 3rd edn, Charles E. Merrill, Columbus, Ohio.

Isaacs, J. D. (1969). The nature of oceanic life. *Scientific American*, **221**, 146–162.

Jannasch, H. W. (1984). Microbes in the oceanic environment. In, *The Microbe 1984*, part II, D. P. Kelly and N. G. Carr, (Eds), Cambridge University Press, Cambridge, pp. 97–122.

Jannasch, H. W., Eimhjellen, K., Wirsen, C. O. and Farmanfarmaian, A. (1971). Microbial degradation of organic matter in the deep sea. *Science (Washington)*, **171**, 672–675.

Jannasch, H. W. and Taylor, C. D. (1984). Deep-sea microbiology. *Annual Review of Microbiology*, **38**, 487–514.

Jannasch, H. W. and Wirsen, C. O. (1973). Deep-sea microorganisms: *in situ* response to nutrient enrichment. *Science (Washington)*, **180**, 641–643.

Jannasch, H. W. and Wirsen, C. O. (1980). Chemosynthetic primary production at East Pacific sea floor spreading centers. *BioScience*, **29**, 592–598.

Jannasch, H. W., Wirsen, C. O., Cuhel, R. L. and Taylor, C. D. (1980). An approach for *in situ* studies of deep-sea amphipods and their microbial gut flora. *Deep-Sea Research*, **27**, 867–872.

Jannasch, H. W., Wirsen, C. O. and Taylor, C. D. (1976). Undecompressed microbial populations from the deep sea. *Applied and Environmental Microbiology*, **32**, 360–367.

Jannasch, H. W., Wirsen, C. O. and Taylor, C. D. (1982). Deep-sea bacteria; isolation in the absence of decompression. *Science (Washington)*, **216**, 1315–1317.

Jannasch, H. W. and Mottl, M. J. (1985). Geomicrobiology of deep sea hydrothermal vents. *Science (Washington)*, **229**, 717–725.

Lin, C. C. and Casida, L. E. (1984). Gelrite as a gelling agent in media for the growth of thermophilic microorganisms. *Applied and Environmental Microbiology*, **47**, 427–429.

Rippka, R., Deruelles, J., Waterbury, J. B., Herdman, M. and Stanier, R. Y. (1979). Generic assignments, strain histories and properties of pure cultures of Cyanobacteria. *Journal of General Microbiology*, **111**, 1–61.

Rowe, G. T. and Gardner, W. D. (1979). Sedimentation rates in the slope water of the northwest Atlantic Ocean measured directly with sediment traps. *Journal of Marine Research*, **37**, 581–600.

Schwartz, J. R. and Colwell, R. R. (1975). Heterotrophic activity of deep-sea sediment bacteria. *Applied Microbiology*, **30**, 639–649.

Schwartz, J. R., Walker, J. D and Colwell, R. R. (1974). Growth of deep-sea bacteria on hydrocarbons at ambient and *in situ* pressure. *Developments in Industrial Microbiology*, **15**, 239–249.

Seki, H., Wada, E., Koike, I. and Hattori, A. (1974). Evidence of high organotrophic potentiality of bacteria in the deep ocean. *Marine Biology*, **26**, 1–4.

Silver, M. W., Shanks, A. L. and Trent, J. D. (1978). Marine snow: microplankton habitat and source of small-scale patchiness in pelagic populations. *Science (Washington)*, **201**, 371–373.

Tabor, P. S. and Colwell, R. R. (1976). Initial investigations with a deep ocean *in situ* sampler. *Ocean '76*, 13D1–13D4.

Tabor, P. S., Deming, J. W., Ohwada, K., Davis, H., Waxman, M. and Colwell, R. R. (1981). A pressure-retaining deep ocean sampler and transfer system for measurement of microbial activity in the deep sea. *Microbial Ecology*, **7**, 51–65.

Taylor, C. D. (1979). Growth of a bacterium under a high-pressure oxyhelium atmosphere. *Applied and Environmental Microbiology*, **37**, 42–49.

Turner, J. T. (1979). Microbial attachment to copepod fecal pellets and its possible ecological significance. *Transactions of the American Microscopic Society*, **98**, 131–135.

Wada, E., Koike, I. and Hattori, A. (1975). Nitrate metabolism in abyssal waters. *Marine Biology*, **29**, 119–124.

Wirsen, C. O. and Jannasch, H. W. (1983). *In situ* studies on deep sea amphipods and their intestinal microflora. *Marine Biology*, **78**, 69–73.

Zobell, C. E. and Johnson, F. H. (1949). The influence of hydrostatic pressure on the growth and viability of terrestrial and marine bacteria. *Journal of Bacteriology*, **57**, 179–189.

Zobell, C. E. and Morita, R. Y. (1959). Deep-sea bacteria. *Galathea Report* **1**, 139–154.

PART 3

Specialized Groups

Methods in Aquatic Bacteriology
Edited by B. Austin
© 1988 John Wiley & Sons Ltd.

9

Anoxygenic Phototrophic Bacteria

J. F. Imhoff

Institut für Mikrobiologie, Meckenheimer Allee 168, 5300 Bonn, FRG

9.1 INTRODUCTION

Anoxygenic phototrophic bacteria are found in various kinds of natural waters including extremes of salinity, pH, and temperature, i.e. freshwater environments and hypersaline brines, alkaline lakes and acid waters, hot springs and arctic lakes. Some species have also been isolated from soil samples.

Phototrophic bacteria may develop wherever anoxic conditions prevail, and light is available as an energy source. The normal habitat is the lower anoxic part of the photic zone in lakes and sediments. The concentrations of sulphide and oxygen, the pH-value, light conditions and stability of the anoxic conditions are of particular importance for the occurrence of different species of phototrophic bacteria (Pfennig, 1967; 1977). Chromatiaceae and Chlorobiaceae representatives are generally found in sulphide containing habitats, where sulphate-reducing bacteria are active. These bacteria form mass developments, predominantly in marine coastal areas, and in lakes with a well-established chemocline in the photic zone. In general, the purple non-sulphur bacteria (Rhodospirillaceae) are less sulphide tolerant and are dominant in sulphide-poor freshwater habitats, but some species also regularly occur together with phototrophic sulphur bacteria in habitats enriched in sulphide (Imhoff, 1982a). The organic matter produced by the phototrophic sulphur bacteria is about 3–5% of the total annual production in lakes poor in sulphide, 9–25% in lakes rich in sulphide, and may reach values of 83% in Solar Lake and Fayetteville Green Lake (Takahashi and Ichimura, 1968; Culver and Brunskill, 1969; Cohen *et al.*, 1977). The contribution to the daily production under stratified conditions is even higher, ranging from 20 to 91%.

The Ectothiorhodospiraceae and Chloroflexaceae are much more restricted in their natural distribution due to their growth requirements (alkalinity, salinity and temperature, see below).

9.2 TAXONOMY

Two major groups are distinguished among the anoxygenic phototrophic bacteria: the phototrophic purple bacteria and the phototrophic green bacteria. In addition, some recent isolates have been obtained, whose taxonomic position is not clear, e.g. *Heliobacterium chlorum* (Gest and Favinger, 1983) and some aerobic heterotrophs that contain bacteriochlorophyll *a*, but are unable to photosynthesize (Nishimura *et al.*, 1981; Sato, 1978; Shiba and Simidu, 1982; Shiba *et al.*, 1979).

9.2.1   The phototrophic purple bacteria

The phototrophic purple bacteria contain bacteriochlorophyll *a* or *b* as antenna pigments. These are located in the cytoplasmic membrane and in various kinds of invagination of it.

1.   The purple sulphur bacteria (Chromatiaceae) grow well under photo-autotrophic conditions. All of them can grow with sulphide and some representatives also use elemental sulphur, sulphite, or thiosulphate as electron donors. During oxidation of sulphide to sulphate, elemental sulphur is deposited inside the cells. Photoheterotrophic, chemoautotrophic and/or chemoheterotrophic growth has been de-monstrated for several species (Gorlenko, 1974; Kondratieva *et al.*, 1976; Kämpf and Pfennig, 1980). Some species require vitamin B_{12} as the only growth factor. Various types of carotenoids and, in most species, bacteriochlorophyll *a* serve as photosynthetic pigments. These are located in intracytoplasmic membranes that are formed as vesicles in all but one species (see Table 9.1).

2.   The Ectothiorhodospiraceae are distinguished from the Chromatiaceae by the intermediate deposition of elemental sulphur outside the cells. There is a characteristic dependence on saline and alkaline growth conditions. Intracytoplasmic membranes are present as lamellar stacks. Inclusion into the Chromatiaceae has been made possible only by emendation of the family characteristics (Pfennig and Trüper, 1971), but significant distinguishing structural and physiological properties later made separation necessary (Imhoff, 1984a; see Table 9.2).

3.   The purple non-sulphur bacteria (Rhodospirillaceae) are the most diverse and best studied group of phototrophic bacteria. Most species are motile, require growth factors, and have various types of carotenoids and bacteriochlorophyll *a* as photosynthetic pigments. These bacteria grow preferably under photoheterotrophic growth conditions and many representatives are aerotolerant and can grow as chemoheterotrophs in the dark. Some species are also able to use sulphide as an electron donor, but only very few oxidize sulphide via

TABLE 9.1 Some characteristics of the family Chromatiaceae. Data were collected from: Eichler and Pfennig, 1986; Trüper and Pfennig, 1981; Madigan, 1986; Gorlenko *et al.* 1979; Schmidt, 1978

Species	Cell shape	Cell size diameter (µm)	Major carotenoid	Growth factor	Motility	Gas vacuoles	G+C content (mol%)	Other significant properties
Chromatium								
C. vinosum	rod	2.0	sp, ly, rh	–	+	–	61.3–66.3	
C. violascens	rod	2.0	rh, rl	–	+	–	61.8–64.3	
C. gracile	rod	1.0–1.3	sp, ly, rh	–	+	–	68.9–70.4	NaCl tolerant
C. minutissimum	rod	1.0–1.2	sp, ly, rh	–	+	–	63.7	
C. purpuratum	rod	1.2–1.7	ok	o	+	–	68.9	NaCl required
C. tepidum	rod	1.2	sp, rv	–	+	–	61.5	thermophilic
C. minus	rod	2.0	ok	–	+	–	52.0–62.2	
C. okenii	rod	4.5–6.0	ok	B_{12}	+	–	48.0–50.0	
C. weissei	rod	3.5–4.0	ok	B_{12}	+	–	48.0–50.0	
C. warmingii	rod	3.5–4.0	ra, ro	B_{12}	+	–	55.1–60.2	
C. buderi	rod	3.5–4.5	ra	B_{12}	+	–	62.2–62.8	NaCl required
Thiodictyon								
T. elegans	rod	1.5–2.0	ra, rh	–	–	+	65.3	nets formed
T. bacillosum	rod	1.5–2.0	ra	–	–	+	66.3	
Lamprocystis								
L. roseopersicina	sphere	3.0–3.5	la, lo	–	+	+	63.8	
Thiocystis								
T. violacea	sphere	2.5–3.5	ra, ro, rh	–	+	–	62.8–67.9	
T. gelatinosa	sphere	3.0	ok	–	+	–	61.3	

(continued overleaf)

TABLE 9.1 (Contd.)

Species	Cell shape	Cell size diameter (μm)	Major carotenoid	Growth factor	Motility	Gas vacuoles	G+C content (mol%)	Other significant properties
Amoebobacter								
A. roseus	sphere	2.0–3.0	sp	B_{12}	–	+	64.3	
A. pendens	sphere	1.5–2.5	sp	–	–	+	65.3	
A. pedioformis	sphere	2.0	sp	(B_{12})	–	+	65.5	platelets formed
Thiopedia								
T. rosea	ovoid	1.0–2.0	ok	o	–	+	o	platelets formed
Thiocapsa								
T. roseopersicina	sphere	1.2–3.0	sp	–	–	–	63.3–66.3	
T. pfennigii	sphere	1.2–1.5	ts	–	–	–	69.4–69.9	bchl b, ICM as tubes
Lamprobacter								
L. modestohalophilus	rod	2.0–2.5	o	B_{12}	+	+	64.0	NaCl required
Thiospirillum								
T. jenense	spiral	2.5–4.5	rh, ly	B_{12}	+	–	45.5	

Abbreviations

o: not determined; bchl; bacteriochlorophyll; ICM: intracytoplasmic membrane system; (B_{12}): vitamin B_{12} strongly enhancing growth, but not absolutely required.

carotenoids: la: lycopenal; lo: lycopenol; ly: lycopene; ok: okenone; ra: rhodopinal; rh: rhodopin; ro: rhodopinol; rv: rhodovibrin; sp: spirilloxanthin; ts: 3,4,3′,4′, tetrahydrospirilloxanthin.

TABLE 9.2 Some characteristics of the family Ectothiorhodospiraceae. Data were collected from: Imhoff, 1984b; Imhoff and Trüper, 1981; Imhoff et al., 1981; Trüper and Imhoff, 1981; Schmidt, 1978

Species	Cell Shape	Cell size diameter (μm)	Colour of culture	Major carotenoid	Major bchl	Major quinone	Gas vacuoles	Optimal salinity	G+C content (mol%)
Ectothiorhodospira									
E. mobilis	rod-spiral	0.7–1.0	red	sp, rh	a	Q-7, MK-7	–	3–15%	62.0–69.9
E. shaposhnikovii	rod-spiral	0.8–0.9	red	sp, rh	a	Q-7, MK-7	–	1–3%	61.2–62.8
E. vacuolata	rod	1.5	red	sp	a	Q-7, MK-7	+	1–6%	61.4–63.6
E. halophila	spiral	0.8–0.9	red	sp	a	Q-8, MK-8	–	15–30%	64.3–69.7
E. halochloris	spiral	0.5–0.6	green	rhg*, rh	b	Q-8, MK-8	–	14–27%	50.5–52.9
E. abdelmalekii	spiral	0.9–1.2	green	**	b	Q-8, MK-8	–	12–20%	63.3–63.8

Abbreviations:
o: not determined; bchl: bacteriochlorophyll; Q: ubiquinone; MK: menaquinone.
Carotenoids: rh: rhodopin; rhg*: rhodopin glucoside and derivatives; sp: Spirilloxanthin; ** most probably similar as in *E. halochloris*.

elemental sulphur to sulphate (Neutzling *et al.*, 1984; Imhoff *et al.*, 1984; Hansen and Imhoff, 1985). Globules of elemental sulphur never appear inside the cells (see Table 9.3).

9.2.2 The phototrophic green bacteria

The green bacteria contain bacteriochlorophyll *c*, *d* or *e* as antenna pigments. These pigments are localized in special organelles, characteristic of all members of this group (except *Heliothrix oregonense*) i.e. the chlorosomes (or chlorobium vesicles), which are firmly attached to the cytoplasmic membrane. The photosynthetic reaction centres containing bacteriochlorophyll *a* are located in the cytoplasmic membrane.

1. The phototrophic green sulphur bacteria require strictly anaerobic conditions for growth, and can assimilate only a limited number of organic compounds. In their sulphide-, CO_2-, and light-dependent metabolism, elemental sulphur globules are formed outside the cells, in the medium. CO_2-assimilation does not proceed via the reductive pentose phosphate cycle but via the reductive tricarboxylic acid cycle (Fuchs *et al.*, 1980a,b). They are dependent on reduced sulphur compounds, do not assimilate sulphate and many strains require vitamin B_{12}. With the exception of *Chloroherpeton thalassium* (gliding motility) they are non-motile. Chain formation is common in many strains; chains are straight in rod-shaped strains and spiral in vibrio-shaped strains. A number of brown and green coloured species are distinguished (see Table 9.4).
2. The multicellular filamentous green bacteria (Chloroflexaceae) are motile by gliding. Both *Chloroflexus* and *Heliothrix* are found in alkaline hot springs. They grow preferably under photoheterotrophic conditions, and are tolerant towards oxygen. This is interesting because oxygen represses the formation of bacteriochlorophyll and of chlorosomes in *Chloroflexus aurantiacus*, but not of the orange coloured carotenoids. *Heliothrix oregonense* does not contain chlorosomes and intracytoplasmic membranes carrying photosynthetic pigments. It is thought that this species represents a phototrophic green bacterium which has lost the chlorosomes. *Chloronema* and *Oscillochloris* are further representatives of this group (see Table 9.4). So far, only *Chloroflexus aurantiacus* bas been isolated in pure culture.

9.3 SELECTIVE ENRICHMENT

As a consequence of their specific growth requirements, phototrophic bacteria are not as widespread in nature as many other bacteria. Therefore,

TABLE 9.3 Some characteristics of the purple non-sulphur bacteria (Rhodospirillaceae). Data were collected from: Trüper and Pfennig, 1981; information on new species by Akiba *et al.*, 1983; Drews, 1981; Eckersley and Dow, 1980; Hansen and Imhoff, 1985; Imhoff, 1983; Kompantseva, 1985; Kompantseva and Gorlenko, 1984; Neutzling *et al.*, 1984; Nissen and Dundas, 1984; Schmidt and Bowien, 1983; and information on quinone composition by Hiraishi and Hoshino, 1984; Imhoff, 1984b; data on carotenoids from Schmidt, 1978

Species	Cell shape	Cell size diameter (μm)	ICM	Major carotenoid	Major quinone	Growth factors	G+C content	Other significant properties
Rhodospirillum								
R. rubrum	spiral	0.8–1.0	V	sp, rv	Q-10, RQ-10	b	63.8–65.8	*'
R. photometricum	spiral	1.2–1.5	S	rv, rh	Q-8, RQ-8	YE	65.8	
R. molischianum	spiral	0.7–1.0	S	ly, rh	Q-9, MK-9	AA	61.7–64.8	
R. fulvum	spiral	0.5–0.7	S	ly, rh	Q-9, MK-9	paba	64.3–65.3	
R. salexigens	spiral	0.6–0.7	L	sp	Q-10, MK-10	glutamate	64.0	NaCl required
R. salinarum	spiral	0.8–0.9	V	sp	Q-10, MK-10	YE	67.4–68.1	NaCl required
R. mediosalinum	spiral	0.8–1.0	V	sp	o	t, paba, n	66.6	NaCl required
Rhodopila								
R. globiformis	sphere	1.6–1.8	V	kts	Q-9, 10; MK-9, 10 RQ-9, 10	b, paba	66.3	acidic pH
Rhodomicrobium								
R. vannielii	ovoid-rod	1.0–1.2	L	rh, ly, sp	Q-10, RQ-10	none	61.8–63.8	peritr. flagella, exospore formation acidic pH
Rhodobacter								
R. capsulatus	rod	0.5–1.2	V	sn, se	Q-10	t, (b, n)	65.5–66.8	
R. veldkampii	rod	0.6–0.8	V	sn, se	Q-10	b, t, paba	64.4–67.5	nonmotile
R. sphaeroides	ovoid-rod	0.7	V	sn, se	Q-10	b, t, n	68.4–69.9	
R. sulfidophilus	rod	0.6–0.9	V	sn, se	Q-10	b, t, n, paba	67.0–71.0	NaCl required
R. euryhalinus	rod	0.7–1.0	V	se	o	b, t, n, paba	62.1–68.6	NaCl required
R. adriaticus	rod	0.5–0.8	V	sn, se	Q-10	b, t	64.9–66.7	NaCl required nonmotile

(continued overleaf)

TABLE 9.3 (Contd.)

Species	Cell shape	Cell size diameter (μm)	ICM	Major carotenoid	Major quinone	Growth factors	G+C content	Other significant properties
Rhodocyclus								
R. purpureus	half circle	0.6–0.7	T	ra, rh	Q-8, MK-8	B_{12}	65.3	nonmotile
R. gelatinosus	rod	0.4–0.5	T	sn, se	Q-8, MK-8	b, t	70.5–72.4	gelatin liquefied
R. tenuis	spiral	0.3–0.5	T	ly, rh, ra	Q-8, MK-8	none	64.8	
Rhodopseudomonas								
R. palustris	rod	0.6–0.9	L	sp, rv, rh	Q-10	paba, (b)	64.8–66.3	
R. viridis	rod	0.6–0.9	L	neu*, ly*	Q-9, MK-9	paba, b	66.3–71.4	bchl b
R. sulfoviridis	rod	0.5–0.9	L	neu, sp	Q-8, 10; MK-7, 8	b, p, paba	67.8–68.4	bchl b
R. blastica	rod	0.6–0.8	L	sn, se	Q-10	B_{12}, b, n, t	65.3	
R. acidophila	rod	1.0–1.3	L	rh, rg, rag	Q-10, MK-10, RQ-10	none	62.2–66.8	acidic pH
R. rutila	rod	0.4–1.0	L	sp, rv	Q-10	none	67.6–69.4	
R. marina	rod	0.7–0.9	L	sp	Q-10, MK-10	o	61.5–63.8	NaCl required, "*"

Abbreviations:
o: not determined; bchl: bacteriochlorophyll; ICM: structure of intracytoplasmic membrane system; "*" characteristic low absorption maximum at 800–805 nm and long wavelength maximum at 880–890 nm.
Membrane system: V: vesicles; L: lamellae; S: stacks; T: tubes
Vitamins: b: biotin; n: niacin; t: thiamine; paba: p-aminobenoic acid; YE: yeast extract; AA: amino acids; vitamins in brackets are required only by some strains
Quinones: Q: ubiquinone; MK: menaquinone; RQ: rhodoquinone (numbers give isoprenoid chain length)
Carotenoids: kts: ketocarotenoids; ly: lycopene; ly*: 1,2 dihydrolycopene; neu: neurosporene; neu*: 1,2 dihydroneurosporene; ra: rhodopinal; rag: rhodopinal glucoside; rg: rhodopin glucoside; rh: rhodopin; rv: rhodovibrin; se: spheroidene; sn: spheroidenone; sp: spirilloxanthin.

TABLE 9.4 Some characteristics of phototrophic green bacteria. Data were collected from: Dubinina and Gorlenko, 1975; Gibson et al., 1984; Gorlenko and Lebedeva, 1971; Gorlenko and Pivovarova, 1977; Pierson et al., 1985; Schmidt, 1978; Trüper and Pfennig 1981

Species	Cell shape	Cell size diameter (μm)	Colour	Major carotenoid	Major bchl	Motility	Gas vacuoles	G+C content (mol%)	Other significant properties
CHLOROBIACEAE									
Chlorobium									
C. limicola	rod	0.7–1.0	green	chl	c or d	–	–	51.0–58.1	
C. vibrioforme	vibrio	0.5–0.7	green	chl	c or d	–	–	52.0–57.1	
C. phaeobacteroides	rod	0.6–0.8	brown	irt, β-irt	e	–	–	49.0–50.0	
C. phaeovibrioides	vibrio	0.3–0.4	brown	irt, β-irt	e	–	–	52.0–53.0	
C. chlorovibrioides	vibrio	0.3–0.4	green	chl	c or d	–	–	o	
Prosthecochloris									
P. aestuarii	sphere	0.5–0.7	green	chl	c	–	–	50.0–56.0	prosthecae
P. phaeoasteroidea	sphere	0.3–0.6	brown	irt	e	–	–	52.2	prosthecae
Ancalochloris									
A. perfilievii	sphere	0.5–1.0	green	o	o	–	+	o	prosthecae, not in pure culture
Pelodictyon									
P. luteolum	ovoid	0.6–0.9	green	chl	c or d	–	+	53.5–58.1	
P. phaeum	vibrio	0.6–0.9	brown	irt	e	–	+	o	
P. clathratiforme	rod	0.7–1.2	green	chl	c or d	–	+	48.5	nets formed, not in pure culture
Chloroherpeton									
C. thalassium	rod	1.0	green	γ-c	c	+	+	45.0–48.2	flexible, gliding, NaCl required

(continued overleaf)

TABLE 9.4 (Contd.)

Species	Cell shape	Cell size diameter (µm)	Colour	Major carotenoid	Major bchl	Motility	Gas vacuoles	G+C content (mol%)	Other significant properties
CHLOROFLEXACEAE									
Chloroflexus C. aurantiacus	filaments	0.6–0.7	green-orange	β-c, γ-c	c	+	–	53.1–54.9	flexible, thermophilic
Heliothrix H. oregonense	filaments	1.5	orange	γ-cg*	*	+	–	o	flexible, thermophilic, no chlorosomes, not in pure culture
Chloronema C. giganteum	rod	2.0–2.5	green	o	d	+	+	o	not in pure culture
Oscillochloris O. chrysea	trichomes	4.5–5.5	green-yellow	o	c	+	+	o	not in pure culture

Abbreviations
Carotenoids: chl: chlorobactene; irt: isorenieratene; β-irt: β-isorenieratene; β-c: β-carotene; γ-c: γ-carotene: γ-cg*, γ-carotene glucoside; γ-carotene and oxygenated derivatives.
* Neither bchl *c*, *d*, or *e* present and chlorosomes as well as intracytoplasmic membranes lacking; o: no data available; bchl: bacteriochlorophyll.

the species composition in the environmental sample is important for the kind of phototrophic bacteria that may develop in any subsequent enrichment culture. Several of the known species are quite rare in nature and are represented by single strains, e.g. *Rhodocyclus purpureus* (Pfennig, 1978), *Rhodopila globiformis* (Pfennig, 1974; Imhoff *et al.*, 1984), and *Rhodopseudomonas sulfoviridis* (Keppen and Gorlenko, 1975). *Heliobacterium chlorum* (Gest and Favinger, 1983), *Chloroherpeton thalassium* (Gibson *et al.*, 1984), and *Chromatium tepidum* (Madigan, 1986) have been isolated for the first time quite recently, and information on their natural distribution is lacking. Other species, for example *Chromatium okenii*, *Thiospirillum jenense* and *Thiopedia rosea*, have been regularly observed in nature, but only a few strains have been isolated because of the difficulties in meeting the specific growth requirements.

Selective enrichment techniques for phototrophic bacteria, were devised initially by Winogradsky (1888), i.e. the 'Winogradsky column'. Variations of this technique have been discussed by Pfennig (1965) and van Niel (1971). Many of the known species of phototrophic bacteria may be selectively enriched and isolated in synthetic media. The enrichment media select primarily between the major physiological groups. Of general importance for the selectivity of media are the concentration of sulphide, the pH, temperature of incubation, the presence of vitamins, the light intensity and type of illumination, and the composition and concentration of nutrients. If samples from marine and hypersaline environments are investigated, the salinity and the mineral composition of the medium is also of importance.

9.3.1 Purple and green sulphur bacteria

Pfennig's medium for phototrophic sulphur bacteria is specifically designed for these bacteria and provides rather non-selective growth conditions for most of the purple and green sulphur bacteria (Pfennig and Trüper, 1981). Pfennig (1965) also gives a number of selective methods for the enrichment of different green and purple sulphur bacteria. Slight variation of this medium and of the culture conditions gives selectivity for various phototrophic sulphur bacteria.

A large number of species is obtained in enrichment cultures, providing that lower sulphide concentrations, modest light intensities, and lower temperatures are adopted. At higher sulphide concentrations sensitive strains are inhibited. Moreover, at high light intensities and high temperatures, fast growing strains (of a single species) may outgrow a variety of other species.

Among the purple sulphur bacteria, continuous illumination at high light intensities (1000–2000 lux) and high temperatures (30 °C) favour fast growing, small celled Chromatiaceae, for example *Chromatium vinosum, C. minus, C. violascens, C. gracile, Thiocapsa roseopersicina* and *Thiocystis violacea.* Low light intensities (100–300 lux), intermittent illumination (with light/ dark cycles of 4/8 or 6/6 h) and low temperatures (15–20 °C) favour large celled and gas-vacuolated species like *Thiospirillum jenense, Chromatium okenii, Thiodictyon elegans, Amoebobacter roseus* and *Thiopedia rosea* (Pfennig, 1965).

The importance of sulphide concentration and illumination in light/dark cycles for competition between large (*Chromatium weissei*) and small cell (*Chromatium vinosum*) was superbly demonstrated in continuous culture experiments by van Gemerden (1974). In continuous illumination, with sulphide as a growth limiting factor, the growth rate of *Chromatium vinosum* exceeds that of *Chromatium weissei* regardless of the sulphide concentration. With intermittant light/dark cycles, however, both organisms showed a balanced coexistence, because sulphide oxidation in *Chromatium weissei* occurred more rapidly.

In the medium for phototrophic sulphur bacteria described by Pfennig (1961, 1965) and Pfennig and Trüper (1981) the Chlorobiaceae have a general advantage over purple bacteria regardless of the light conditions and the temperature of incubation. The only proviso is that the cultures should not be incubated with an infrared filter that absorbs light at wavelength below 800 nm. Under these conditions, the green bacteria, which have absorption maxima at 700–750 nm, may not develop. Light filters, transmitting only radiation above 900–1000 nm, may also be used for selective enrichment of bacteria containing bacteriochlorophyll *b* e.g. *Thiocapsa pfennigii, Rhodopseudomonas viridis, Rhodopseudomonas sulfoviridis, Ectothiorhodospira halochloris* and *Ectothiorhodospira abdelmalekii,* which absorbs above 1000 nm (Table 9.5).

Chlorobiaceae were selectively enriched by van Niel (1932) in a medium for purple sulphur bacteria, in which the pH was reduced (Pfennig recommends pH 6.6–6.9) and the concentration of sulphide was increased to 0.1–0.2%. (A number of purple sulphur bacteria are inhibited by concentrations of sulphide of $\geqslant 0.1\%$ $Na_2S \cdot 9H_2O$). The addition of thiosulphate favours thiosulphate utilizing Chlorobiaceae (Pfennig, 1965). Very low light intensities (5–50 lux) are also selective for phototrophic green bacteria due to their efficient light harvesting machinery, that enables growth at light intensities not supportive of phototrophic purple bacteria (Biebl and Pfennig, 1978).

Representatives of the green sulphur bacteria, for example, *Chlorobium limicola* and *Prosthecochloris aestuarii,* are also favoured at high light intensities, high sulphide concentration and high temperature. Others, e.g. *Pelodictyon clathratiforme,* prefer low sulphide concentration (1–2 mM), low light intensities (100–500 lux) with diurnal light/dark cycles, and low incubation temperatures (15–20 °C).

TABLE 9.5   Characteristic absorption maxima of different bacteriochlorophylls in living cells.

Bacteriochlorophyll	Absorption maxima (nm)
bchl *a*	375, 590, 805, 830–890
bchl *b*	400, 605, 840, 1020–1040
bchl *c*	745–755
bchl *d*	725–745
bchl *e*	710–725
bchl *g*	770–790

9.3.2 Ectothiorhodospiraceae

Ectothiorhodospira spp. have been isolated from marine sources and salt lakes from many parts of the world. Alkaline soda lakes show a natural abundance of *Ectothiorhodospira* spp., which is taken as proof of their successful adaptation to these environments. For the extremely halophilic species, the tolerance of, and the dependence on, high salinity and alkalinity are strongly selective conditions for their enrichment. Marine strains of *Ectothiorhodospira* can selectively be enriched under photoauto-trophic conditions with sulphide as the electron donor and in saline and alkaline media (3% (w/v) NaCl and pH 8.5–9.0), even in the presence of high proportions of *Chromatium* spp. within the sample. Using a medium (Imhoff and Trüper, 1977) based on the mineral composition of the soda lakes of the Wadi Natrun (Jannasch, 1957; Imhoff *et al.*, 1979), many *Ectothiorhodospira* strains have been isolated from various locations. Dependent on the species composition of the sample and the salt concentration of the media (from 3 to 25%), various *Ectothiorhodospira* spp. can be isolated. With many environmental samples, isolation can be achieved in agar dilution series without prior enrichment.

9.3.3 Purple non-sulphur bacteria (Rhodospirillaceae)

For selective enrichment of purple non-sulphur bacteria, media have been used with lowered sulphate concentration and containing organic sub-strates not used by sulphate-reducing bacteria (to avoid production of sulphide). Some species of the purple non-sulphur bacteria, however, tolerate or even require sulphide as a reduced sulphur source and/or photosynthetic electron donor, for example, *Rhodobacter sulfidophilus* (Hansen and Veldkamp, 1973), *Rhodobacter adriaticus* (Neutzling *et al.*, 1984), and *Rhodopseudomonas sulfoviridis* (Keppen and Gorlenko, 1975). In many cases and in particular with marine samples, small celled Chroma-tiaceae successfully compete with purple non-sulphur bacteria even in the complete absence of sulphate in the enrichment media. This is attributable to the nutritional versatility of these Chromatiaceae. Frequent species in natural habitats, e.g. *Rhodocyclus gelatinosus* and *Rhodopseudomonas palustris*, are often outcompeted by fast growing species, e.g. *Rhodobacter capsulatus* and *Rhodospirillum fulvum*. This competition can be avoided by direct inoculation of samples into agar dilution series, using suitable media. A few improved methods for selective enrichment of purple non-sulphur are described below.

A succinate-mineral medium without growth factors and with an initial pH of 5.2 is highly selective for *Rhodopseudomonas acidophila* and *Rhodomicrobium vannielii* (Pfennig, 1969).

Rhodopseudomonas palustris is a very common species in nature, and many enrichment methods end up with the development of this species. Selective carbon sources include benzoate and formate (Quadri and Hoare, 1968; Stammel, 1977).

Enrichments, under photoautotrophic conditions, using hydrogen as the electron donor favour *Rhodobacter capsulatus*, which grows faster under these conditions than most other phototrophic bacteria (Klemme, 1968).

Pelargonate and caproate (not more than 0.04% (w/v) at pH 7.5) are highly selective for *Rhodospirillum fulvum* and *Rhodospirillum molischianum* (Pfennig, 1967; van Niel, 1944).

Citrate has been proved to be useful for selective enrichment of *Rhodocyclus gelatinosus*, although a few strains of other species, namely *Rhodopseudomonas palustris*, *Rhodopseudomonas viridis*, *Rhodobacter capsulatus* and *Rhodobacter sphaeroides*, may also grow.

9.3.4 Chloroflexaceae

The only species of this group that has been isolated in pure culture is *Chloroflexus aurantiacus*. However, selective enrichment techniques are not available.

9.4 ISOLATION PROCEDURES

Environmental samples are inoculated into suitable enrichment media, contained in 20 ml capacity screw-cap test tubes or 50 ml volume bottles, which are completely filled with the medium. For isolation, agar dilution series are prepared using enrichment cultures or promising environmental samples. Selectivity is not necessary for media used for isolation. Actually, a non-selective medium is of advantage to allow as many species as possible to develop as separated colonies in the agar. This method may also be used to determine the composition of natural populations. In particular, slow growing strains that are rapidly outgrown by enrichment techniques can be easily isolated in this way. If the expected cell density in a sample is low, the cells may be concentrated by centrifugation for agar dilution series, and by membrane filtration for inoculation on agar plates.

9.4.1 Preparation of agar dilution series

Dissolve and distribute purified agar (1.8% w/v) in amounts of 3 ml into cotton plugged test tubes. Sterilize by autoclaving. Use immediately or store until used. Dissolve agar in boiling water and keep at 50 °C in a water bath until use. Warm up suitable medium in the same water bath, and add 6 ml of the prewarmed medium in each test tube. Mix thoroughly and keep at 50 °C. Use six to eight tubes for each dilution series. Inoculate the first tube with an environmental sample or enrichment culture, mix carefully, transfer approximately 0.5–1.0 ml to second tube, mix carefully and continue the procedure up to the last tube. Cool the tubes immediately in a cold water bath. After the agar has solidified, seal with a paraffin mixture (1 part paraffin and 3 parts paraffin oil) and keep in the dark for several hours before incubating them in the light. After the cells have grown to form visible colonies, the paraffin layer is removed by melting and well-separated colonies are picked up with a pasteur pipette (the tip is drawn out to a small capillary) and transferred to a second dilution series. In general, three to four such dilution series are necessary to obtain pure cultures. When pure cultures have been obtained, single colonies are inoculated into liquid medium.

For agar dilution series, non-selective media are of advantage. For phototrophic sulphur bacteria, Pfennig's medium with the addition of 0.05% (w/v) Na-acetate is recommended (Pfennig and Trüper, 1981, see below). Incubation should be at 20–28 °C, and 200–500 lux. For isolation of purple non-sulphur bacteria, use the medium given below. This is supplemented with succinate, malate, fumarate, pyruvate or acetate (0.1% w/v) as substrates, with the addition of 0.05% (w/v) yeast extract, and—depending on the sample and desired species—0.4–1.0 mM sulphide. *Chloroflexus aurantiacus* can also be isolated by this technique.

9.4.2 Cultivation on agar plates

In particular, purple non-sulphur bacteria have been successfully isolated on agar plates. When high numbers of phototrophic bacteria are present in the sample, streaking by conventional methods is appropriate. Samples containing low numbers of phototrophic bacteria can be concentrated on membrane filters (cellulose acetate or cellulose nitrate with 0.4 μm pore size; Biebl, 1973; Biebl and Drews, 1969; Swoager and Lindstrom, 1971; West-macott and Primrose, 1975). The plates may be incubated in an anaerobic jar in which the air is replaced by an oxygen-free mixture of nitrogen with 5% carbon dioxide and 3% hydrogen. Remaining traces of oxygen are removed by reduction with hydrogen over a palladium catalyst. Alternatively, the GasPak system or comparable systems may be used. A more detailed description of the methods is given by Biebl and Pfennig (1981).

A method to grow phototrophic sulphur bacteria on agar plates has been described by Irgens (1983). A medium without added sulphide, and the GasPak system, is used. As a source of sulphide, a test tube with 0.05-0.1 g thioacetamide and 1 ml of 0.5 N HCl is placed in the anaerobic jar. Thioacetamide is not stable under acidic conditions and decomposes to ammonia, hydrogen sulphide and acetic acid. The gas is released over a period of at least 1 week. As indicators, methylene blue and a strip with lead acetate are included. This is a convenient method to obtain viable counts of purple sulphur bacteria. It has been successfully used for the isolation of *Amoebobacter, Chromatium, Ectothiorhodospira, Lamprocystis, Thiocapsa* and *Thiocystis* spp. (Irgens, 1983).

9.5 MEDIA FOR PHOTOTROPHIC BACTERIA

Due to the enormous physiological variability of phototrophic bacteria and their greatly varying growth requirements, a number of different media have to be used for their cultivation and isolation. Basic media for phototrophic green and purple sulphur bacteria (Pfennig, 1961; 1965; Pfennig and Trüper, 1981), for *Ectothiorhodospira* spp. (Raymond and Sistrom, 1969; Imhoff and Trüper, 1977; Trüper and Imhoff, 1981), and for purple non-sulphur bacteria (van Niel, 1944; Drews, 1965; Imhoff and Trüper, 1976; Pfennig, 1969; Biebl and Pfennig, 1981; Imhoff, 1982b) have been described. These media, with a number of modifications, are suitable for the cultivation of most phototrophic bacteria.

9.5.1 Media for phototrophic green and purple sulphur bacteria

By variation of the sulphide concentration, the pH and the illumination, most purple and green phototrophic sulphur bacteria grow well in 'Pfennig's medium'. Two modifications of this medium are described by Pfennig and Trüper (1981). The first is suitable also for the species that are most difficult to grow; the second is suitable for the most common phototrophic sulphur bacteria, and its preparation is easier. The medium is prepared as follows (Biebl and Pfennig, 1978; Pfennig and Trüper, 1981; Trüper, 1970). Preparation should be done in a 2-l capacity Erlenmeyer flask with an outlet near the bottom which is connected with tubing to a bell for easy aseptic distribution of the prepared medium to culture vessels.

Solution 1

Distilled water	950 ml
Trace elements (solution 2)	1 ml
KH_2PO_4	1.0 g
NH_4Cl	0.5 g
$MgSO_4 \cdot 7 H_2O$	0.4 g
$CaCl_2 \cdot 2 H_2O$	0.05 g

For the cultivation of marine strains, 20 g NaCl are added and the amount of $MgSO_4 \cdot 7H_2O$ is increased to 3 g.

Solution 2 (trace element solution SL 8)

Distilled water	1000 ml
$EDTA - Na_2$	5.2 g
$FeCl_2 \cdot 4 H_2O$	1500 mg
$ZnCl_2$	70 mg
$MnCl_2 \cdot 4 H_2O$	100 mg
H_3BO_3	62 mg
$CoCl_2 \cdot 6 H_2O$	190 mg
$CuCl_2 \cdot 2 H_2O$	17 mg
$NiCl_2 \cdot 6 H_2O$	24 mg
$Na_2MoO_4 \cdot 2 H_2O$	36 mg

The salts are dissolved in the given order, and the solution is stored in a refrigerator (4 °C). After cooling the autoclaved solution 1, the sterile solutions 3 to 5 are added with gentle stirring by a magnetic stirrer.

Solution 3 (vitamin B_{12})
2 mg vitamin B_{12} in 100 ml distilled water

The stock solution is sterilized by filteration (0.22 μm porosity filters), and 1 ml is added.

Solution 4
5% (w/v) $NaHCO_3$ in distilled water

The solution is filter-sterilized, and 40 ml are added.

Solution 5
5% $Na_2S \cdot 9 H_2O$ in distilled water

6 ml of a freshly autoclaved solution is added for purple sulphur bacteria, and 12 ml for green sulphur bacteria.

After complete mixing of solutions 1–5, the pH is adjusted with sterile H_2SO_4 or Na_2CO_3 (2 mol/l each) to pH 6.8 in the case of green sulphur bacteria, and to 7.2 for purple sulphur bacteria. The medium is immediately distributed into sterile screw cap culture bottles (50 or 100 ml capacity bottles are recommended). These bottles are filled completely, leaving only a small (pea sized) air bubble to compensate for possible pressure changes. The concentration of sulphide in this medium is kept low, because some species are inhibited by higher concentrations of sulphide and most species will not tolerate concentrations higher than 0.2% of sodium sulphide. Because sulphide is the only electron source in these media, high population densities can only be achieved by 'feeding' (repeated addition of small amounts of sulphide: Pfennig and Trüper, 1981).

Feeding solution
   3 g $Na_2S \cdot 9H_2O$ in 100 ml distilled water

The solution is autoclaved in a tightly closed screw cap bottle, after replacement of the air by dinitrogen. A known amount of this solution is partially neutralized to pH 8.0 by dropwise addition of sterile H_2SO_4 (2 M). This partially neutralized solution is immediately used for feeding the cultures. Depending on the density of the culture, 1–2 ml per 100 ml are used for Chromatiaceae, and 2–3 ml for Chlorobiaceae. A convenient method to prepare this sulphide solution is described by Siefert and Pfennig (1984).

Acetate solution
   2.5% NH_4-acetate
   2.5% Mg-acetate

To increase the growth yield of pure cultures of phototrophic green and purple bacteria, 1 ml of this (autoclaved) solution is added to 100 ml culture medium. Addition of acetate is also recommended for media used in agar dilution series.

9.5.2 Media for *Ectothiorhodospira* species

The medium described for the cultivation of *Ectothiorhodospira* spp. by Imhoff and Trüper (1977) has been modified several times. Investigations on the vitamin requirement of a number of *Ectothiorhodospira* isolates have shown that vitamins are not required by any species, although vitamin B_{12} sometimes shows some growth stimulation. Addition of vitamins to the medium was, therefore, omitted. Although solutions of trace elements and sulphide were sterilized separately, the filtration procedure used formerly yielded less reproducible growth, presumably because of some interaction of the medium constituents with the filters. The reproducibility was much better with the preparation method described here.

   The basal medium has the following composition (amounts per litre for a medium with 15% salinity):

1 ml trace elements solution SLA (see below)
0.8 g KH$_2$PO$_4$
2.0 g Na-acetate
1.0 g Na$_2$S$_2$O$_3 \cdot$ 5 H$_2$O
130.0 g NaCl
200 ml 1 M Na-carbonate; pH 9.0

The components are dissolved in 600 ml of distilled water and the carbonate buffer is added. The volume is adjusted to 980 ml, and this solution is autoclaved. The salinity is adjusted to the desired value by variation of the NaCl content (assuming a contribution of 2% by the other medium constituents).

The remaining salts are sterilized separately and, after they have cooled down, are added with gentle stirring.

2% MgCl$_2 \cdot$ 7 H$_2$O (5 ml/l)
1% CaCl$_2 \cdot$ 2 H$_2$O (5 ml/l)
5% Na$_2$S $\cdot$ 9 H$_2$O (5–10 ml/l)
20% NH$_4$Cl (4 ml/l)

The medium is then immediately placed into sterilized culture vessels, which are filled completely with only a small air bubble left. Before inoculation, they are preincubated at the incubation temperature to achieve volume expansion, and then inoculated with 10–20% of a fresh preculture.

The addition of Na$_2$S$_2$O$_3$ is not necessary for the two green coloured *Ectothiorhodospira* spp. as these can not use thiosulphate. In the red coloured species, it is suitable as an additional electron donor and makes the feeding with sulphide unnecessary. Standard media for *E. halochloris* and *E. abdelmalekii* contain, in exchange for the same amount of NaCl (in g per litre), Na$_2$SO$_4$ (10% of the total salinity). For photoheterotrophic growth of *E. mobilis*, reduced sulphur compounds are omitted, and 0.1% (w/v) Na-ascorbate are added to achieve anoxic conditions. Under these conditions sulphate is used as the sole sulphur source.

Trace element solution SLA (Imhoff and Trüper, 1977)
1800 mg FeCl$_2 \cdot$ 4 H$_2$O
250 mg CoCl$_2 \cdot$ 6 H$_2$O
10 mg NiCl$_2 \cdot$ 6 H$_2$O
10 mg CuCl$_2 \cdot$ 2 H$_2$O
70 mg MnCl$_2 \cdot$ 4 H$_2$O
100 mg ZnCl$_2$
500 mg H$_3$BO$_3$
30 mg Na$_2$MoO$_4 \cdot$ 2 H$_2$O
10 mg Na$_2$SeO$_3 \cdot$ 5 H$_2$O

Dissolve separately in a total of 900 ml of distilled water, adjust pH with 1N HCl to about 2–3 and bring volume up to 1000 ml. This trace element solution has been also successfully used for the cultivation of all purple sulphur and non-sulphur bacteria.

9.5.3 Medium for purple non-sulphur bacteria (Rhodospirillaceae)

A number of media, which differ in their mineral composition only slightly, are used in various laboratories. The following AT-medium (Imhoff and Trüper, 1976; Imhoff, 1982b) has been used over many years for the isolation and cultivation of most freshwater and marine species of purple non-sulphur bacteria:

> Dissolve in 1000 ml distilled water:
> 1 ml trace element solution SLA (see above)
> 1 ml vitamin solution VA (see below)
> 1.0 g KH_2PO_4
> 0.5 g $MgCl_2 \cdot 6H_2O$
> 0.1 g $CaCl_2 \cdot 2H_2O$
> 1.0 g NH_4Cl
> 3.0 g $NaHCO_3$
> 0.7 g Na_2SO_4
> 1.0 g NaCl
> 1.0 g Na-acetate or other carbon source; pH 6.9

This medium, containing acetate, pyruvate, malate, succinate or fumarate as carbon sources, is well suited for the isolation and enumeration of purple non-sulphur bacteria in agar dilution series or on agar plates. For the cultivation of *Rhodopseudomonas acidophila* and *Rhodomicrobium vannielii*, the pH is adjusted to 5.5. For marine species, 30 g NaCl are added for a saline AT-medium (SAT-medium). If desired, other carbon sources may be used. Yeast extract stimulates growth of most known representatives of Rhodospirillaceae and is required as a growth factor by some *Rhodospirillum* spp. It may be added at a concentration of 0.05% (w/v). For some species, e.g. the brown coloured *Rhodospirillum* spp. addition of 0.01% (w/v) Fe-citrate stimulates growth. Although a number of purple non-sulphur bacteria are aerotolerant, others, for example the brown pigmented *Rhodospirillum* spp., *Rhodopseudomonas viridis*, *Rhodopila globiformis* and *Rhodocyclus purpureus*, are more sensitive to oxygen. In most cases, it is sufficient to add 0.05% (w/v) sodium ascorbate to achieve anoxic conditions. Some species, e.g. *Rhodopseudomonas sulfoviridis*, *Rhodobacter adriaticus*, *Rhodobacter veldkampii* and *Rhodopila globiformis* (Keppen and Gorlenko, 1975; Neutzling *et al.*, 1984; Hansen and Imhoff, 1985; Pfennig, 1974) require reduced sulphur compounds and some, namely *Rhodobacter sulfidophilus, Rhodobacter adriaticus, Rhodobacter veldkampii* and *Rhodopseudomonas marina,* tolerate sulphide at low or intermediate concentrations, and can use it as a photosynthetic electron donor (Hansen and Imhoff, 1985; Hansen and van Gemerden, 1972; Imhoff, 1982a; Imhoff, 1983; Neutzling *et al.,* 1984). For these species sodium sulphide is added at concentrations of 0.4–2.0 mM.

Vitamin solution VA (modified after Imhoff and Trüper, 1977)
This vitamin solution meets the vitamin requirement of all known purple non-sulphur bacteria. For more specific use, the vitamins may be used singly or in combinations (at the same concentration). The vitamins are dissolved in 100 ml of distilled water, sterilized by filtration, and refrigerated (4 °C). For use, add 1 ml per litre of medium:

    10 mg biotin
    35 mg niacin
    30 mg thiamine dichloride
    20 mg p-aminobenzoic acid
    10 mg pyridoxolium hydrochloride
    10 mg Ca-panthothenate
     5 mg vitamin B_{12}

9.5.4 Medium for *Rhodopila globiformis* (modified after Pfennig, 1974)

Dissolve in 1000 ml of distilled water:
1.5 g mannitol
0.5 g gluconate
0.4 g KH_2PO_4
0.4 g NaCl
0.05 g $CaCl_2 \cdot 2H_2O$
0.4 g $MgCl_2 \cdot 6H_2O$
0.4 g NH_4Cl
0.2 g Na-thiosulphate

Add 5 ml 0.1% (w/v) Fe-citrate solution, 1 ml VA (see above), and 1 ml SLA (or equivalent trace element solution), adjust pH to 4.9.

9.5.5 Media for halophilic *Rhodospirillum* species

For *Rhodospirillum salexigens* and *Rhodospirillum salinarum* the media described by Drews (1981) and Nissen and Dundas (1984) are suitable. All strains grew well on these complex media.

The complex medium (modified after Nissen and Dundas, 1984):
Dissolve in 1000 ml of distilled water:
3.5 g $MgCl_2 \cdot 6H_2O$
0.3 g KH_2PO_4
100 NaCl (depending on desired salinity)
1.5 g yeast extract (Difco)
1.5 g proteose peptone (Difco)
10 mM Na-malate
1 ml trace element solution SLA; pH 7.0

Synthetic medium
For special biochemical and physiological work, synthetic media are preferable. A recipe for a synthetic medium is given below. It is sterilized by filtration. A medium of 10% salinity contains per litre:
1 ml vitamin solution VA (see above)
1 ml trace element solution SLA (see above)
3.9 g $NaHCO_3$
0.5 g KH_2PO_4
1.0 g KCl
0.05 g $CaCl_2 \cdot 2H_2O$
3.5 g $MgCl_2 \cdot 6H_2O$
1.0 g Na_2SO_4
100 g NaCl (dependent on desired salinity)
5.0 mM proline
2.0 mM acetate
2.0 mM pyruvate
10.0 mM K_3-citrate
5.0 mM Na-glutamate; pH 7.0

9.5.6 Medium for *Chloroflexus aurantiacus* (according to Castenholz and Pierson, 1981)

Medium D for *Chloroflexus aurantiacus* is prepared as a 20-fold concentrated stock solution. The pH is adjusted with 2 M NaOH after dilution and addition of supplements to pH 8.2 (after autoclaving and cooling the pH is at 7.5–7.6):
Dissolve in 1000 ml of distilled water:
0.1 g nitrilotriacetic acid (NTA)
0.06 g $CaSO_4 \cdot 2H_2O$
0.10 g $MgSO_4 \cdot 7H_2O$
0.008 g NaCl
0.07 g Na_2HPO_4
0.036 g KH_2PO_4

0.5 g glycylglycine
0.5 g yeast extract
1.0 ml FeCl$_3$ solution (0.29 g/l)
0.5 ml trace element solution (see below)

Trace element solution
Dissolve in 1000 ml distilled water:
0.5 ml H$_2$SO$_4$ (conc.)
2.28 g MnSO$_4 \cdot$H$_2$O
0.50 g ZnSO$_4 \cdot$7H$_2$O
0.5 g H$_3$BO$_3$
0.025 g CuSO$_4 \cdot$5H$_2$O
0.025 g Na$_2$MoO$_4 \cdot$2H$_2$O
0.045 g CoCl$_2 \cdot$6H$_2$O

For liquid cultures, 0.2 g l^{-1} NH$_4$Cl as nitrogen source and 0.4 g Na$_2$S$\cdot$9H$_2$O as reducing agent and a reduced sulphur source may be added.

9.5.7 Media for *Heliobacterium chlorum* (Gest and Favinger, 1983)

Heliobacterium chlorum may be grown either on a complex medium (ATCC medium 112) or a synthetic medium.

Complex medium
1.0 g K$_2$HPO$_4$
0.5 g MgSO$_4 \cdot$7H$_2$O
10.0 g yeast extract; pH 7.0

A reducing agent should be added to completely remove oxygen from the medium.

Synthetic medium (PLB) according to Gest and Favinger (1983). To about 700 ml of distilled water, 2.1 ml lactic acid (85%) is added and partially neutralized with NaOH (to pH 5.0). Then the following components are added:
0.1 ml 7.5% CaCl$_2 \cdot$2H$_2$O
2.0 ml 20% MgSO$_4 \cdot$7H$_2$O
2.0 ml 1.0% Na$_2$-EDTA
1.0 ml trace element solution
1.5 ml chelated iron solution
10.0 ml 10% (NH$_4$)$_2$SO$_4$
10.0 ml 0.5 M K$_2$HPO$_4$
15 µg biotin
2.2 g Na-pyruvate

Trace element solution
Dissolve in 250 ml distilled water:
2.5 g Na$_2$-EDTA
20 mg CoCl$_2 \cdot$ 6 H$_2$O
50 mg NiCl$_2 \cdot$ 6 H$_2$O
10 mg CuCl$_2 \cdot$ 2 H$_2$O
200 mg MnCl$_2 \cdot$ 4 H$_2$O
50 mg ZnCl$_2$
100 mg H$_3$BO$_3$
100 mg Na$_2$MoO$_4 \cdot$ 2 H$_2$O
5 mg Na$_2$SeO$_3 \cdot$ 5 H$_2$O
5 mg NaVO$_3$

Chelated iron solution
Dissolve in 500 ml of distilled water
0.5 g FeCl$_2 \cdot$ 4 H$_2$O
1.0 g Na$_2$-EDTA
1.5 ml concentrated HCl.
The pH is adjusted to 6.8, and the volume brought up to 1 litre. Just before inoculation, sterilized solutions of Na-ascorbate (final concentration 0.05%) and NaHCO$_3$ (final concentration 0.02%) are added. Ti^{3+}-citrate (0.05%) or Na-thioglycolate (0.04%) may be used instead of Na-ascorbate. Instead of the vitamin solution used by the authors only biotin is added, because other vitamins are not required (Gest and Favinger, 1983). Carbon sources used by *Heliobacterium chlorum* have not yet been determined. It should be possible to use either lactic acid or pyruvic acid. It may also be possible to replace trace element solution, iron solution and EDTA by the trace element solution SLA, described for the Ectothiorhodospiraceae.

9.6 IDENTIFICATION

Both morphological and physiological properties are very important for the identification of phototrophic bacteria. A first selection is usually achieved from the nature of the medium and the culture conditions (for enrichment and isolation). With some experience a number of species may be tentatively identified from size and consistency of the colonies on agar plates, colour of cell suspensions or colonies, and microscopic examination. In general, however, many morphological and physiological properties have to be used in identification. During the last few years, a number of additional characteristics have been utilized, for example the nature of quinones, fatty acids, and polar lipids (Imhoff *et al.*, 1982; Imhoff, 1984b; Imhoff *et al.*, 1984). Also DNA–DNA hybridization studies are very useful in distinguishing between otherwise similar species (DeBont *et al.*, 1981). Of course, for the description of a new species, the base ratio of the DNA has to be determined.

9.6.1 Microscopical studies

During the purification of a new isolate, useful data may be readily obtained. Careful microscopic studies will give information on shape and size of the cells, formation of cell aggregates, mode of cell division, motility, and presence or absence of slime capsules, sulphur globules and gas vacuoles. The measurement of cell dimensions should be done, however, with pure cultures grown under optimal conditions. The quality of photomicrographs not only depends on the quality of optical lenses but also on the patience of the photographer.

The technique is quite simple. Purified washed agar, 2 g, is dissolved in 100 ml of distilled water. The hot agar is evenly distributed in 2 ml portions on cleaned slides and dried. Using a coverslip, a small amount of a culture (10–15 μl) is applied to the dried agar. If the correct volume is used, healthy cells are fixed in one plane between agar and cover slip. If the volume is too large, cells will be swimming in different planes, and out of focus. If the volume is too small, cells are lysed by the pressure created by the swelling agar. An overall magnification of 2000× is sufficient in most cases. Photomicrographs of the bacterial cells may be used for exact determination of size.

Electron microscopic studies are necessary to demonstrate the type of flagellation and ultrastructure of the cells. Negative staining (with 1% (w/v) solutions of phosphotungstic acid or uranyl acetate at pH 6.7) may be used to demonstrate the flagella (with whole cells) and the type of intracytoplasmic membranes (with isolated membranes). Ultrathin sections are, however, required for studies of cell wall structures and for proof of the size and location of the membranes within the cell.

9.6.2 Pigments

The colour of colonies or cell suspensions and the measurement of absorption spectra give valuable information on the kind of pigments present. The absorption maxima of the various types of bacteriochlorophyll, found in phototrophic bacteria, are diagnostic (see Table 9.5). Only in a very small number of species does the colour of cell suspensions represent the exact colour of the bacteriochlorophyll-protein complexes. In most species, the colour is determined by the carotenoids. Within the Chlorobiaceae, green coloured species possess chlorobactene and brown pigmented species contain isorenieratene or its derivatives as major components (the only exception is *Chloroherpeton thalassium*). Within the phototrophic purple bacteria, spirilloxanthin gives rise to the pink or red colouration and increasing amounts of additional rhodopin give brown-red colour. Rhodopin without significant amounts of spirilloxanthin is brown, whereas okenone is purple-red and rhodopinal purple-violet. Carotenoids of the spheroidenone series give colours from yellowish-brown to brownish-red (depending on the content of oxygenated deriva-tives formed in the presence of oxygen) and greenish-brown under strong reducing conditions. Definite identification of the structures of carotenoids and bacteriochlorophylls needs much more sophisticated analyses. Quantitative analyses of pigments generates important information about the various groups of phototrophic bacteria present in the environment. A sensitive HPLC method has been described, separating and quantitating major carotenoids and (bacterio)chlorophylls from phototrophic microbes (Korthals and Steenbergen, 1985).

For identification of phototrophic bacteria, absorption spectra of whole cells are measured using cell suspensions washed twice in medium or appropriate salt solutions, and then suspended in 60% sucrose solution (Biebl and Drews, 1969). Better results are often achieved by using isolated intracytoplasmic membranes, suspended in buffer. For this purpose, it is sufficient to break the cells by ultrasonication or with a French Press, and to separate whole cells and large cell fragments from the intracytoplasmic membranes by centrifugation at 15000 g.

9.6.3 Physiological properties

To define the optimal growth conditions, it is necessary to determine the requirement for and/or the tolerance of sulphide as well as its oxidation products, the optimal pH and temperature, the optimal salinity, the tolerance towards oxygen and dark growth, and the requirement for growth factors. Also, the utilization of sulphur, nitrogen and carbon sources have to be determined. Some characteristic properties are given in Tables 9.1–9.4. Details on substrate utilization are found in the respective chapters of *Bergey's Manual of Systematic Bacteriology* (in press). The physiological properties, together with data on morphology and pigments, allow a definite identification in most cases.

9.6.4 Diagnostic features

As already discussed previously, cells growing under autotrophic condi-tions, in the presence of sulphide, and forming globules of elemental sulphur inside their cells undoubtedly belong to the family Chromatiaceae. Further differentiation is possible by the properties given in Table 9.1.

Growth factors are generally not required by many representatives of the Chromatiaceae. In fact, only vitamin B_{12} is necessary for some species. Intracytoplasmic membranes are vesicles in all known species except *Thiocapsa pfennigii*.

For species of the genus *Ectothiorhodospira*, the presence of rod to spiral shaped cells with intracytoplasmic membranes as lamellar stacks that grow under alkaline (pH 7.5–9.0) and saline (1–30% salinity) conditions are characteristic. Sulphide is used by all species as the photosynthetic electron donor, but elemental sulphur is deposited outside the cells and appears in the medium. Growth factors are not required. The distinguishing features are presented in Table 9.2.

The majority of the purple non-sulphur bacteria may be distinguished from the families Chromatiaceae and Ectothiorhodospiraceae by higher sensitivity to sulphide. With the exception of *Rhodobacter veldkampii* and *Rhodobacter adriaticus*, the species that oxidize sulphide either form elemental sulphur only or do not form elemental sulphur at all. Most species require one or more growth factors.

The green bacteria are clearly recognized by their content of different bacteriochlorophyll and carotenoid structures, depicted in their absorption spectra. Chlorobium vesicles can be seen by electron microscopy. The Chlorobiaceae are strictly autotrophic, and depend on the presence of reduced sulphur compounds as photosynthetic electron donor and sulphur source. They assimilate only a limited number of organic carbon compounds. Cells are non-motile or, as with *Chloroherpeton thalassium*, motile by gliding. The Chloroflexaceae have multicellular filaments and move by gliding. Only one species, *Chloroflexus aurantiacus*, has been described on the basis of pure culture studies, to date.

9.6.5 Analyses of polar lipids, fatty acids and quinones

Chloroform soluble membrane components are extracted in a mixture of chloroform/methanol/water, i.e. 1/2/0.8 (Imhoff *et al.*, 1982). Cells are first evenly suspended in water (or salt solution), and chloroform and methanol are then added to the desired ratio. By addition of further amounts of chloroform and water (or salt solution), the ratio of the mixture is brought up to 1/1/0.9 (chloroform/methanol/water) and forms a lower chloroform phase and an upper methanol/water phase. The components to be analysed dissolve in the lower chloroform phase that can be easily separated by the use of a separatory funnel. This 'lipid' solution is concentrated in a rotary evaporator and may be used, after methanolysis, for the analysis of fatty acids, or may be separated by acetone precipitation into polar lipids (containing phospholipids and other polar lipids) and non-polar lipids (containing pigments, quinones and other nonpolar lipids).

9.6.5.1 Analysis of fatty acids

Aliquots of the chloroform dissolved lipids are brought to dryness in a heating block under nitrogen flux. Methanol and toluene (4 ml each) and 0.2 ml concentrated sulphuric acid are added to the dry lipids, mixed thoroughly and heated in closed tubes at 70 °C overnight. Methyl esters are brought into hexane, concentrated to a small volume, quantitatively applied on silica gel TLC plates, and purified by development of the plates in appropriate solvent systems (e.g. hexane/diethylether/acetic acid = 80/ 20/1, or hexane/diethylether = 95/5 Mangold, 1967). Fatty acid methyl esters are detected by their yellow fluorescence under UV-light after spraying with 0.2% 2', 7'-dichlorofluorescein in ethanol. They are scraped off the plates, eluted from the silica gel with hexane, brought to dryness and redissolved in a suitable solvent (CS_2) for gas chromatographic analysis. Separation and quantitation is achieved by suitable gas chromatographic systems.

Major fatty acids of phototrophic bacteria are straight chain saturated and monounsaturated fatty acids with 14, 16 and 18 carbon atoms. Various groups of phototrophic bacteria can be distinguished by the proportions of these fatty acids. Chlorobiaceae have only small amounts of C 18:1, but high proportions of C 14:0, C 16:0, and C 16:1. Chromatiaceae have high proportions of C 16:0, C 16:1, and C 18:1 (Kenyon, 1978). In *Chloroflexus aurantiacus* C 16:0, C 18:0 and C 18:1 are dominant. Among the purple non-sulphur bacteria only in *Rhodocyclus* species C-16 fatty acids are dominant, all others have C 18:1 as major fatty acid, but various proportions of C 18:0, C 16:1 and C 16:0, so that different groups can be distinguished (Kato *et al.*, 1985; Imhoff, unpublished).

9.6.5.2 Analysis of quinones

The acetone-soluble fraction is concentrated to a small volume and the quinones are separated on silica gel thin-layer plates (Sil G-25 with UV indicator at 254 nm) with petrol ether/diethylether = 85/15 according to their ring structure. Ubiquinones, menaquinones and rhodoquinones are well separated in this system. Quinones are detected under UV light and eluted with acetone. On reverse phase thin-layer plates (nano Sil-C_{18}-100-F_{254}) with acetone/acetonitrile = 60/40, ubiquinones and rhodoquinones are separated according to the isoprenoid chain length. For menaquinones, a solvent system with acetone/water = 97/3 (or 95/5) may be used. For identification on thin-layer plates appropriate standards are used. Analysis of the chain length and quantitation can be performed on similar systems with HPLC (Collins and Jones, 1981; Imhoff, 1984b). Identification of the quinone ring structure is possible by their UV absorption spectra (Collins and Jones, 1981).

The quinone composition is quite different in the various groups of phototrophic bacteria. So far, it appears that Chromatiaceae have Q-8 and MK-8 as major components and *Ectothiorhodospira* spp. have either Q-8 and MK-8 or Q-7 and MK-7. The Rhodospirillaceae show a great variation in their quinone composition (Hiraishi and Hoshino, 1984; Hiraishi *et al.*, 1984; Imhoff, 1984b; see Tables 9.2 and 9.3). In Chlorobiaceae a special quinone, the 'chlorobium quinone' has been found (see Collins and Jones, 1981).

9.6.5.3 *Analysis of polar lipids*

After precipitation with ice cold acetone, the polar lipids are redissolved in chloroform/methanol (4/1). This solution is applied on silica gel thin layer plates. For comparison of different strains, a one-dimensional solvent system is used (Chloroform/methanol/acetic acid/water = 85/15/10/3.5). Appropriate standards are applied together with the lipid samples. For complete resolution, the plates are developed in two dimensions with chloroform/methanol/water (65/25/4) in the first dimension and (after the plates have been carefully dried) with the solvent system used for one dimensional plates in the second dimension. For complete detection of the lipids, the plates may be sprayed with 40% sulphuric acid in ethanol, and charred at about 200 °C, or iodine vapour may be used. Group specific reagents can be used to detect phospholipids, glycolipids, lipids with amino groups and phosphatidylcholine (Mangold, 1967).

Great variation occurs in the polar lipid composition of phototrophic bacteria. Chlorobiaceae contain characteristic glycolipids; Chromatiaceae have a number of different glycolipids; and Ectothiorhodospiraceae lack significant amounts of glycolipids and ornithine lipids. The latter are characteristic for the purple non-sulphur bacteria (Kenyon, 1978; Imhoff *et al.*, 1982). Often the lipid composition is species specific, and in some species, characteristic components are found that are not present in other species (Imhoff *et al.*, 1982). Species of the genus *Rhodobacter* can be distinguished on the basis of their polar lipid composition (Hansen and Imhoff, 1985).

9.7 REFERENCES

Akiba, T., Usami, R. and Horikoshi, K. (1983). *Rhodopseudomonas rutila*, a new species of nonsulfur purple photosynthetic bacteria. *International Journal of Systematic Bacteriology*, **33**, 551–556.
Biebl, H. (1973), *Die Verbreitung der schwefelfreien Purpurbakterien im Plussee und anderen Seen Ostholsteins*. Ph. D. Thesis, University of Freiburg.
Biebl, H. and Drews, G. (1969). Das in-vivo-Spektrum als taxonomisches Merkmal bei Untersuchungen zur Verbreitung von Athiorhodaceae. *Zentralblatt für*

Bakteriologie, Hygiene und Parasitenkunde, Originale Abteilung, II, 123, 425–452.
Biebl, H. and Pfennig, N. (1978). Growth yields of green sulfur bacteria in mixed
cultures with sulfur and sulfate reducing bacteria. Archives of Microbiology, 117,
9–16.
Biebl, H. and Pfennig, N. (1981). Isolation of members of the family Rhodospiril-
laceae. In The Prokaryotes. A Handbook on Habitats, Isolation, and Identification of
Bacteria, M. P. Starr, H. Stolp, H. G. Trüper, A. Balows and H. G. Schlegel (Eds),
Springer, Heidelberg and New York, chapter 14.
Castenholz, R. W. and Pierson, B. K. (1981). Isolation of members of the family
Chloroflexaceae. In The Prokaryotes. A Handbook on Habitats, Isolation, and
Identification of Bacteria, M. P. Starr, H. Stolp, H. G. Trüper, A. Balows and H. G.
Schlegel (Eds.), Springer, Heidelberg and New York, chapter 17.
Cohen, Y., Krumbein, W. E. and Shilo, M. (1977). Solar Lake (Sinai). II.
Distribution of photosynthetic microorganisms and primary production. Limnolo-
gy and Oceanography, 22, 609–620.
Collins, M. D. and Jones, D. (1981). Distribution of isoprenoid quinone structural
types in bacteria and their taxonomic implications. Bacteriological Reviews, 45,
316–354.
Culver, D. A. and Brunskill, G. J. (1969). Fayetteville Green Lake, New York. V.
Studies of primary production and zooplankton in a meromictic lake. Limnology
and Oceanography, 14, 862–873.
De Bont, J. A. M., Scholten, A. and Hansen, T. A. (1981). DNA–DNA
hybridization of Rhodopseudomonas capsulata, Rhodopseudomonas sphaeroides, and
Rhodopseudomonas sulfidophila strains. Archives of Microbiology, 128, 271–274.
Drews, G. (1965). Die Isolierung schwefelfreier Purpurbakterien. Zentralblatt für
Bakteriologie, Parasitenkunde, Infektionskrankheiten und Hygiene, I. Abteilung, Supple-
ment, 1, 170–178.
Drews, G. (1981). Rhodospirillum salexigens, spec. nov., an obligatory halophilic
phototrophic bacterium. Archives of Microbiology 130, 325–327.
Dubinina, G. A. and Gorlenko, V. M. (1975). New filamentous photosynthetic
green bacteria containing gas vacuoles. Microbiologiya, 44, 452–458.
Eckersley, K. and Dow, C. S. (1980). Rhodopseudomonas blastica sp. nov.: a member
of the Rhodospirillaceae. Journal of General Microbiology, 119, 465–473.
Eichler, B. and Pfennig, N. (1986). Characterization of a new platelet-forming
purple sulfur bacterium, Amoebobacter pedioformis sp. nov. Archives of Microbiolo-
gy, 146, 295–300.
Fuchs, G., Stupperich, E. and Jaenchen, R. (1980a). Autotrophic CO_2 fixation in
Chlorobium limicola. Evidence against the operation of the Calvin cycle in growing
cell. Archives of Microbiology, 128, 56–63.
Fuchs, G., Stupperich, E. and Eden, G. (1980b). Autotrophic CO_2 fixaton in
Chlorobium limicola. Evidence for the operation of a reductive tricarboxylic acid
cycle in growing cells. Archives of Microbiology, 128, 64–71.
Gest, H. and Favinger, J. L. (1983). Heliobacterium chlorum, an anoxygenic
brownish-green photosynthetic bacterium containing 'new' form of bacterio-
chlorophyll. Archives of Microbiology, 136, 11–16.
Gibson, J., Pfennig, N. and Waterbury, J. B. (1984). Chloroherpeton thalassium gen.
nov. et spec. nov., a non-filamentous, flexing and gliding green sulfur bacterium.
Archives of Microbiology, 138, 96–101.
Gorlenko, V. M (1974). The oxidation of thiosulfate of Amoebobacter roseus in the
dark under microaerophilic conditions. Mikrobiologiya, 43, 729–731.
Gorlenko, V. M., Krasilnikova, E. N., Kikina, O. G. and Tatarinova, N. Ju. (1979).

The new motile purple sulphur bacteria *Lamprobacter modestohalophilus* nov. gen., nov. spec. with gas vacuoles. *Biological Bulletin of the Academy of Science, USSR*, **6**, 631–642.

Gorlenko, V. M. and Lebedeva, E. V. (1971). New green sulfur bacteria with apophyses. *Mikrobiologiya*, **40**, 746–747.

Gorlenko, V. M. and Pivovarova, T. A. (1977). On the belonging of blue-green alga *Oscillatoria coerulescens* Gicklhorn, 1921 to a new genus of Chlorobacteria *Oscillochloris* nov. gen. *Izvestiya Akademii Nauk SSSR, Series Biology*, **3**, 396–409.

Hansen, T. A. and Imhoff, J. F. (1985). *Rhodobacter veldkampii*, a new species of phototrophic purple nonsulfur bacteria. *International Journal of Systematic Bacteriology*, **35**, 115–116.

Hansen, T. A. and van Gemerden, H. (1972). Sulfide utilization by purple nonsulfur bacteria. *Archives of Microbiology*, **86**, 49–56.

Hansen, T. A. and Veldkamp, H. (1973). *Rhodopseudomonas sulfidophila* nov. spec., a new species of the purple nonsulfur bacteria. *Archives of Microbiology*, **92**, 45–58.

Hiraishi, A. and Hoshino, Y. (1984). Distribution of rhodoquinone in Rhodospirillaceae and its taxonomic implications. *Journal of General Microbiology*, **30**, 435–448.

Hiraishi, A., Hoshino, Y. and Kitamura, H. (1984). Isoprenoid quinone composition in the classification of Rhodospirillaceae. *Journal of General and Applied Microbiology*, **30**, 197–210.

Imhoff, J. F. (1982a). Response of phototrophic bacteria to mineral nutrients. In *Handbook of Biosolar Resources*, vol. 1, *Basic Principles*, part 2, A. Mitsui and C. C. Black (Eds), CRC Press, Boca Raton, pp. 135–146.

Imhoff, J. F. (1982b). Occurrence and evolutionary significance of two sulfate assimilation pathways in the Rhodospirillaceae. *Archives of Microbiology*, **132**, 197–203.

Imhoff, J. F. (1983). *Rhodopseudomonas marina* sp. nov., a new marine phototrophic purple bacterium. *Systematic and Applied Microbiology*, **4**, 512–521.

Imhoff, J. F. (1984a). Reassignment of the genus *Ectothiorhodospira* Pelsh 1936 to a new family Ectothiorhodospiraceae fam. nov., and emended description of the Chromatiaceae Bavendamm 1924. *International Journal of Systematic Bacteriology*, **34**, 338–339.

Imhoff, J. F. (1984b). Quinones of phototrophic purple bacteria. *FEMS Microbiology Letters*, **25**, 85–89.

Imhoff, J. F. and Trüper H. G. (1976). Marine sponges as habitats of anaerobic phototrophic bacteria. *Microbial Ecology*, **3**, 1–9.

Imhoff, J. F. and Trüper H. G. (1977). *Ectothiorhodospira halochloris* sp. now., a new extremely halophilic phototrophic bacterium containing bacteriochlorophyll *b*. *Archives of Microbiology*, **114**, 115–121.

Imhoff, J. F. and Trüper H. G. (1981). *Ectothiorhodospira abdelmalekii* sp. nov., a new halophilic and alkaliphilic phototrophic bacterium. *Zentralblatt für Bakteriologie und Hygiene, I. Abteilung Originale*, **C2**, 228–234.

Imhoff, J. F., Sahl, H. G., Soliman, G. S. H. and Trüper, H. G. (1979). The Wadi Natrun: Chemical composition and microbial mass developments in alkaline brines of eutrophic desert lakes. *Geomicrobiology Journal*, **1**, 219–234.

Imhoff, J. F., Tindall, B., Grant, W. D. and Trüper, H. G. (1981). *Ectothiorhodospira vacuolata* sp. nov., a new phototrophic bacterium from soda lakes. *Archives of Microbiology*, **130**, 238–242.

Imhoff, J. F., Kushner, D. J., Kushwaha, S. C. and Kates, M. (1982). Polar lipids in phototrophic bacteria of the Rhodospirillaceae and Chromatiaceae families. *Journal of Bacteriology*, **150**, 1192–1201.

Imhoff, J. F., Trüper, H. G. and Pfennig, N. (1984). Rearrangement of the species and genera of the phototrophic 'Purple Non-sulfur Bacteria'. *International Journal of Systematic Bacteriology*, **34**, 340–343.

Irgens, R. L. (1983). Thioacetamide as a source of hydrogen sulfide for colony growth of purple sulfur bacteria. *Current Microbiology*, **8**, 183–186.

Jannasch, H. W. (1957). Die bakterielle Rotfärbung der Salzsen des Wadi Natrun (Ägypten). *Archives of Hydrobiology*, **53**, 425–433.

Kämpf, C. and Pfennig N. (1980). Capacity of Chromatiaceae for chemotrophic growth. Specific respiration rates of *Thiocystis violacea* and *Chromatium vinosum*. *Archives of Microbiology*, **127**, 125–135.

Kato, S.-I., Urakami, T. and Komagata, K. (1985). Quinone systems and cellular fatty acid composition in species of Rhodospirillaceae genera. *Journal of General and Applied Microbiology*, **31**, 381–398.

Kenyon, C. N. (1978). Complex lipids and fatty acids of photosynthetic bacteria. In *The Photosynthetic Bacteria*, R. K. Clayton and W. R. Sistrom (Eds), Plenum Press, New York and London, chapter 14.

Keppen, O. I. and Gorlenko, V. M. (1975). A new species of purple budding bacteria containing bacteriochlorophyll *b*. *Microbiology*, **44**, 224–229.

Klemme, J.-H. (1968). Untersuchungen zur Photoautotrophie mit molekularem Wasserstoff bei neuisolierten schwefelfreien Purpurbakterien. *Archiv für Mikrobiologie*, **64**, 29–42.

Kompantseva, E. I. (1985). *Rhodobacter euryhalinus* new-species; a new halophilic purple bacterial species. *Mikrobiologyia*, **54**, 974–982.

Kompantseva, E. I. and Gorlenko, V. M. (1984). A new species of moderately halophilic purple bacterium *Rhodospirillum mediosalinum* sp. nov. *Mikrobiologiya*, **53**, 775–781.

Kondratieva, E. N., Zhukov, V. G., Ivanovsky, R. N., Petushkova, Yu. P. and Monosov, E. Z. (1976). The capacity of phototrophic sulfur bacterium *Thiocapsa roseopersicina* for chemosynthesis. *Archives of Microbiology*, **108**, 287–292.

Korthals, H. J. and Steenbergen, C. L. M. (1985). Separation and quantification of pigments from natural phototrophic microbial populations. *FEMS Microbiology and Ecology*, **31**, 177–185.

Madigan, M. T. (1986). *Chromatium tepidum* sp. now., a thermophilic photosynthetic bacterium of the family Chromatiaceae. *International Journal of Systematic Bacteriology*, **36**, 222–227.

Mangold, H. K. (1967). Aliphatische Lipide. In *Dünnschichtchromatographie*, 2nd ed., E. Stahl (Ed.) Springer, Heidelberg and New York, pp. 350–404.

Neutzling, O., Imhoff, J. F. and Trüper, H. G. (1984). *Rhodopseudomonas adriatica* sp. nov., a new species of the Rhodospirillaceae, dependent on reduced sulfur compounds. *Archives of Microbiology*, **137**, 256–261.

Nishimura, Y., Shimadzu, M. and Iizuka, H. (1981). Bacteriochlorophyll formation in radiation resistant *Pseudomonas radiora*. *Journal of General and Applied Microbiology*, **27**, 427–430.

Nissen, H. and Dundas I. D. (1984). *Rhodospirillum salinarum* sp. nov., a halophilic photosynthetic bacterium isolated from a Portuguese saltern. *Archives of Microbiology*, **138**, 251–256.

Pfennig, N. (1961). Eine vollsynthetische Nährlösung zur selektiven Anreicherung einiger Schwefelpurpurbakterien. *Naturwissenschaften*, **48**, 136–137.

Pfennig, N. (1965). Anreicherungskulturen für rote und grüne Schwefelbakterien. *Zentralblatt für Bakteriologie, Parasitenkunde, Infektionskrankheiten und Hygiene, I. Abteilung Supplement*, **1**, 179–189.

Pfennig, N. (1967). Photosynthetic bacteria. *Annual Review of Microbiology*, **21**, 285–324.

Pfennig, N. (1969). *Rhodopseudomonas acidophila*, sp. n., a new species of the budding purple nonsulfur bacteria. *Journal of Bacteriology*, **99**, 597–602.

Pfennig, N. (1974). *Rhodopseudomonas globiformis*, sp. n., a new species of the Rhodospirillaceae. *Archives of Microbiology*, **100**, 197–206.

Pfennig, N. (1977). Phototrophic green and purple bacteria: a comparative systematic survey. *Annual Review of Microbiology*, **31**, 275–290.

Pfennig, N. (1978). *Rhodocyclus purpureus* gen. nov. and sp. nov., a ring-shaped vitamin B_{12}-requiring member of the family Rhodospirillaceae. *International Journal of Systematic Bacteriology*, **28**, 283–288.

Pfennig, N. and Trüper H. G. (1971). Conservation of the family name Chromatiaceae Bavendamm 1924 with the type genus *Chromatium* Perty 1852, request for an opinion. *International Journal of Systematic Bacteriology*, **21**, 15–16.

Pfennig, N. and Trüper, H. G. (1981). Isolation of members of the families Chromatiaceae and Chlorobiaceae. In *The Prokaryotes. A Handbook on Habitats, Isolation, and Identification of Bacteria*, M. P. Starr, H. Stolp, H. G. Trüper, A. Balows, and H. G. Schlegel (Eds), Springer, Heidelberg and New York, chapter 16.

Pierson, B. K., Giovannoni, S. J., Stahl, D. A. and Castenholz, R. W. (1985). *Heliothrix oregonenese*, gen. nov., spec. nov., a phototrophic filamentous gliding bacterium containing bacteriochlorophyll a. *Archives of Microbiology*, **142**, 164–167.

Quadri, S. M. H. and Hoare, D. S. (1968). Formic hydrogen lyase and the photoassimilation of formate by a strain of *Rhodopseudomonas palustris*. *Journal of Bacteriology*, **95**, 2344–2357.

Raymond, J. C. and Sistrom W. R. (1969). *Ectothiorhodospira halophila*: A new species of the genus *Ectothiorhodospira*. *Archives für Mikrobiology*, **69**, 121–126.

Sato, K. (1978). Bacteriochlorophyll formation by facultative methylotrophs, *Protaminobacter ruber* and *Pseudomonas AM 1*. *FEBS Letters*, **85**, 207–210.

Schmidt, K. (1978). Biosynthesis of carotenoids. In *The Photosynthetic Bacteria*. R. K. Clayton, and W. R. Sistrom (Eds). Plenum New York and London, chapter 19.

Schmidt, K. and Bowien, B. (1983). Notes on the description of *Rhodopseudomonas blastica*. *Archives of Microbiology*, **136**, 242.

Shiba, T. and Simidu, U. (1982). *Erythrobacter longus* gen. nov., sp. nov., an aerobic bacterium which contains bacteriochlorophyll a. *International Journal of Systematic Bacteriology*, **32**, 211–217.

Shiba, T., Simidu, U. and Taga, N. (1979). Distribution of aerobic bacteria which contain bacteriochlorophyll a. *Applied and Environmental Microbiology*, **38**, 43–45.

Siefert, E. and Pfennig, N. (1984). Convenient method to prepare neutral sulfide solution for cultivation of phototrophic sulfur bacteria. *Archives of Microbiology*, **139**, 100–101.

Stammel, B. (1977). *Neuisolierung formiatverwertender phototropher Bakterien, sowie Isolierung und Anreicherung der Formiatdehydrogenase von* Rhodopseudomonas palustris. Diploma Thesis, University of Bonn.

Swoager, W. C. and Lindstrom E. S. (1971). Isolation and counting of Athiorhodaceae with membrane filters. *Applied Microbiology*, **22**, 683–687.

Takahashi, M. and Ichimura, S. (1968). Vertical distribution and organic matter production of photosynthetic sulfur bacteria in Japanese lakes. *Limnology and Oceanography*, **13**, 644–655.

Trüper, H. G. (1970) Culture and isolation of phototrophic sulfur bacteria from the marine environment. *Helgoländer wissenschaftliche Meeresuntersungen*, **20**, 6–16.

Trüper, H. G. and Imhoff J. F. (1981) The genus *Ectothiorhodospira*. In *The Prokaryotes. A Handbook on Habitats, Isolation and Identification of Bacteria*, M. P. Starr, H. Stolp, H. G. Trüper, A. Balows and H. G. Schlegel (Eds), Springer, Heidelberg and New York, chapter 15.

Trüper, H. G. and Pfennig, N. (1981). Characterization and Identification of the anoxygenic phototrophic bacteria. In *The Prokaryotes. A Handbook on Habitats, Isolation and Identification of Bacteria*, M. P. Starr, H. Stolp, H. G. Trüper, A. Balows, and H. G. Schlegel (Eds), Springer, Heidelberg and New York, chapter 18.

van Gemerden, H. (1974) Coexistence of organisms competing for the same substrate: An example among the purple sulfur bacteria. *Microbial Ecology*, **1**, 104–119.

van Niel, C. B. (1932). On the morphology and physiology of the purple and green sulfur bacteria. *Archiv für Mikrobiologie*, **3**, 1–112.

van Niel, C. B. (1944). The culture, general physiology, morphology and classification of the nonsulfur purple and brown bacteria. *Bacteriological Reviews*, **8**, 1–118.

van Niel, C. B. (1971). Technics for the enrichment, isolation, and maintenance of photosynthetic bacteria. In *Methods in Enzymology*, vol. XXIII, part A, S. P. Collowick and N. V. Kaplan (Eds), Academic Press, New York and London, pp. 3–28.

Westmacott, D. and Primrose S. B. (1975). An anaerobic bag for photoheterotrophic growth of some Rhodospirillaceae in Petri dishes. *Journal of Applied Bacteriology*, **38**, 205–207.

Winogradski, S. N. (1888). *Zur Morphologie und Physiologie der Schwefelbakterien*, Felix-Verlag, Leipzig.

Methods in Aquatic Bacteriology
Edited by B. Austin
© 1988 John Wiley & Sons Ltd.

10

Cyanobacteria: Isolation, Interactions and Ecology

M. J. Daft

Department of Biological Sciences, University of Dundee, Dundee DD1 4HN, Scotland

10.1 INTRODUCTION

Fossil records show that the cyanobacteria are an ancient group of photosynthetic micro-organisms. Brock (1973) and Schopf and Walter (1982) consider that they may have been present for some 2.5–3.0 billion years varying in their dominance. The Rhynia deposits in Aberdeenshire contain recognizable heterocystous, and hence presumably, nitrogen-fixing forms that resemble members of present-day *Nostoc* species (Croft and George, 1959). In more modern times they have become less important, although in recent decades they have attracted considerable attention due to their success in colonizing fresh waters subjected to eutrophication. As photosynthetic micro-organisms, and with many forms being capable of fixing atmospheric nitrogen, the cyanobacteria can be found in many diverse habitats, the major limiting factor being that of illumination. It has been estimated that approximately 70% of the biosphere is composed of lakes and oceans, and one of the most widespread micro-organisms is a *Synechococcus* species probably capable of fixing atmospheric nitrogen (Waterbury *et al.*, 1979).

Freshwater aquatic habitats such as ponds, lakes, lochs, reservoirs, drinking water filters and drainage ditches have been investigated for the ecological contributions and impact made by the indigenous cyanobacteria. Under certain conditions the cyanobacteria can reach very high densities and cause blooms. These blooms contain mainly species of *Anabaena*, *Aphanizomenon*, *Coelosphaerium*, *Gloeotrichia*, *Microcystis* and *Oscillatoria* (Figs 10.1–10.3). Considerable fluctuations in the presence and amounts of cyanobacteria may occur in different habitats. In some African lakes the

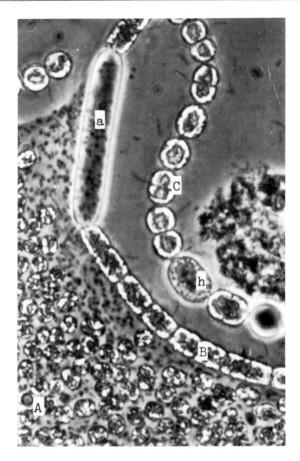

Fig. 10.1 Sample of a natural bloom from Long Loch containing (A) *Microcystis*, (B) *Aphanizomenon* and (C) *Anabaena* species (× 1600). Note akinete (a) and heterocysts (h) in the filament.

bloom may be permanent. In subtropical zones they may be seasonable and in temperate regions somewhat irregular growths are produced. Blooms may be very detrimental to the value of the water particularly by causing deoxygenation and tainting of the water when the blooms decay rapidly (Stewart and Daft, 1977). Some freshwater isolates can produce toxic compounds (Carmichael, 1984). Conversely, *Spirulina platensis* may be eaten by man. The collapse of a bloom, where the cyanobacteria no longer dominates the habitats, may be due to light, oxygen or predators.

Marine environments have received less attention than freshwater habitats. Waterbury and Stanier (1981) have outlined the marine habitats

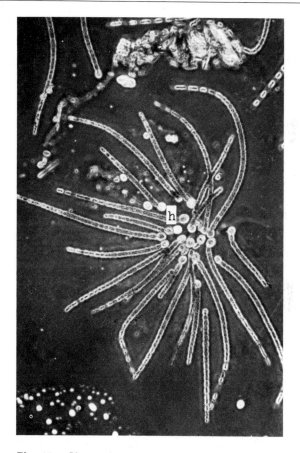

Fig. 10.2 *Gloeotrichia* sp. from a bloom in Monikie Reservoir (×1500). Note terminal heterocysts (h).

exploited by cyanobacteria, namely the warmer temperate and tropical regions, the intertidal and subtidal, open ocean and coral reefs. An interesting source of *Phormidium corallyticum* is the black band disease of the Atlantic coral reefs (Taylor, 1984). Brock (1976) isolated halophilic cyanobacteria, and showed that this habitat has a limited flora. Strains of *Aphanothece* can grow in salinities close to 3 M NaCl (Waterbury and Stanier, 1981).

Mud, either wet or dried, from temperate or tropical regions is a rich source of cyanobacteria. Sugar cane fields have been found to contain numerous nitrogen-fixing species in Central and Southern Taiwan (Chung *et al.*, 1981). Strains of *Anabaena*, *Nostoc*, *Cylindrospermum*, *Calothrix*, *Saytonema* and *Hapalosiphon* were isolated and purified as axenic cultures from rice fields by Huang and Chow (1983).

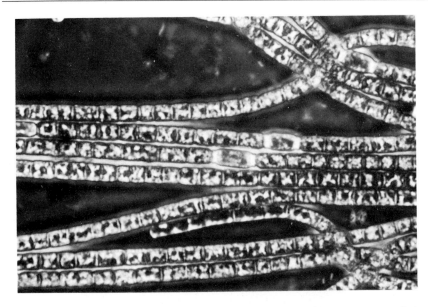

Fig. 10.3 *Aphanizomenon* sp. showing the aggregation of filaments into 'rafts' (×1700).

Environments with extreme conditions can support the growth of cyanobacteria. Hot springs at temperatures of about 70 °C, have been shown to contain several thermophilic genera (Castenholz, 1981). At the other end of the temperature range, cyanobacteria have been isolated from the Antarctic (Fogg *et al.*, 1973).

Perhaps the most specialized habitats are those in which the cyanobacteria have entered into symbiotic associations with plants, animals and fungi.

10.2 ISOLATION

In the previous section, the wide range of habitats were outlined in which cyanobacteria may be found. It is from these diverse habitats that isolations are made and hence a variety of isolation techniques must be employed in order to obtain either uni-cyanobacterial or axenic cultures. Isolation techniques fall into four main categories: (1) simple enrichment, (2) continuous culture, (3) filtration or (4) direct isolation (Fig. 10.4).

Material collected from the natural habitat should be examined under the light microscope in order to ascertain the relative abundance of the types of cyanobacteria present in the sample. If too low, the sample can be

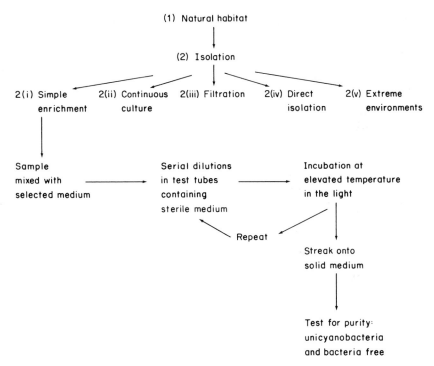

Fig. 10.4. Enrichment for cyanobacteria.

concentrated by centrifugation or, if too high, diluted with a suitable medium. Many different media have been used for enrichment in liquid culture and growth of the isolate on solid media. The composition of five media are given in Tables 10.1 and 10.2. These are suitable for isolates taken from freshwater, marine, halophilic and hot spring habitats.

10.2.1 Simple enrichment

In the simple enrichment method, the inoculum is prepared by first mixing the sample with the selected medium and then serial dilutions made in test tubes containing a similar sterile medium. If heterocystous cyanobacteria predominate in the original sample, then the nitrogen sources in the medium can be omitted in order to encourage the growth of the nitrogen-fixing species. The test tubes are incubated in the light (150 μW cm^{-2}), either continuously or with a dark cycle, at an elevated temperature of about 30°C. Higher temperatures suggested by Allen and Stanier (1968) may be of use for certain isolations. Such elevated temperatures depress the growth of many contaminating green algae.

TABLE 10.1   Trace element solutions used for the isolation of cyanobacteria

Ingredients	Amount ($g\,l^{-1}$) in medium				
	Modified Castenholz D[1]	BG-11[2]	MN[3]	ASM[4]	Difco algae broth (AB medium)[5]
NaCl	160.0	–	–	–	–
NaNO$_3$	0.7	1.5	0.75	0.017	1.0
Na$_2$HPO$_4$	0.11	–	–	0.0141	–
Na$_2$CO$_3$	–	0.02	0.02	–	–
KNO$_3$	0.1	–	–	–	–
K$_2$HPO$_4 \cdot 3H_2O$	–	0.04	0.02	0.0174	0.25
MgSO$_4 \cdot 7H_2O$	0.1	0.095	0.038	0.049	0.513
MgCl$_2 \cdot 6H_2O$	–	–	–	0.041	–
CaCl$_2 \cdot 2H_2O$	–	0.036	0.018	0.029	0.058
CaSO$_4 \cdot 2H_2O$	0.06	–	–	–	–
NH$_4$Cl	–	–	–	–	0.05
FeCl$_3 \cdot 6H_2O$	0.0003	0.006	0.003	0.003	0.003
Citric acid	–	0.006	0.003	–	–
Ferric ammonium citrate	–	–	–	–	–
EDTA (Na Mg salt)	–	0.001	0.0005	0.008	–
Nitrilotriacetic acid	0.1	–	–	–	–
Seawater (ml)	–	–	750	–	–
Distilled water (ml)	1000	1000	250	1000	1000
pH after autoclaving	7.5	7.1	8.5	7–7.6	6.8

[1] Brock (1976).
[2] Stanier et al. (1971).
[3] Waterbury and Stanier (1978).
[4] Gorham et al. (1964).
[5] Burnham et al. (1976).

TABLE 10.2 Trace element solutions used for the isolation of cyanobacteria

| | g 1000 ml^{-1} distilled water | | |
Ingredients	BG-ll/MN	Modified Castenholz D	ASM
H_3BO_3	0.00286	0.5	0.0025
$MnCl_2 \cdot 4H_2O$	0.00181	–	0.0014
$ZnSO_4 \cdot 7H_2O$	0.000222	0.5	–
$Na_2MoO_4 \cdot 2H_2O$	0.000390	0.025	–
$CuSO_4 \cdot 5H_2O$	0.000079	0.025	–
$Co(NO_3)_2 \cdot 6H_2O$	0.0000494	0.045	–
Concentrated H_2SO_4 (ml)	–	0.5	–
$MnSO_4 \cdot H_2O$	–	2.28	–
$(NH_4)_6MO_7O_{24} \cdot 4H_2O$	–	–	0.001

Daily inspection of the tubes under a binocular microscope will indicate when the next step may be undertaken. The tube selected should contain sufficient growth to allow either a repeat of the simple enrichment stage in tubes or to streak onto the same medium solidified with 1.5% (w/v) agar. If the petri dishes are deep-poured, they may be incubated under the same environmental conditions without drying out. When single colonies develop they may be picked off, re-streaked or cultured in more liquid medium and finally regrown on the solid medium to obtain a unicyano-bacterial culture which is bacteria-free. Selective usage of antibiotics, ultraviolet light or micromanipulation may aid removal of the contaminating bacteria. However, some cyanobacteria have close associations with the bacteria (Daft *et al.* 1975) and other bacteria may grow within the mucilagenous sheath surrounding the cyanobacterial cells making their removal very difficult. Obtaining bacteria-free cyanobacteria in these situations is most time-consuming. Other bacteria, under certain conditions may act as saprophytes or pathogens of cyanophycean blooms (Stewart and Daft, 1977) whereas 'bactericized' cultures can grow better than axenic cultures (Fitzsimons and Smith, 1984).

10.2.2 Continuous enrichment cultures

In freshwater ecosystems, particular cyanobacteria may overwinter in the water column, either as spores or akinetes. Other forms overwinter in the sediments as vegetative cells. Cyanobacteria, in their resting stages contained in water samples taken in winter, may be concentrated either by centrifugation or filtration. Mud samples taken from the lake or loch are diluted in surface water and then added to filtered and sterilized water collected at the same site. This water is contained in a culture vessel, and

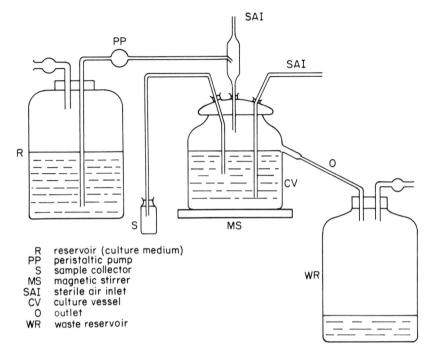

R   reservoir (culture medium)
PP   peristaltic pump
S   sample collector
MS   magnetic stirrer
SAI   sterile air inlet
CV   culture vessel
O   outlet
WR   waste reservoir

Fig. 10.5 Continuous culture apparatus.

incubated in the light at room temperature. A diagrammatic representation of the apparatus is shown in Fig. 10.5. The culture vessel is supplied via a peristaltic pump with sterile lake water contained in the reservoir. Flow rates can be adjusted as required for each system. Sterile air is pumped into the vessel, and the culture stirred by a magnet. Samples are collected by closing the overflow tube which, in turn, increases the pressure in the culture vessel and hence the culture fluid passes into the sampling bottle. the culture vessel is illuminated at a suitable intensity and, if required, the culture vessel may be placed in a cooled or heated water bath in order to control the incubation temperature.

Using this continuous culture apparatus, Fallowfield (1981) followed the changes in algal and bacterial cell counts over a period of 25 days. The source of the inoculum was mud, taken from a Scottish loch. After 8 days, the culture contained the green algae *Senedesmus quadricauda*, *Ankistrodesmus falactus*, a *Sphaerocystis* species, the cyanobacterium *Anabaena spiroides* plus the diatom *Melosira granulata*. At the termination of the experiment (18 days), *A. spiroides* and the *Sphaerocystis* species were the only micro-organisms detected in the culture.

10.2.3 Filtration

Fitzsimons and Smith (1984) described a method for the isolation and the axenic culture of planktonic cyanobacteria. Essentially, their method entails filtration through various sized filters by means of the application of gentle suction onto the sample which contains cyanobacteria plus contaminating bacteria. The pore size of the filters is selected to allow contaminating bacteria to pass through while retaining the planktonic cyanobacteria. If the original sample has clumped, then gentle sonication may be used to separate the cells or filaments of the cyanobacterium. Using sterile medium, the filtration and resuspension is repeated 10 times. The final concentrate is then diluted serially (up to 10^{-9}) and incubated in continuous light (12–15 $\mu E\,m^{-2}\,s^{-1}$). From each of these dilution tubes, contaminating bacteria are detected by using 16 different test media. Tubes that show cyanobacterial growth and are free of bacterial contamination may also be examined using phase contrast microscopy. This technique, modified by Fitzsimons and Smith, brings about an alteration in the cyanobacterium:bacteria ratio. For example, in a natural sample of *Aphanizomenon flos-aquae*, the ratios were 0.3:12 and 14:0.2, before and after filtration, respectively.

10.2.4 Direct isolation

This technique depends on the original sample containing an accumulation of the cyanobacterium that easily fragments when suspended in the minimal amount of liquid medium. If this can not be done by shaking, then the sample may require homogenization or sonication before streaking onto solid media. Table 10.1 gives the composition of media that have been found useful for direct isolation. Rippka *et al.* (1981) recommended that the streaked plates are incubated in low light intensities and at comparatively lower temperatures ($>30\,°C$) than those used in the simple enrichment technique described in 10.2.1. The plates need to be examined frequently, and when colonies appear, they are removed and re-streaked as often as needed in order to obtain pure cultures. Purity of the cultures may be determined by inoculating poured-plates containing a range of media with the culture, as suggested by Fitzsimons and Smith (1984).

10.2.5 Isolation of cyanobacteria from extreme environments

Thermophilic cyanobacteria may be enriched and isolated at temperatures below or at their normal levels. If cultured at subnormal, i.e. $<45\,°C$, it may be necessary to add cycloheximide to reduce the growth of eukaryotic algae. At temperatures above $45\,°C$, few green algae will survive for long periods. Castenholz (1981) suggested that a high proportion of native water should be used in the initial stage of enrichment, in order to reduce shock to the cyanobacterium under investigation.

Some modifications to the media will enhance the growth of specific types, i.e. omission of combined nitrogen or the supplementation of the medium with ammonium chloride or sodium sulphide for tolerant and sulphide utilizing isolates. Direct isolation is used to obtain axenic cultures. To test for purity, with regard to contaminating bacteria, the usual plating out techniques are used at a range of temperatures.

10.3 INTERACTIONS

Growth of primary cyanobacterial producers are limited by light and the availability of nutrients. In temperate climates, the concentration of nitrogen within the photic zone decreases in the spring and summer. The sources of phosphate are mainly water runoff, domestic sewage, and agricultural wastes. Phosphate is probably the most important ion in the determination of the total yield of the biomass. However, this biomass is potentially available for exploitation by other micro-organisms. These microbes may range from specific obligate parasites through to those that can utilize the extracellular products of the primary producers. Interactions amongst the prey (cyanobacterial primary producers) and their predators have been discussed by Daft et al. (1985a, b). Yields of primary producers in different parts of the world are reduced by the effects of grazing by rotifers, crustaceans, benthic invertebrates, insects, fish, crocodiles and mammals (Frost, 1980; Goldman and Horne, 1983). Amoebae are now known to play a significant role in reducing cyanobacterial blooms (Yamamoto, 1981).

Several types of microbes will parasitize cyanobacteria to varying degrees. Bacteria may be pathogenic (Shilo, 1970; Daft and Stewart, 1971; Burnham, et al. 1981), utilize extracellular products (Fallowfield, 1981) or enhance the growth of cyanobacteria (Fitzsimons and Smith, 1984). Some of the major interactions of cyanobacteria with other organisms are listed in Table 10.3.

The isolation and partial purification of a cyanophage was first carried out by Safferman and Morris (1963). This phage infects species of *Lyngbya, Plectonema and Phormidium* (LPP). Different cyanophages have now been isolated that infect unicellular, colonial, non-heterocystous and heterocystous cyanobacteria. The shape of the cyanophage head, morphology of the tail, host range, physical and chemical composition were all used in their classification (Stewart and Daft, 1977).

TABLE 10.3 Interactions of cyanobacteria with other organisms

Type of interaction	Mode of action	Examples	References
Pathogenic	(i) Internal	Cyanophages	Stewart and Daft (1977)
			Saffermann and Morris (1963)
			Burnham et al. (1976)
		Bdellovibrios	Grilli-Caiola and Pellegrini (1984)
			Canter (1972)
	(ii) Contact	Fungi (lower)	Daft and Stewart (1971)
		Lysobacter CP-I	Shilo (1970)
		FP-I	Burnham et al. (1981)
		Myxococcus sp.	Daft et al. (1985b)
			Yamamoto (1981)
		Amoebae	Huang and Wu (1982)
			Bader et al. (1976)
		Ciliates	Stewart and Brown (1969)
	(iii) Extracellular	*Cytophaga* sp. N-5	Redhead and Wright (1980)
	lytic compounds	Fungi (higher)	Al-Tai et al. (1986)
		Actinomycetes	Saffermann and Morris (1965)
Saprophytic	(iv) Utilization of	*Lysobacter* CP-I	Fallowfield (1981)
	extracellular	Ensheathed bacteria	Caldwell (1979)
	products from	Non-specific	Daft and Fallowfield (1977)
	cyanobacteria		
Symbiotic	(v) Contact with	Diatoms	Mague et al. (1974)
	tissues of the	Lichens	Stewart et al. (1983)
	host	Cycads	Grilli-Caiola (1975)
		Ferns	Peters (1975); Duckett et al. (1975)
		Liverworts	Rodgers and Stewart (1977)
		Angiosperms	Silvester and Macnamara (1976)

10.3.1 Internal pathogens

Cyanophages show the most specific relationships with their particular hosts. Frequently phages are even strain specific. Experimentally, specificity can be shown in batch culture (Cowlinshaw and Mrsa, 1975) and in chemostats containing a cyanobacteria–cyanophage system (Barnet *et al.*, 1981). The latter authors showed reciprocal oscillations in both the numbers of *Plectonema boryanum* cells and the LPP-DUN1 cyanophage particles over a period of 65 days. A similar pattern of results was found with *Aphanothece stagnina* and cyanophage AS-1. Table 10.4 gives some modified data on wild type *P. boryanum*/LPP-DUN1 and *P. boryanum*/ mutant D_m. After the experimental period, the chemostat contained resistant *P. boryanum* strains and mutant phages. The development of resistance in the host and mutation of the cyanophage may be a factor in the maintenance of each component in nature. Total loss of the wild type host can only be brought about by the effect of a mutant phage (Phage D_m). Another component that may protect the cyanobacteria is the influence of silt. In nature, storms or overturn will add colloidal substances such as clays and particulate material to the water column. Cyanobacteria are removed from suspension by bentonite (Avnimelech *et al.*, 1982) and bacteria may become covered colloids (Santoro and Stotzky, 1968; Marshall, 1969). Phage particles absorb onto colloids (Sykes and Williams, 1978). Barnet *et al.* (1984) showed the effect of particulate material on a host/prey system. The latter authors showed that the addition of silt prevented the lysis of *Plectonema boryanum* by the cyanophage LPP-DUN1 in batch culture and damped down oscillations of the host in continuous culture. Silt, therefore, appeared to protect the cyanobacterium, in part, from lysis by the cyanophage. Data from these experiments showed an increase in cyanophage numbers with the particulate material due, perhaps, to the higher numbers of host cells available for cyanophage multiplication.

TABLE 10.4 Influence of wild type cyanophage LPP-DUN1 and mutant phage D_m on numbers of wild type *Plectonema boryanum* cells in a chemostat culture

	Time (days)						
	0	4	6	8	10	14	25
Numbers of *P. boryanum* cells ml^{-1}							
LPP-DUN1 (wild type)	8×10^7	4×10^6	5×10^5	6×10^4	2×10^4	6×10^2	6×10^2
Mutant phage D_m	4×10^7	5×10^4	2×10^6	4×10^4	4×10^4	3×10^7	3×10^7

The antimicrobial effects of two strains of *Cellvibrio* on blue-green algae have been reported by Granhall and Berg (1972). Each strain produces extracellular compounds during the stationary phase of culture. The mode of action of these compounds appears to be one of inhibition of the host cell wall synthesis; primarily in the walls of the vegetative cells. Lytic properties were attributed to a mixture of heat resistant, low molecular weight substances. These substances were active against an *Anabaena* sp. This particular cyanobacterium formed part of the annual waterbloom in a river. Burnham *et al.* (1976) also showed lysis of a cyanobacterium, specifically *Phormidium luridum* by *Bdellovibrio bacteriovorus*. Cells of *B. bacteriovorus* and a heat resistant factor from the culture medium lysed viable cells of the test cyanobacterium. Changes in the appearance of the *P. luridum* could be seen after 16 h under a light microscope. Samples of a natural bloom, containing *Microcystis aeruginosa* collected from a lake by Grilli-Caiola and Pellegrini (1984), were examined under a transmission electron microscope. Some cells of *M. aeruginosa* contained bacteria similar to *Bdellovibrio* spp. Lysis of the cyanobacterium was attributed to the disruption of certain sites within the cell wall. It was also suggested that these lysed areas were caused by contact between the bacterium and cyanobacterium host both externally and internally, permitting entry and exit of the pathogen. Hence, this disease could be another cause of sudden bloom disappearance coupled with the extracellular lysis discussed above.

Canter (1972) described many lower fungi, mainly chytrids, that attack cyanobacteria. Great difficulties are encountered in studies on these pathogens due to their sporadic appearance and rapid disappearance in nature. However, there is a positive correlation between the abundance of both the fungal parasites and their specific cyanobacterial hosts. Often maximal numbers of the parasites occur just before mass death of the hosts. Pathogenicity of the fungi is dependent on an adequate oxygen supply, pH (6.4–8.0 units) and carbohydrate alkalinity.

10.3.2 Contact lysis

Bacteria that bring about lysis by establishing contact with the cyanobacteria may all belong to the genus *Lysobacter* (Christensen and Cook, 1978). Two groups have been studied extensively, the FP-1 group by Shilo (1970) and the CP isolates by Daft *et al.* (1975). These bacteria are opportunist parasites in that they can lyse the host cyanobacteria within 20 min or utilize the extracellular products of the same types of cyanobacteria (Fallowfield, 1981). A lytic mechanism that requires polar contact between bacterium and cyanobacterium would ensure that maximum efficiency was maintained, as opposed to the widespread production of extracellular lytic compounds. A cell-free extract of CP-1 causes lysis of several cyanobacteria

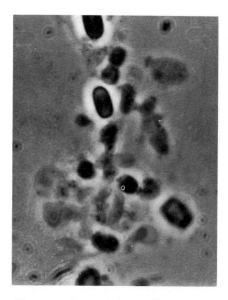

Fig. 10.6 General lysis of vegetative cells of *N. ellipsosporum* filaments by a cell-free preparation of CP-l plus ethylenediamine tetraacetic acid. The heterocysts appear unaffected (×1400).

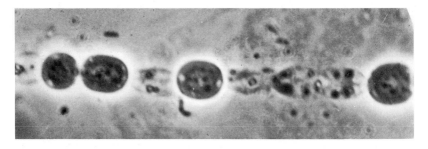

Fig. 10.7 Filament of *Nostoc ellipsosporum* showing lysis of individual vegetative cells by bacterium CP-l and four intact cells (×1500).

only in the presence of chelating agents (Daft & Browning, unpublished data). This suggests that more than one enzyme is passed across the point of contact between prey and predator. The chelating agents may substitute for part of the lytic system and the remainder is a stable complex from the lytic bacterium (Figs 10.6 and 10.7). It would appear that the role of bacteria such as CP-1 is complex within a cyanophycean bloom. The relative concentrations of the cyanobacterium may influence the mechanisms exhibited by the lytic bacteria at any given time during the development of the bloom.

The *Myxococcus* species manifest various forms of morphology depending on the growth conditions (Daft *et al.*, 1985a). In flask culture, the bacterium produces spherules (Burnham *et al.*, 1981). These spherules lyse cyanobacteria efficiently at both high and low concentrations (Burnham *et al.*, 1984). In continuous culture, the bacterium adheres to the vessel walls or to any substrata in the culture vessel. When adhered to the walls of the vessel, it produces long colonial 'feathery' strands. These may cover the entire surface when grown on glass beads. These three morphological forms probably exist in nature, the spherules being free living, the long strands attached to plant stems and small stones providing a substratum for the uniform growth form. The mechanism of lysis of the cyanobacteria is by means of entrapment and then enzymatic degradation of the host cells. The highly lytic enzymes are not lost to the environment because the cyanobacterium is confined within the colonial *Myxococcus* sp. The predator can be maintained indefinitely by supplying viable cells of the prey (Burnham *et al.*, 1984). Maintenance of lytic efficiency over a long period and at low prey densities makes this type of bacterium an attractive one for control purposes.

10.3.3 Ingestion by animals

In Japan, Yamamoto (1981) isolated amoebae belonging to the genera *Nuclearia* and *Acanthamoeba*. The technique used was that of mixing contaminated lake or river water with a susceptible cyanobacterium in soft agar, and then overlaying this mixture on prepoured petri dishes. Algophorus amoebae were then isolated from plaques produced on the host cyanobacterium. Microscopic observations ensure that the plaque is caused by an amoeba rather than other lytic agents. Rates of consumption of the cyanobacteria depend on temperature, type of prey and the generation times of the amoebae. Generation times of the predator also vary with temperature and type of prey. This is shown in Table 10.5 for *Phormidium luridum* and *Anabaena flos-aquae* at 5, 17 and 24 °C. Ingestion of the cyanobacterial filaments is followed by the production of food vacuoles which are typical of the feeding mechanism of amoebae. The vegetative cells are digested in preference to the heterocysts which are frequently ejected apparently unaffected by the digestion processes.

Grazing of cyanobacteria by ciliates is another biological system that may reduce, to some degree, the growth of blooms. In laboratory studies, Bader *et al.* (1976) found steady states of coexistence were produced in chemostat cultures, between the ciliate and cyanobacterial host. This differs in that

TABLE 10.5 Influence of temperature, time and prey on the generation times of algophorus amoebae (data from Daft and Yamamoto, unpublished).

Amoebae	Temperature (°C)	Generation time (h) when prey is:	
		Phormidium luridum	*Anabaena flos-aquae*
Nuclearia sp.			
(4–1)	5	132	43
(Japan)	24	17	22
Acanthamoeba			
castellanii (pH2)	5	310	166
(Scotland)	24	82	73

other prey–predator systems often show marked oscillations of each protagonist, e.g. reciprocal oscillations of cyanobacterium/cyanophage (Barnet *et al.*, 1981).

10.3.4 Non-specific associations

Daft and Fallowfield (1977) found that two different pseudomonads, one white and the other pale yellow in colour, multiplied rapidly during the development of high concentrations of diatoms followed by a cyanophycean bloom in late summer and early autumn. These bacteria appear to be non-specific in their association with the dominant photosynthetic organism growing at any given time. The pale yellow type appears in March, peaks in April, mid-May and late July, whereas the white type appears in mid-June, peaks in late July and disappears along with the other in October. Ensheathed bacteria pose another situation both in purification techniques and the relationship between the cyanobacteria and the bacterium (Caldwell, 1979).

10.3.5 Beneficial associations

Diatoms, both marine and freshwater pennate and centric forms, are associated with clearly recognizable cyanobacteria or cyanobacteria-like inclusion bodies. In *Rhizosoneia*, short filaments with basal heterocysts, similar to that of a *Calothrix* species, occur within the cell and such cells possess nitrogenase activity (Mague *et al.*, 1974, 1977). In the northern Pacific, *Rhizosoneia* is a frequent component of the marine phytoplankton. Unusual cytoplasmic inclusion bodies in *Phopalodia* have been described by Drum and Pankratz (1965). In these diatoms, the bodies have the appearance of unicellular cyanobacteria.

Some lichens contain cyanobacteria, but their numerical contribution to the group is small. Stewart *et al.* (1983) quoted a figure of 8% for both bipartite and tripartite forms. In the bipartite lichens such as *Peltigera canina* or *Lichina confinis*, the symbiosis or 'controlled parasitism' is between a cyanobacterium and a fungus. In the tripartite *P. aphthosa*, the third component is a green eukaryotic alga. The eukaryotic alga is the major source of photosynthate and the cyanobacterium the source of fixed nitrogen in the tripartite forms. In the bipartite forms, the cyanobacterium supplies both fixed carbon and fixed nitrogen. Hitch and Millbank (1975) reported great variability amongst lichens in their ability to fix atmospheric nitrogen. They also found that there was a greater proportion of heterocysts in the tripartite types compared with the bipartite types. Lichens are widespread in their distribution. This distribution can be explained, in part, by geological history and the long-distance dispersal mechanisms possessed by lichens. Lichens can obtain water from dew, fog or melting snow and, given a suitable temperature, these sources may be important ecological strategies. However, the comparative longevity of lichens may prove disastrous to grazing animals as seen in Lapland after the nuclear accident at Chernobyl, USSR in 1986.

The cycads develop coralloid root nodules which contain specific cyanobacteria. These phycobionts have been studied by Grilli-Caiola (1975), who has shown that they are localized in intercellular spaces with abundant mucilage. Some morphological differences were found between the cyanobacterium growing in the living coralloids and those growing on solid medium. This same author in 1980 also compared cycads from five Italian Botanical Gardens. All the phycobionts had feature characteristics of *Nostoc*. However, the *Nostoc* isolates did vary in some aspects with regard to interactions amongst the hosts.

Ferns such as *Azolla* possess cavities in their leaves which contain nitrogen fixing *Anabaena azollae*. The fixed nitrogen is transferred to the host but the fern can grow without the symbiont if there is a source of combined nitrogen in the medium (Duckett *et al.*, 1975).

In two publications, Rodgers and Stewart (1977) and Stewart and Rodgers (1977) studied the morphology, physiology, nitrogen fixation and interchange of nitrogen and carbon in the cyanophyte–hepatic symbioses. *Anthoceros punctatus* and *Blasia pusilla* contain a species of *Nostoc* probably *N. sphaericum*. The gametophyte generation houses the cyanobacterium within cavities on the undersurface of the thallus. Nitrogen fixed by the *Nostoc* contributes significantly to the nitrogen contents of the thalli. Carbon fixation by the *Nostoc* is insignificant and transfer of photosynthate is considered to be from the liverworts to the phycobionts.

The association between *Gunnera*, an angiosperm, and *Nostoc*, a cyanophyte, is unique. All species of *Gunnera* studied so far are symbiotic with the cyanophycean endophytes. Both components of this association have been grown separately and resynthesized by Silvester and McNamara (1976). The isolated cyanophyte was probably *N. punctiforme*, but had considerably lower acetylene reduction activity than when in association with the host. The initiated nodules contain the cyanobacterium which is enclosed within membranes of the host. Within the nodule the phycobiont has a modified thylakoid distribution and high numbers of heterocysts compared with the vegetative cells.

10.3.6 Extracellular lytic compounds

Stewart and Brown (1969, 1970) worked with a species of *Cytophaga* (coded N-5) which was originally isolated from sewage. N-5 produces extracellular products which lyse certain cyanobacteria and green algae. It can grow in a chemically defined medium or on autoclaved or living algae, on agar or in liquid media. In addition, Gram-positive and Gram-negative bacteria were lysed or inhibited by N-5. However, lysis of cyanobacteria, using the double-agar overlay technique, took up to 2 weeks for the plaques to become visible. This contrasts markedly with cyanophages and the 'contact lytic' bacteria that give plaques within 5–7 days.

A second group of micro-organisms that produce extracellular inhibitory compounds, probably antibiotics, are septate fungi, as described by Redhead and Wright (1980). Similar fungi have been isolated from a reservoir near Oxford but these lost their lytic or inhibitory properties after storage and several subcultures (Daft, 1981).

During a survey on antagonistic effects of actinomycetes isolated from waste stabilization ponds, Safferman and Morris (1963) found that several cyanobacteria were susceptible to degradation. The actinomycetes produce antibiotics and other extracellular products that inhibit the growth or lyse the hosts. Their culture filtrates were particularly efficient on *Anacystis nidulans*. Soils have yielded a range of actinomycetes that are active against cyanobacteria. Mud from a salt marsh in England (Pearson and Jawad, 1982) and agricultural soils from India contained agents that produced lytic extracellular metabolites (Sharma *et al.*, 1984) Subsequently Whyte *et al.* (1985) found that *Streptomyces achromogenes* was the main species in soil that lysed *Anabaena cylindrica* and *Tolypothrix tenuis*. An actinomycete, coded AN6, originally isolated from Iraqi soil has a very wide host range (Al-Tai *et al.*, 1986). Cyanobacteria, fungi, bacteria and green algae are all susceptible to degradation by the extracellular products of AN6. This may indicate that AN6 produces extracellular compounds that are multi-target inhibitors. Superimposed on the effects of the extracellular products on the vegetative and heterocysts is the entrapment mechanism of AN6. Each AN6 colony is discrete, rounded, up to 1 mm in diameter and pale pink in

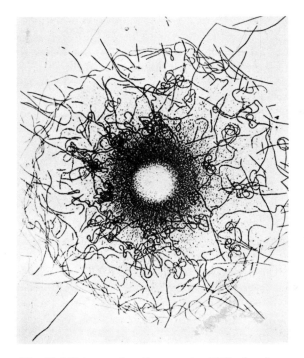

Fig. 10.8 Colony of actinomycete AN6 showing entrapped filaments of *Phormidium luridum* around the colony (×70).

colour. Each colony shows a less dense central area surrounded by a deeper pink radiating zone. The outermost layer is composed of a network of hyphae, being least dense at the periphery of the sphere of AN6. A high proportion of the prey filaments become entrapped in the outermost layer of the spheres (Fig. 10.8). The cyanobacterial filaments become coiled towards the central zone of the actinomycete colony. Filaments of the cyanobacterium are then surrounded by the hyphae of the actinomycete with cells of both prey and predator in direct contact. Lysis of the prey species can be detected after 24 h but this period varies with the type of cyanobacterium. Resistant cyanobacterial cells were not found in the treated cultures such as those found in cyanobacteria/cyanophage interactions (Barnet *et al.*, 1981).

10.3.7 Utilization of extracellular products

Lysobacter CP-1 has minimum growth requirements of a single carbohydrate, acetate, plus glutamic acid, histidine, asparagine, cysteine, tryptophan and arginine. Each of these are produced as extracellular products by cyanobacteria (Daft and Fallowfield, 1977). Hence, it is possible that the bacteria, such as CP-1, can be maintained by the waste products of the cyanophycean blooms. With their ability to lyse the cyanobacteria and to grow in the lysate, such bacteria have several strategies for their maintenance in nature. The dynamic balance between these two types of micro-organisms under natural conditions requires more research.

10.4 ECOLOGY

10.4.1 Introduction

The distribution of aquatic photosynthetic micro-organisms (such as members of the cyanophyceae), both spatially and temporally, is controlled mainly by oxygen and light. Within the photic zone, oxygen levels can be at a constant high, a constant low or alternate between aerobic and anaerobic conditions. Cyanobacteria can adapt to any of these three conditions. In part this is due to the cyanobacteria possessing photosystems I and II and that under microaerophilic conditions photosystem I can utilize hydrogen sulphide as the electron donor. Hence the two extremes with regard to oxygen availability can be exploited.

Light passing through the photic zone varies in its intensity and spectral composition. These changes induce the development of layers composed of different species within the water column. At the surface and at the highest intensity of illumination the bloom-forming cyanobacteria can withstand conditions of photo-oxidation. Benthic organisms, on the other hand, are subjected to a continuous precipitation of silt and organic debris which can reduce their exposure to light. However, many benthic cyanobacteria exhibit gliding movements which enable them to reposition themselves for efficient photosynthesis. Another adaptation of benthic cyanobacteria is the change of the vegetative cell wall from being hydrophobic to that of reproductive cells being hydrophilic. This permits the progeny to be dispersed over a wide area. On settling, the reproductive cells then change into hydrophobic vegetative cells and begin their benthic development (Shilo, 1982; Fattom and Shilo, 1984).

The position of cyanobacteria in the phytoplankton can be influenced by the possession of gas vacuoles. Gas vacuoles control the buoyancy of the cells and their collapse results in the disappearance of such species from the surface waters. Accumulation of nutrients at various interfaces in the water column is also associated with the abundance of certain cyanobacteria.

10.4.2 Distribution

The contribution of the cyanobacteria to primary production is difficult to assess. It is probable that they are responsible for much larger figures than have been considered previously. For example, Waterbury *et al.* (1979) found a marine chroococcalean belonging to the genus *Synechococcus* that is widespread throughout different parts of the world. This cyanobacterium can reach concentrations of 10^5 cells ml^{-1} in surface waters and be detected at a depth of 400 m.

Some examples of the widespread distribution and the great variety of habitats exploited by cyanobacteria are given below. When the Peel-Harvey Estuary in Australia received an increased nutrient run-off from the surrounding agricultural land this resulted in marked growths of macroalgae and increased fish catches. However, the subsequent blooms of the cyanobacterium, *Nodularia spumigena* reduced water clarity and also the fish catches were reduced (Lenanton *et al.* 1985). Laboratory studies on this bloom-forming cyanobacterium showed that phosphorus is the limiting nutrient and this nutrient within the blooms was derived from sediment release and nutrient recycling (Hamel and Huber, 1985). Nutrients introduced by land drainage into the Great Barrier Reef Lagoon are closely associated with the seasonal occurrence of cyanobacteria, such as an *Oscillatoria* sp. (Revelante and Gilmartin, 1982). After the cyanophycean bloom period within the lagoon, the diversity of the subsequent diatom growths was greatly reduced suggesting a residual inhibition effect caused by the same *Oscillatoria* sp. The chemical composition of the water in lowland Iraqi marshes is characterized as being slightly saline, highly calcareous and mesotrophic. Blooms of cyanobacteria are produced in September and form felts containing several species (Maulood *et al.* 1981). In the Mozhaisk reservoir in Russia, there were seasonal differences in the qualitative and quantitative composition of the primary producers. During maximum blooms the cyanobacteria dominated (Kanikovskaya and Sadchikov, 1985). Comparatively high salinity exists in the Don Juan Pond, an antarctic pond, which supports a population of *Oscillatoria nigrans* and *Chroococcus minutus*. In experimental systems, the cyanobacteria grow and withstand this stressed environment for up to 40 days. Ott and Waber (1982) consider that the cyanobacteria may be the major primary producers. Dense blooms of cyanobacteria that develop in Lake Hartbees-Port Dam in South Africa are covered by a crust of photo-oxidized cells. These 'hyper-scums' can persist for more than 100 days and may be anoxic, aphotic and have low pH values (Zohary, 1985). Thick algal mats can be formed on salt marsh soils, such as those described by Zedler (1982). These soils were covered by more cyanobacteria during the warmer periods. Coastal waters of the South Baltic produced cyanophycean blooms due to increased nutrients in the inflowing waters of the estuaries. To some degree the development of the blooms can be predicted mathematically (Plinski, 1984). In southeast Scotland *Rivularia atra* grows in the upper littoral zone of a sheltered bay. Here the source of organic phosphorus to the cyanobacterium is the decaying seaweeds. Khoja *et al.* (1984) consider that this nitrogen-fixing *R. atra* shows adaptation to its unstable substratum in that nitrogenase activity is greatly reduced on the exclusion of light.

10.4.3 Beneficial effects

Although a great deal of work has been done on understanding factors that may be involved in the development of cyanophycean blooms, with a view to reducing their deleterious effects, there are some situations in which cyanobacteria can be useful. An obvious example is in agriculture where atmospheric nitrogen fixation is exploited for plant nutrition. Attempts have been made in biotechnology to produce ammonia from light, air and water by N_2-fixing cyanobacteria. Musgrave *et al.* (1983) immobilized a cyanobacterium (*Anabaena sp.*) in calcium alginate gels and compared several methods for ammonium production. L-methionine-DL-sulphoximine was added to the culture medium in order to inhibit glutamine synthetase with the subsequent accumulation of NH_4^+ in the medium.

As mentioned previously (Lenanton *et al.*, 1985), higher productivity of fisheries may be linked with dense blooms of cyanobacteria. In the supratidal zone (between high and high spring tides) cyanobacteria fix and bind the quartz sand flats. Protection from erosion and desiccation of the cyanobacteria made this habitat, in North Germany, a rich source of food for marine and terrestrial invertebrates (Gerdes *et al.*, 1985). Reclamation properties of cyanobacteria in sodic soils has been shown by Subhashini and Kaushik (1981). Soils supporting the cyanobacteria produced (a) lower pH values, (b) reduced electrical conductivity and (c) less exchangeable sodium. Coupled with an increased nitrogen content of the soil, cyanobacteria may be exploited for the reclamation of certain soils.

10.4.4 Adaptation

As with other micro-organisms, cyanobacteria show a wide range of adaptabilities and tolerances to different environments. *Lyngbya* strains can develop a high tolerance to calcium and lead.

This tolerance is enhanced by the availability of inorganic phosphorus (Begchi *et al.*, 1985). Diesel fuel is more readily biodegraded by myxobacteria in the presence of cyanobacteria. Evidently, the extracellular products of the cyanobacteria stimulate the interactions between bacteria and the hydrocarbons (Gusev *et al.* 1982). Conversely water soluble fractions of lubricating oils can inhibit photosynthesis of *Anacystis marina*. In particular, used lubricating oil appears to be particularly toxic to phytoplankton (Bate and Crafford, 1985).

10.4.5 Control mechanisms

Microbial pathogens of cyanobacteria, along with grazing agents, probably reduce the growth rates of the hosts considerably. Whether any one predatory organism can completely eliminate a natural bloom is questionable. However, manipulation of certain parasites such as cyanophages may be exploited under certain circumstances. The use of mutant cyanophages, that completely lyse a wild-type population, is an example of a potential control method (Barnet *et al*, 1981). Interactions amongst predators and prey species discussed in Section 10.3 indicate some of the biological systems that might be exploited. A single agent may not possess all the attributes for complete control but a mixture of agents that combine different modes of attack on the host may offer better chances of success.

Management of the flow rate through a body of water is an effective control procedure. Mass culture of *Synechocystis* in a pond receiving swine waste could be controlled by the rate of waste loading. This cyanobacterium had a competitive advantage over other algal species particularly in the summer and autumn (Lincoln *et al.*, 1984). An increase in the flow rate of water into Lake Washington, in the USA, resulted in a decrease in phosphorus content and a reduction in the predicted total algal biomass (Welch and Patmont, 1980). Dilution water containing low levels of macronutrients was added to the lake on three occasions in 1977, and once in 1978. Average phosphorus levels flowing into the lake were reduced by 30% and the proportion of *Aphanizomenon* within the bloom from 96% to 68%. This change in species composition was attributed by the authors to the excretory products of the predominating cyanobacteria. Maintenance of the chemical quality flowing into large bodies of water and a greater understanding of microbial interactions may reduce some of the deleterious effects of cyanobacteria.

10.5 REFERENCES

Allen, M. M. and Stanier, R. Y. (1968). Selective isolation of blue-green algae from water and soil. *Journal of General Microbiology*, **51**, 203–209.

Al-Tai, A., Daft, M. J. and Stewart, W. D. P. (1986). Lysis of cyanobacteria, green algae and fungi by an actinomycete AN6 (in press).

Avnimelech, Y., Troeger, B. W. and Reed, L. W. (1982). Mutual flocculation of algae and clay evidence and implications. *Science*, **216**, 63–65.

Bader, F. G., Tsuchiya, H. M. and Fredrickson, H. (1976). Grazing of ciliates on blue-green algae: effects of ciliate encystment and related phenomenon. *Biotechnology and Bioengineering*, **XVIII**, 311–331.

Bagchi, S. N., Karamchandai, A. and Bisen, P. S. (1985). Isolation and preliminary characterisation of cadmium and lead tolerant strains of the cyanobacterium *Lyngbya* sp., lu 487. *FEMS Microbiology Letters*, **29**, 65–68.

Barnet, Y. M., Daft, M. J. and Stewart, W. D. P. (1981). Cyanobacteria–cyanophage interactions in continuous culture. *Journal of Applied Bacteriology*, **51**, 541–552.

Barnet, Y. M., Daft, M. J. and Stewart, W. D. P. (1984). The effect of suspended particulate material on cyanobacteria–cyanophage interactions in liquid culture. *Journal of Applied Bacteriology*, **56**, 109–115.

Bate, G. C. and Crafford, S. D. (1985). Inhibition of phytoplankton photosynthesis by the water-soluble fractions of used lubricating oil. *Marine Pollution Bulletin*, **16**, 401–404.

Brock, T. D. (1973). Evolutionary and ecological aspects of the cyanophytes, In *The Biology of Blue-Green Algae*, Botanical Monographs 9, N. G. Carr and B. A. Whitton (Eds). University of California Press, Berkeley, pp. 487–500.

Brock, T. D. (1976). Halophilic blue-green algae. *Archiv für Mikrobiologie*, **107**, 109–111.

Burnham, J. C., Collart, S. A. and Highison, B. W. (1981). Entrapment and lysis of the cyanobacterium *Phormidium luridum* by aqueous cultures of *Myxococcus xanthus* PCO2. *Archives of Microbiology*, **129**, 285–294.

Burnham, J. C., Collart, S. A. and Daft, M. J. (1984). Myxobacterial predation of the cyanobacterium *Phormidium luridum* in aqueous environments. *Archives of Microbiology*, **137**, 220–225.

Burnham, J. C., Stetak, I. and Locher, G. (1976). Extracellular lysis of the blue-green alga *Phormidium luridum* by *Bdellovibrio bacteriovorus*. *Journal of Phycology*, **12**, 306–313.

Caldwell, D. E. (1979). Associations between photosynthetic and heterotrophic prokaryotes in plankton. In *III International Symposium on Photosynthetic Prokaryotes.*, J. M. Nicols (Ed.), vol. 14, Oxford University Press, Oxford.

Canter, H. M. (1972). A guide to the fungi occurring on planktonic blue-green algae. In *Taxonomy and Biology of Blue-Green Algae*, T. V. Desikachary (Ed.), University of Madras, pp. 145–158.

Carmichael, W. W. (1984). Isolation culture and toxicity of freshwater cyanobacteria blue-green algae. First American Symposium on animal, plant and microbial toxins. *Toxicon*, **23**, 25–26.

Castenholz, R. W. (1981). Isolation and cultivation of thermophilic cyanobacteria. In *The Prokaryotes. A Handbook on Habitats, Isolation and Identification of Bacteria*, vol. 1, M. P. Starr, H. Stolp, H. G. Trüper, A. Balows and H. G. Schlegel (Eds), Springer-Verlag, Berlin, Heidelberg and New York, pp. 236–246.

Christensen, P. and Cook, F. D. (1978). *Lysobacter*, a new genus of nonfruiting, gliding bacteria with high base ratio. *International Journal of Systematic Bacteriology*, **28**, 367–393.

Chung, Y. T., Huang, C-M. and Chen, W-P. (1981). Studies on the cultivation of nitrogen fixing microorganisms. 1. Isolation of blue-green algae from sugarcane fields and measurements of their nitrogen fixing activities. *Report of the Taiwan Sugar Research Institute*, **92** 29–42.

Cowlinshaw, J. and Mrsa, M. (1975). Co-evolution of a virus-alga system. *Applied Microbiology*, **29**, 234–239.

Croft, W. N. and George, E. A. (1959). Blue-green algae from the Middle Devonian of Rhynie (Aberdeenshire). *Bulletin of British Museum (Natural History) Geology*, **3**, 341–353.

Daft, M. J. (1981). The use of natural pathogens to control nuisance algal blooms. *Report to the Water Research Council*.

Daft, M. J., Burnham, J. C. and Yamamoto, Y. (1985a). Lysis of *Phormidium luridum* by *Myxococcus fulvus* in continuous flow cultures. *Journal of Applied Bacteriology*, **59**, 73–80.

Daft, M. J., Burnham, J. C. and Yamamoto, Y. (1985b). Algal blooms: Consequences, and potential cures. *Journal of Applied Bacteriology Symposium Supplement*, 175S–186S.

Daft, M. J. and Fallowfield, H. J. (1977). Seasonal variations in algal and bacterial populations in Scottish freshwater habitats. In *The Oil Industry and Microbial Ecosystems*, K. W. A Chater and H. J. Sommerville (Eds), Heyden et Sou Ltd, pp. 41–50.

Daft, M. J., McCord, S. B. and Stewart, W. D. P. (1975). Ecological studies on algal-lysing bacteria in freshwater. *Freshwater Biology*, **5**, 577–596.

Daft, M. J. and Stewart, W. D. P. (1971). Bacterial pathogens of fresh water blue-green algae. *New Phytologist*, **70**, 819–829.

Drum, R. W. and Pankratz, S. (1965). Fine structure of an unusual cytoplasmic inclusion in the diatom genus *Rhopalodia*. *Protoplasma*, **60**, 141–149.

Duckett, J. G., Toth, R. and Soni, S. L. (1975). An ultrastructural study of the *Azolla, Anabaena azollae* relationship. *New Phytologist*, **75**, 111–118.

Fallowfield, H. J. (1981). Microbiological studies of three Scottish freshwater habitats. Ph. D. Thesis. University of Dundee.

Fattom, A. and Shilo, M. (1984). Hydrophobicity as an adhesion mechanism of benthic cyanobacteria. *Applied and Environmental Microbiology*, **47**, 135–143.

Fitzsimons, A. G. and Smith, R. V. (1984). The isolation and growth of axenic cultures of planktonic blue-green algae. *British Phycology Journal*, **19**, 157–162.

Fogg, G. E., Stewart, W. D. P., Fay, P. and Walsby, A. E. (1973). *The Blue-green Algae*, Academic Press, London, New York.

Frost, B. W. (1980). Grazing. In *The Phycological Ecology of Phytoplankton*, I. Morris (Ed.), University of California Press, Berkeley, pp. 465–491.

Gerdes, G., Krumbein, W. E. and Reineck, H. E. (1985). The depositional record of sandy vericoloured tidal flats Mellum Island, Southern North Sea, West Germany. *Journal of Sedimental Petrology*, **55**, 265–278.

Goldman, C. R. and Horne, A. J. (1983). *Limnology*, McGraw-Hill, New York.

Gorham, P. R., McLachlan, J. J., Hammer, U. T. and Kim, W. K. (1964). Isolation and culture of toxic strains of *Anabaena flos-aquae* (Lyngb) de Breb. *Verhandlungen der Internationalen Vereinging Theoretische und Angewandle Limnologie*, **15**, 796–804.

Granhall, U. and Berg, B. (1972). Antimicrobial effects of *Cellvibrio* on blue-green algae. *Archiv für Mikrobiologie*, **84**, 234–242.

Grilli-Caiola, M. G. (1980). On the phycobionts of the cycad coralloid-roots. *New Phytologist*, **85**, 537–544.

Grilli-Caiola, M. G. (1975). A light and electron microscope study of blue-green algae growing in the coralloid-roots of *Encephalartos altensteinii* and in culture. *Phycologia*, **14**, 25–33.

Grilli-Caiola, M. G. and Pellegrini, S. (1984). Lysis of *Microcystis aeruginosa* (Kutz) by *Bdellovibrio*-like bacteria. *Journal of Phycology*, **20**, 471–475.

Gusev, M. V., Koronelli, T. V., Lin'kova, M. A. and Il'inskii, V. V. (1982). The effect of excretions and cell biomass of cyanobacteria on hydrocarbon oxidising myxobacteria. *Mikrobiologiya*, **54**, 152–155.

Hamel, K. S. and Huber, A. L. (1985). Relationship of cellular phosphorus in the cyanobacterium *Nodularia* to phosphorus availability in the Peel-Harvey Estuarine system, Western Australia. *Hydrobiologia*, **124**, 57–64.

Hitch, C. J. B. and Millbank, J. W. (1975). Nitrogen metabolism in lichens. VII. Nitrogenase activity and heterocyst frequency in lichens with blue-green phycobiants. *New Phytologist*, **75**, 239–244.

Huang, T-C. and Chow, T-J. (1983). Axenic culture of the nitrogen fixing cyanobacteria isolated from rice field. *Proceedings of the National Science Council Republic of China, Part B*, **4**, 438–445.

Huang, T-C. and Wu, H-Y. (1982). Predation of amoebae on the filamentous blue-green algae. *Botanical Bulletin of the Academy of Science (Taipei)*, **22**, 63–70.

Kanikovskaya, A. A. and Sadchikov, A. P. (1985). Study of seasonal changes in the interrelationship of phytoplankton and bacterioplankton in Mozhaisk reservoir Russian-SFRS USSR 1. Seasonal changes in the population and biomass of plankton as a function of principal hydrobiological characteristics. *Biologia Naukii (Moscow)*, **7**, 55–62.

Khoja, T. M., Livingstone, D. and Whitton, B. A. (1984). Ecology of marine *Rivularia* population. *Hydrobiologia*, **108**, 65–74.

Lenanton, R. C. J., Loneregan, N. R. and Potter, I. C. (1985). Blue-green algal blooms and the commercial fishery of a large Australian estuary. *Marine Pollution Bulletin*, **16**, 477–482.

Lincoln, E. P., Koopman, B. and Hall, T. W. (1984). Control of a unicellular blue-green alga *Synechocystis* sp. in mass algal culture. *Aquaculture*, **42**, 349–358.

Mague, T. H., Mague, F. C. and Holm-Hansen, O. (1977). Physiology and chemical composition of nitrogen fixing phytoplankton in the central North Pacific Ocean. *Marine Biology*, **41**, 213–227.

Mague, T. H., Weare, N. M. and Holm-Hansen, O. (1974). Nitrogen fixation in the North Pacific Ocean. *Marine Biology*, **24**, 109–119.

Marshall, K. C. (1969). Studies by microelectrophoretic and microscope techniques of the sorption of illite and montmorillonite to rhizobia. *Journal of General Microbiology*, **56**, 301–306.

Maulood, B. K., Hinton, G. C. F., Whitton, B. A. and Al-Saadi, H. A. (1981). Algal ecology of the lowland Iraqi marches. *Hydrobiologia*, **80**, 269–276.

Musgrave, S. C., Kerby, N. W., Codd, G. A. and Stewart, W. D. P. (1983). Structural features of calcium alginate entrapped cyanobacteria modified for ammonia production. *European Journal of Applied Microbiology and Biotechnology*, **17**, 133–136.

Ott, J. A. R. and Waber, J. (1982). Survival of blue-green algae in a simulated antartic pond. *Environmental and Experimental Botany*, **22**, 9–14.

Pearson, H. W. and Jawad, A. L. M. (1982). Interactions between blue-green algae and other microorganisms isolated from mud cores taken from the River Dee salt marsh in Cheshire, England, U. K. *British Phycological Journal*, **17**, 237.

Peters, G. A. (1975). The *Azolla-Anabaena* relationship, III. Studies on metabolic capabilities and a further characterisation of the symbiont. *Archives of Microbiology*, **103**, 113–122.

Plinksi, M. (1984). Predictive model of cyanophyta invasion in coastal waters of the South Baltic. *Polish Archives of Hydrobiology*, **30**, 177–188.

Redhead, K. and Wright, S. J. L. (1980). Lysis of the cyanobacterium *Anabaena flos-aquae* by antibiotic producing fungi. *Journal of General Microbiology*, **119**, 95–101.

Revelante, M. and Gilmartin, M. (1982). Dynamics of phytoplankton in the Great Barrier Reef Lagoon, Australia. *Journal of Plankton Research*, **4**, 47–76.

Rippka, R., Waterbury, J.B. and Stanier, R. Y. (1981). Isolation and purification of cyanobacteria: Some general principles. In *The Prokaryotes. A handbook on Habitats, Isolation and Identification of Bacteria*, vol. 1, M. P. Starr, H. Stolp, H. G. Trüper, A.

Balows and H. G. Schlegel (Eds), Springer-Verlag, Berlin, Heidelberg and New York, pp. 212–220.

Rodgers, G. A. and Stewart, W. D. P. (1977). The cyanophyte–hepatic symbiosis 1. Morphology and physiology. *New Phytologist*, **78** 441–458.

Safferman, R. S. and Morris, M. E. (1963). Algal virus isolation. *Science, New York*, **140**, 679–680.

Santoro, T. and Stotzky, G. (1968). Sorption between microorganisms and clay minerals as determined by the electrical sensing zone particle analyser. *Canadian Journal of Microbiology*, **14**, 299–307.

Schopf, J. W. and Walter, M. R. (1982). Origin and early evolution of cyanobacteria: The geological evidence. In *The Biology of Cyanobacteria*, N. G. Carr and B. A. Whitton (Eds), Blackwell Scientific, Oxford.

Sharma, P., Henriksson, E., Rosswall, T. and Vadehra, D. V. (1984). Soil colloidal particles as carriers of inhibitory agents against the cyanobacterium *Anabaena* in an Indian Soil. *OIKOS* **43**, 235–240.

Shilo, M. (1970). Lysis of blue-green algae by *Myxobacter*. *Journal of Bacteriology*, **104**, 453–461.

Shilo, M. (1982). Photosynthetic microbial communities in aquatic ecosystems. *Transactions of the Royal Society London, B. Biological Science*, **297**, 565–574.

Silvester, W. B. and McNamara, P. J. (1976). The infection process and ultrastructure of the *Gunnera–Nostoc* symbiosis. *New Phytologist*, **77**, 135–141.

Stanier, R. Y., Kunisawa, R., Mandel, M. and Cohen-Bazire, G. (1971). Purifications and properties of unicellular blue-green algae (order Chroococcales). Bacteriological Reviews, **35**, 171–205.

Stewart, J. R. and Brown, R. M. (1969). *Cytophaga* that kills or lyses algae. *Science, New York*, **164**, 1523–1524.

Stewart, J. R and Brown, R. M. (1970). Killing of green and blue-green algae by a non-fruiting myxobacterium *Cytophaga* N-5. *Bacteriological Proceedings*, **18**.

Stewart, W. D. P. and Daft, M. J. (1977). Microbial pathogens of cyanophycean blooms. *Advances in Aquatic Microbiology*, **1**, 177–219.

Stewart, W. D. P. and Rodgers, G. A. (1977). The cyanophyte-hepatic symbiosis II. Nitrogen fixation and the interchange of nitrogen and carbon. *New Phytologist*, **78**, 459–471.

Stewart, W. D. P., Rowell, P. and Rai, A. N. (1983). Cyanobacteria-eukaryotic plant symbiosis. *Annales de Microbiologie (Institut Pasteur)*, **134B**, 205–228.

Subhashini, D. and Kaushik, B. D. (1981). Amelioration of sodic soils with blue-green algae. *Australian Journal of Soil Research*, **19**, 361-366.

Sykes, I. K. and Williams, S. T. (1978). Interactions of actinophage and clays. *Journal of General Microbiology*, **108**, 97–102.

Taylor, D. L. (1984). The black band disease of Atlantic reef corals. 2. Isolation cultivation and growth of *Phormidium-corallyticum*. *Marine Ecology*, **4**, 321–328.

Waterbury, J. B. and Stanier, R. Y. (1978). Patterns of growth and development in pleurocapsalean cyanobacteria. *Microbiological Reviews*, **42**, 2–44.

Waterbury, J. B. and Stanier, R. Y. (1981). Isolation and growth of cyanobacteria from marine hyper saline environments. In *The Prokaryotes. A Handbook on Habitats, Isolation and Identification of Bacteria*, vol. 1, M. P. Starr, H. Stolp, H. G. Trüper, A. Balows and H. G. Schlegel (Eds), Springer-Verlag, Berlin, Heidelberg and New York, pp. 221–223.

Waterbury, J. B., Watson, S. W., Guillard, R. R. L. and Brand, L. E. (1979). Widespread occurrence of a unicellular, marine planktonic, cyanobacterium. *Nature* **277**, 293–294.

Welch, E. B. and Patmont, C. R. (1980). Lake restoration by dilution Moses Lake, Washington, U.S.A. *Water Research*, **14**, 1317–1326.

Whyte, L. G., Maule, A. and Cullimore, D. R. (1985). Method for isolating cyanobacteria-lysing streptomycetes from soil. *Journal of Applied Bacteriology*, **58**, 195–197.

Yamamoto, Y. (1981). Observations on the occurrence of microbial agents which cause lysis of blue-green algae in Lake Kasumigaura. *Japanese Journal of Limnology*, **42**, 20–27.

Zedler, J. B. (1982). Salt marsh algal mat composition spatial and temporal comparisons. *Bulletin of Southern Californian Academy of Science*, **81**, 41–50.

Zohary, T. (1985). Hyperscums of the cyanobacterium *Microcystis aeruginosa* in a hypertrophic Lake Hartees-poort Dam, South Africa. *Journal of Plankton Research*, **7**, 399–410.

Methods in Aquatic Bacteriology
Edited by B. Austin
© 1988 John Wiley & Sons Ltd.

11

Sulphate-Reducing Bacteria

N. S. Battersby

Water Research Centre, Medmenham Laboratory, Henley Road, Medmenham, PO Box 16, Marlow, Buckinghamshire SL7 2HD, England

11.1 INTRODUCTION

The sulphate-reducing bacteria (SRB) are a diverse group of bacteria, unified by their use of sulphate as a terminal electron acceptor during anaerobic respiration. The group contains several morphological types of differing metabolic capabilities, indicating that genetically unrelated taxa have acquired the capacity of dissimilatory sulphate reduction (Pfennig *et al.*, 1981). Although this suggests that the ability to use sulphate as an electron acceptor is plasmid-coded, only two out of eleven marine SRB examined to date possess cryptic plasmids (Postgate *et al.*, 1984; Battersby and Stewart, unpublished results). SRB can be found in many anaerobic environments (e.g. soils, sewage sludge digesters, sediments, rumen). However, their principal habitat is the marine environment, where the concentration of sulphate in seawater is high and fairly constant (approximately 29 mM at a salinity of 35‰).

SRB are strict anaerobes and require a low redox potential of around $-100\,mV$ before growth can occur (Postgate, 1984). However, the possession of oxygen-protective enzymes such as superoxide dismutase and catalase (Hardy and Hamilton, 1981) enable them to survive in oxygenated sea water at levels of 10^1–$10^2\,ml^{-1}$ (Kimata *et al.*, 1955; Hardy, 1981). High oxygen consumption within a sediment and hydrographic conditions which prevent the exchange of water can lead to the formation of anoxic bodies of seawater where higher numbers of SRB can occur. Schneider (1977) reported from 10^2 to 10^3 SRB ml^{-1} in water samples from the Baltic Sea, and SRB activity has been recorded in stagnant basins off the Californian and Venezuelan coasts (Fonselius, 1976). The ubiquity of SRB in seawater has led to major problems of corrosion, toxicity and spoilage

269

in the offshore oil and gas industries, where large volumes of seawater are used as ballast and in the maintenance of well pressure (Battersby et al., 1985c).

SRB are most active within the upper 20 cm of coastal and shelf sediments where sulphate is replenished from the overlying seawater and heterotrophic bacteria maintain reducing conditions. Here, acetate, lactate and propionate-oxidizing SRB occur at populations of around $10^2-10^3 \, ml^{-1}$ (Laanbroek and Pfennig, 1981). As sulphate is generally non-limiting in such sediments (approximately $>2 \, mM \, SO_4^{2-}$, Goldhaber and Kaplan, 1974), SRB can outcompete methanogenic bacteria for their common substrates, acetate and hydrogen (Nedwell, 1982). Under such conditions, sulphate respiration is the predominant means of anaerobic oxidation and can account for 50% of organic carbon mineralization (Jørgensen, 1977b). In coastal sediments, SRB can oxidize as much organic matter to CO_2 as aerobic respiration (Jørgensen, 1982). The substrates (electron donors) used for dissimilatory sulphate reduction in marine sediments are fermentation end-products (e.g. short-chain fatty acids and H_2) produced by other sedimentary bacteria. Data by Sørensen et al. (1981) and Christensen (1984) indicate that the quantitative contribution of these substrates to sulphate reduction is: acetate (40–65%), propionate (10–20%), butyrate (5–20%), isobutyrate (8%) and hydrogen (5–10%). If the oxidation of butyrate and propionate is incomplete, however, then acetate oxidation can account for almost two-thirds of the sulphate reduction occurring in the sediment. SRB therefore have four important roles in anaerobic, marine food chains: acetate oxidation, diverse substrate oxidations (see below), hydrogen production (under conditions of sulphate limitation) and hydrogen utilization (Peck and Odom, 1984).

The relative importance of sulphate reduction in marine sediments decreases with increasing water depth and distance from land. The general lowering of the metabolic rate in such sediments leads to a downwards migration of the oxic zone and the supply of more recalcitrant organic matter to the SRB. However, sulphate reduction can occur in oxic marine sediments in reduced microsites of 50–100 μm diameter within faecal pellets and organic aggregates (Jørgensen, 1977a; Battersby et al., 1985b).

SRB play a key role in the oceanic sulphur cycle by removing much of the approximately $10^{11} \, kg$ sulphate which enters the sea each year as river run-off via terrestrial erosion. Sulphate reduction and precipitation with Ca^{2+} maintains the concentration of sulphate in seawater at around 29 mM. This is reflected in the sulphate:chlorine (a conservative element) ratio being constant at $0.14 \pm 0.4\%$ (Wilson, 1975). SRB have also been identified as the principal methylators of mercuric ions (Hg^{2+}) in anoxic, estuarine sediments leading to the formation of toxic and biomagnification-prone methylmercury compounds (Compeau and Bartha, 1985). A knowledge of the types and numbers of SRB present in anoxic marine environments is therefore of prime importance to microbiologists, geochemists, petroleum engineers and toxicologists.

11.2 ISOLATION METHODS

11.2.1 Exposure to oxygen

As SRB are obligately anaerobic organisms, the exposure of sediment and water samples to the atmosphere is undesirable. Although SRB can survive in oxic conditions (see above), the degree of oxygen tolerance is strain specific and can vary from <3 hours for *Desulfococcus multivorans* and vegetative *Desulfotomaculum* spp. (Cypionka *et al.*, 1985) to 72 hours for *Desulfovibrio vulgaris* (Hardy and Hamilton, 1981). Samples from sediment cores can be removed with minimal exposure to the air by the use of truncated, plastic syringes inserted through pre-drilled holes in the core tube (Hines and Buck, 1982). The holes are sealed with waterproof tape during sampling. Small sediment cores and anoxic water samples can be handled within an anaerobic cabinet under an atmosphere of N_2 or $N_2/CO_2/H_2$. Any manipulations of SRB media should be performed in an anaerobic cabinet or under a stream of nitrogen or other inert gas using the syringe needle gassing technique of Hungate (1969).

11.2.2 Temperature

Most marine SRB grow best at between 25 and 35 °C, with an upper temperature limit of around 40 °C (Widdel, 1980). They are usually handled at room temperature and cultivated at 30 °C. However, Trüper *et al.* (1969) have reported the isolation of several strains of *Desulfovibrio*, from hot brine deeps in the Red Sea, that grew optimally at 40 °C. Although SRB activity has been observed at low temperatures, reports of psychrophilic SRB in the literature are rare (e.g. Barghoorn and Nichols, 1961). ZoBell (1958) reported that *Desulfovibrio* strains isolated from cold, marine environments had optimum growth temperatures of 20–30 °C. However, little is known about the temperature characteristics of non-lactate-utilizing SRB, such as *Desulfobacter* sp. (see below).

11.2.3 Growth substrate

As SRB are differentiated by their growth substrates and the extent by which these substrates are oxidized, the choice of substrate is the most important selective factor for determining what types of SRB are isolated from a particular environment (Pfennig et al., 1981). Until quite recently, the use of media based on lactate, and reduced by thioglycollate, had led to the regular isolation of only one genus of SRB (Desulfovibrio) from marine samples (Trüper et al., 1969; Hardy, 1981). However, the use of modified cultural techniques (low phosphate, CO_2–bicarbonate buffered, Na_2S reduced media) has led to the isolation of a range of morphologically and physiologically diverse SRB from marine and brackish sediments (Widdel and Pfennig, 1981, 1982; Widdel et al., 1983). There is growing evidence that it is these novel SRB which are most active in situ and not Desulfovibrio spp. (Sørensen et al., 1981; Taylor and Parkes, 1985). A range of media suitable for the cultivation of the main nutritional types of SRB is given in Section 11.5.1.

11.3 ENRICHMENT OF SRB

11.3.1 Batch culture enrichment

Traditionally, SRB have been isolated using batch culture enrichment techniques. Samples of sediment or seawater are incubated in a selective liquid medium until SRB growth is evident. This necessitates the cultivation of samples in media that contain a substrate concentration far in excess of that found in the environment. Selection of bacteria in batch cultures is dependent on maximum specific growth rate (μ_{max}). SRB which have a high substrate affinity (K_s) but low μ_{max} are not enriched (Veldkamp, 1976). As such organisms are probably the most active in the marine environment (Bull and Brown, 1979), batch culture enrichments may yield an inaccurate picture of the active population of SRB in situ (Parkes and Taylor, 1983).

11.3.2 Continuous culture enrichment

In continuous culture (chemostat) enrichments, growth medium is pumped into a culture vessel at the same rate (f) as culture is removed. The culture volume in the system remains constant (v) and under steady-state conditions the dilution rate (D) is equal to the specific growth rate of the bacterial culture ($D = f/v = \mu$). This growth rate is determined by the concentration of the growth-limiting substrate (e.g. acetate, lactate), which is in turn determined by the dilution rate. As sulphate reduction in marine sediments is usually limited by the availability of electron donors (Nedwell, 1982), chemostat enrichments of SRB are usually carried out under conditions of carbon limitation. The use of continuous culture enrichments enables bacteria to grow at a submaximal rate (as is usually the case in nature), under low nutrient concentrations. The theory and application of continuous culture methods and chemostat enrichments is discussed in greater detail by Veldkamp (1976) and Bull and Brown (1979).

Anaerobic chemostat enrichments have been used to isolate acetate-oxidizing SRB from estuarine sediments (Keith *et al.*, 1982). Descriptions of anaerobic chemostats for the cultivation of pure cultures of SRB can be found in Leban *et al.* (1966), Ware and Postgate (1971), Vosjan (1975), Postgate (1984) and Wardell *et al.* (1986).

A simple chemostat for the enrichment of SRB is shown in Fig. 11.1. The unit is based on a 1 l 'Quickfit' glass vessel (Scientific Glass Laboratories, Stoke-on-Trent) as sulphide corrodes stainless steel. Anaerobic conditions in the system are maintained by gassing the chemostat and medium reservoir with 10 ml min^{-1} oxygen-free nitrogen. The culture is kept at 30 °C by means of a circulating water bath and stirred magnetically. Butyl

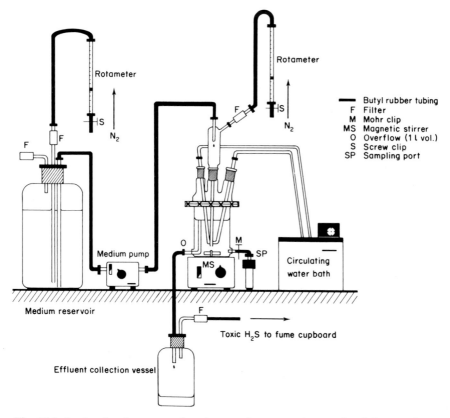

Fig. 11.1 A simple chemostat for the continuous culture of sulphate-reducing bacteria.

rubber tubing is used throughout, as silicone rubber tubing is permeable to air and can allow the diffusion of oxygen into the culture (Postgate, 1984).

The pH of the culture can be maintained by the use of buffered media, by varying the proportion of CO_2 in a N_2/CO_2 gas phase (Ware and Postgate, 1971; Ingvorsen et al., 1984) or by the use of a pH electrode and acid/alkali dosing pumps. However, the high levels of sulphide encountered in SRB enrichments can react with the silver ions of the reference electrolyte to form black Ag_2S. This blocks the porous ceramic plug of the liquid junction between the culture sample and the reference electrode leading to inaccurate pH readings (Bühler and Baumann, 1986).

The redox potential of the medium may be poised to approximately $< -100\,mV$ prior to inoculation with sterile sodium sulphide solution (final concentration 5 mM). The medium in the reservoir is often left unpoised, however, as the high level of sulphide produced during SRB growth acts as a redox buffer and can inhibit SRB growth at concentrations $>10\,mM$ (Brown et al., 1973). Full redox control is possible by the use of a platinum Eh electrode coupled with nitrogen sparging (raises Eh) and a Na_2S supply which lowers Eh (Brown et al., 1973). However, nitrogen sparging as described above should minimize any problems of growth inhibition due to accumulated sulphide in the culture.

Published dilution rates for continuous cultures of marine SRB range from $0.01\,h^{-1}$ for Desulfobacter postgatei (Ingvorsen et al., 1984) to $0.80\,h^{-1}$ for a Desulfovibrio desulfuricans strain (Vosjan, 1975). A suitable dilution rate for the enrichment of SRB from marine sediments is $0.035\,h^{-1}$ (Keith et al., 1982; Battersby and Stewart, unpublished results). The vessel is inoculated by filling it to the overflow with a suspension of approximately 50 g of sediment suspended in seawater or medium. Alternatively, the vessel is filled with seawater. These amounts are for a 1 l chemostat. Medium is then pumped into the vessel at a rate which will give the desired dilution rate (e.g. $35\,ml\,h^{-1}$ into a 1 l pot for $D = 0.035\,h^{-1}$). The contents of the chemostat are sampled at 2-day intervals and SRB enumerated by agar-shake or roll-tubes (see below).

11.4 ISOLATION OF PURE CULTURES OF SRB

Pure cultures of SRB are commonly isolated from enrichment cultures by three methods: roll-tubes (Hungate, 1969), agar-shake tubes and by plating out on agar plates. The roll-tube method is superior to the other two techniques as it requires relatively inexpensive equipment, has a high success rate and is not subject to artifacts caused by the mixotrophic growth of SRB on H_2 (see below). SRB colonies are also clearly visible (cf. agar-shake tubes).

11.4.1 Roll-tubes

The medium is removed from the autoclave as soon as the pressure is down to atmospheric and held at 45 °C under a stream of oxygen-free nitrogen. Aliquots of 4 ml are dispensed into 25 ml roll-tube bottles (Astell Scientific, London), using either the Hungate technique, or an anaerobic dispenser based on a gas-tight, pre-set syringe (Herbert and Gilbert, 1984). The roll-tubes are closed with rubber stoppers (Astell Scientific, London), while removing the gassing needle. Resazurin-containing medium should be used to enable the rejection of any tubes that contain pink (oxidized) medium.

Enrichment (0.5 ml) is injected into one bottle, the contents mixed and 0.5 ml from this bottle inoculated into a second bottle. This is repeated for a maximum of eight bottles. The resealing ability of stoppers, after puncturing by syringe needle, can be extended by coating the surface of the stopper with a layer of 'Silcoset 151' silicone rubber sealing compound (Ambersil Ltd, Basingstoke). After inoculation, the bottles are placed on a water-cooled roller (Astell Scientific, London) which coats the inside of each bottle with a thin layer of agar. The bottles are incubated at 30 °C until discrete SRB colonies are visible in the agar of roll-tubes containing high dilutions of the enrichment culture. Well-separated colonies are picked from the agar using suction through a sterile Pasteur pipette, which has had its capillary bent through 90°. This is most easily performed using a dissection microscope. The colonies are inoculated into liquid medium and incubated until growth occurs.

Cultures which produce hydrogen sulphide in a medium containing sulphate and a suitable substrate contain SRB. These should be checked for purity by microscopical examination, the absence of aerobic growth and the production of only black colonies when grown in tubes of tryptone-soya agar supplemented with NaCl and 0.05% $(NH_4)_2Fe(SO_4)_2 \cdot 6H_2O$ (Postgate, 1984). Impure cultures should be passed through further sets of roll-tubes.

11.4.2 Agar-shake tubes

Test tubes containing 10 ml of medium are held at 45 °C under anaerobic conditions. Serial 'Pasteur pipette dilutions' are carried out on six to eight tubes, using a few drops of enrichment culture as inoculum. The tubes are stoppered and the contents mixed by inverting the tubes twice. They are then placed in cold water to solidify the agar and plugged with 2 cm of agar to prevent the ingress of air into the culture. Tubes are incubated in the air until SRB colonies are visible (see below). SRB colonies are obtained by breaking a tube and picking out the colonies from the agar using suction and a Pasteur pipette. Reinoculation and purity checks are performed as detailed above. This method however, often results in contamination problems during tube breakage and SRB recovery.

11.4.3 Plating out

Within the O_2-free atmosphere of an anaerobic cabinet, enrichment cultures of SRB can be plated out using standard techniques. However, the success of this method is dependent on strict anaerobiosis within the chamber and reduced agar plates (e.g. pour plates and allow to cool in the anaerobic cabinet). Under these conditions, the technique yields pure cultures of SRB at a high success rate. However, the use of nutritionally selective media is not possible, as many SRB can utilize the H_2 present in the atmospheres of most commercial anaerobic cabinets (e.g. 80% N_2/10% CO_2/10% H_2) as an electron donor (see below). Media containing $FeSO_4 \cdot 7H_2O$ as an indicator of SRB growth should be used.

11.5 MEDIA FOR SRB

11.5.1 Fatty acid and benzoate-oxidizing SRB

The use of modified cultural techniques by Widdel and co-workers has led to the isolation of a wide range of nutritionally diverse SRB. The medium used to enrich and isolate these organisms is a CO_2–bicarbonate buffered basal medium, to which are added reductants, substrate and growth-promoting factors (Pfennig et al., 1981).

Mineral salts

KH_2PO_4	0.2	g
KCl	0.3	g
NH_4Cl	0.3	g
$CaCl_2 \cdot 2H_2O$	0.15	g
$MgCl_2 \cdot 6H_2O$	3.0	g
Na_2SO_4	3.0	g
NaCl	20.0	g
Distilled water	97.0	ml

This mineral salts base is autoclaved at 121 °C for 15 min and cooled under 90% N_2/10% CO_2. The following solutions are then added from sterile, stock solutions under anaerobic conditions:

1 ml trace element solution

HCl (25% v/v)	6.5 ml

FeCl$_2$·4H$_2$O	1.5 g (dissolve in acid)
Distilled water	993 ml
CuCl$_2$·2H$_2$O	15 mg
Na$_2$MoO$_4$·2H$_2$O	25 mg
NiCl$_2$·6H$_2$O	25 mg
H$_3$BO$_3$	60 mg
ZnCl$_2$	70 mg
MnCl$_2$·4H$_2$O	100 mg
CoCl$_2$·6H$_2$O	120 mg

(Autoclave.)

1 ml selenite solution

Na$_2$SeO$_3$	3 mg
NaOH	0.5 g
Distilled water	1 l

(Autoclave.)

30 ml CO$_2$–NaHCO$_3$ solution

NaHCO$_3$	8.5 g
Distilled water	100 ml

(Saturate with CO$_2$ and filter-sterilize into a gas-tight container.)

3 ml sodium sulphide solution

Na$_2$S·9H$_2$O	12 g (wash crystals with distilled water and blot dry with filter paper)
Distilled water	100 ml

(Autoclave in a gas-tight bottle under N$_2$.)

1 ml vitamin solution

Biotin	1 mg
p-aminobenzoic acid	5 mg
B$_{12}$	5 mg
Thiamin	10 mg
Distilled water	100 ml

(Filter-sterilize.)

The pH of the medium is adjusted to 7.2 using sterile 2M Na$_2$CO$_3$ or HCl. The basal medium is then amended with autoclaved substrate to the final concentration shown below:

Acetate-oxidizing SRB:	20 mM CH$_3$ COONa·3H$_2$O
Benzoate-oxidizing SRB:	4 mM C$_6$H5 COONa
Butyrate-oxidizing SRB:	10 mM CH$_3$CH$_2$CH$_2$COONa
Lactate-oxidizing SRB:	20 mM CH$_3$CH(OH) COONa

Palmitate-oxidizing SRB: $2\,mM\ CH_3(CH_2)_{14}\ COOH$
(use $10\,ml\,l^{-1}\,0.2\,M$ solution, neutralized with $0.78\,g$ NaOH. Clear solution by heating before use.)

Propionate-oxidizing SRB: $10\,mM\ C_2H_5COONa$

As an aid to SRB growth, $1\,ml\,l^{-1}$ of sterile 3% (w/v) sodium dithionite can be added to the complete medium prior to inoculation. The dithionite solution is made up in O_2-free distilled water and anaerobically filter-sterilized into N_2-gassed bottles. The solution is dispensed anaerobically. The medium can be solidified with $10\,g\,l^{-1}$ Agar No. 1 (Oxoid). As this medium does not contain ferrous sulphate as an indicator of sulphide production (see below), SRB colonies can only be identified tentatively by shape, colour and microscopical examination of cell suspensions (Laanbroek and Pfennig, 1981). The high level of c-type cytochromes in many SRB leads to characteristically salmon-pink colonies in this medium (R. Herbert, personal communication). Black SRB colonies can be obtained in this medium by adding $66\,\mu g\ FeSO_4\cdot7H_2O$ to $9\,ml$ of complete medium (Taylor and Parkes, 1985).

11.5.2 Filamentous gliding SRB (*Desulfonema* spp.)

Desulfonema spp. can be enriched from marine sediments using either $15\,mM$ acetate or $4\,mM$ benzoate (*Desulfonema magnum*) as substrates. The medium used is based on that described above, with increased concentrations of $MgCl_2\cdot6H_2O$ ($5\,g\,l^{-1}$) and $CaCl_2\cdot2H_2O$ ($1\,g\,l^{-1}$) for *Desulfonema magnum* (Widdel *et al.*, 1983). $1\,ml\,l^{-1}$ of the organic acid mixture of Widdel *et al.* (1983) is also added to the medium:

Isobutyric acid	$5\,g$
n-Valeric acid	$5\,g$
Isovaleric acid	$5\,g$
2-Methylbutyric acid	$5\,g$
n-Caproic acid	$2\,g$
n-Heptanoic acid	$2\,g$
n-Octanoic acid	$2\,g$
Succinic acid	$45\,g$
Distilled water	$1\,l$

(Neutralize with NaOH and autoclave.)

Growth of filamentous SRB is stimulated by the use of artificial sediments, which support gliding movement (Widdel, 1980). Widdel *et al.* (1983) used either a viscous medium $2\,g\,l^{-1}$ agar) or a light sediment of precipitated aluminium phosphate ($5 \, \text{ml} \, l^{-1}$ $0.2 \, \text{M}$ $AlCl_3 \cdot 6H_2O$; pH adjusted with $1.6 \, \text{ml}$ $1 \, \text{M}$ Na_2CO_3). A detailed description of the purification techniques for filamentous SRB is given by Widdel (1983). Enrichment culture (1 ml) is filtered through an electron microscope grid, washed with 20 ml of liquid medium and resuspended in fresh medium. The filaments of *Desulfonema magnum* approximately $6-8 \, \mu\text{m}$ diameter) are visible under the dissection microscope and can be serially washed in tubes of fresh media using a finely drawn Pasteur pipette. The filaments are then disrupted (to increase cell numbers) and further purified by passage through shake-tubes of soft agar ($0.7 \, \text{g} \, l^{-1}$ agar) supplemented with $6 \, \text{mM}$ $Na_2S_2O_4$ (to protect the cells from the effects of Br_2, see below). The surface layer of agar, in tubes showing irregular, fluffy colonies of filamentous SRB, is removed to within 8 mm of the SRB colony and the surface sterilized with bromine vapour for 1 min. After removal of the Br_2 with a stream of N_2, the colony is transferred into fresh medium.

11.5.3 Simple media for *Desulfovibrio* and *Desulfotomaculum* spp.

Prior to the studies of Widdel and co-workers, the SRB were differentiated into two unrelated genera: *Desulfovibrio* and *Desulfotomaculum* (see section 11.7.1). These organisms (with the exception of *Desulfotomaculum acetoxidans*) grow well in simple lactate based media as described by Pankhurst (1971), Postgate (1984) and Herbert and Gilbert (1984). Such media usually contain $0.5 \, \text{g} \, l^{-1}$ $FeSO_4 \cdot 7H_2O$ as an indicator of SRB growth; the sulphide produced by SRB growth forming a precipitate of FeS, which blackens liquid media and yields black colonies in agar tubes and plates. This means that sodium sulphide cannot be used as a reducing agent. It is generally replaced by $0.1 \, \text{g} \, l^{-1}$ ascorbate plus $0.1 \, \text{g} \, l^{-1}$ thioglycollate. However, the former is only a weak reducing agent, while the latter can be inhibitory to many SRB strains (Khosrovi and Miller, 1975; Pfennig *et al.*, 1981). Nevertheless, such media give acceptable, percentage recoveries of around 80% for pure cultures of marine *Desulfovibrio* spp. (Battersby *et al.*, 1985a).

The high levels of inorganic salts present in these media often result in the formation of precipitates. These are beneficial to SRB growth, as they provide reduced micro-environments where SRB can attach and grow (Pankhurst, 1971; Herbert and Gilbert, 1984; Battersby *et al.*, 1985a). SRB growth is also stimulated by the inclusion of $1 \, \text{g} \, l^{-1}$ yeast extract in the medium. The constituent amino acids of yeast extract are thought to increase the solubility of Fe^{2+} by chelation (see below).

A suitable liquid medium for the cultivation of marine *Desulfovibrio* spp. is medium B of Postgate (1984), supplemented with $0.4 \, \text{M}$ NaCl and containing 30 mM lactate:

KH_2PO_4	0.5 g
NH_4Cl	1.0 g
$CaSO_4$	1.0 g
$MgSO_4 \cdot 7H_2O$	2.0 g
NaCl	23.4 g
$CH_3CH(OH)COONa$ (70% w/v soln.)	3.5 ml
Yeast extract	1.0 g
$FeSO_4 \cdot 7H_2O$	0.5 g
Na thioglycollate	0.1 g
Na ascorbate	0.1 g
Resazurin (1 mg ml^{-1} soln.)	1 ml
Distilled water	1 l

This medium can also be made up with seawater (Postgate, 1984) or 75% sea/source water + 25% distilled water (Herbert and Gilbert, 1984). The pH of the medium is adjusted to 7.5 with 2 M NaOH prior to autoclaving at 121 °C for 15 min (final pH approximately 7.2). To prevent oxidation of the medium, containers should be tightly closed or gassed with nitrogen, as soon as the autoclave pressure is down to atmospheric.

A suitable solid medium for the growth of *Desulfovibrio* and *Desulfotomaculum* spp. is medium E of Postgate (1984) with 20 mM sulphate, 30 mM lactate and 0.4 M NaCl:

KH_2PO_4	0.5 g
NH_4Cl	1.0 g
Na_2SO_4	2.9 g
$CaCl_2 \cdot 6H_2O$	1.0 g
$MgCl_2 \cdot 6H_2O$	2.0 g
NaCl	23.4 g
$CH_3CH(OH)COONa$ (70% w/v soln.)	3.5 ml
Yeast extract	1.0 g
$FeSO_4 \cdot 7H_2O$	0.5 g
Na thioglycollate	0.1 g
Na ascorbate	0.1 g
Resazurin (1 mg ml^{-1} soln.)	1 ml
Agar No. 1 (Oxoid)	10 g
Distilled water	1 l

The pH of the medium is adjusted to 7.6 with 2 M NaOH, prior to autoclaving as described above.

The original media of Postgate (1984) have been amended by the addition of 1 mg l^{-1} resazurin as a redox indicator. Medium which is colourless is at an Eh of $< -110\,mV$ (Costilow, 1981) and can be used. However, medium which shows a pink colouration has been oxidized and should be discarded.

The use of API medium (Anon., 1975), which is available commercially in a dehydrated form (Difco) cannot be recommended. The redox potential of this medium is poorly poised with $0.1\,g\,l^{-1}$ ascorbatle and weakly buffered by K_2HPO_4 $(0.01\,g\,l^{-1})$. The Eh of the medium is $> -110\,mV$, even immediately after autoclaving. If used, API medium should be prepared with 75% seawater, and supplemented with $1.0\,g\,l^{-1}$ ascorbate and a higher level of phosphate $(0.5\,g\,l^{-1}\,KH_2PO_4;$ Hardy, 1981).

The high level of iron and precipitation in Postgate's medium B precludes its use for continuous cultures of SRB. Postgate (1984) recommends a medium containing low amounts of Ca^{2+}, Mg^{2+} and Fe^{2+}, with 1 mM citrate as a chelating agent (medium C). In the absence of chelation, the iron in SRB cultures is present mainly as solid FeS and growth can be limited by the rate of dissolution of this solid phase. Although citrate decreases the concentration of Fe^{2+} in solution, the availability of Fe^{2+} is increased due to the instantaneous adjustment of the equilibrium (Spencer, 1957):

$$\text{complexed } Fe^{2+} \rightleftharpoons \text{citrate} + Fe^{2+}$$

Medium C, with lactate replaced by acetate, has been used in chemostat enrichments of *Desulfobacter* spp. from estuarine sediments (Keith *et al.*, 1982). However, care must be taken when using chelated media as bacteria vary in their tolerance to chelating agents (Bridson and Brecker, 1970). Several strains of *Desulfovibrio desulfuricans* isolated from a Scottish sea-loch could not grow in Postgate's medium C. Citrate interferred with respiratory growth on lactate plus sulphate but had no effect on fermentative growth with pyruvate in the absence of sulphate (Battersby, 1983).

The composition of a carbon-limited version of medium C is shown below:

KH_2PO_4	0.5 g
NH_4Cl	1.0 g
Na_2SO_4	2.9 g
$CaCl_2 \cdot 6H_2O$	60 mg
$MgSO_4 \cdot 7H_2O$	60 mg
NaCl	23.4 g
$CH_3CH(OH)COONa$ (70% w/v soln.)	3.5 ml
Yeast extract	10 mg
$FeSO_4 \cdot 7H_2O$	4 mg
Trisodium citrate	0.3 g
Trace elements (see 11.5.1)	1 ml

| Resazurin (1 mg ml^{-1} soln.) | 1 ml |
| Distilled water | 1 l |

The pH of the medium is adjusted to 7.6 prior to autoclaving at 121 °C for 15 min (90 min for 20 l batches for chemostat use). The medium is cooled under a stream of nitrogen (see Fig. 11.1). Medium C is sometimes cloudy after autoclaving but should clear on cooling. The pH of the medium during SRB growth reportedly remains steadier if Na_2SO_4 is replaced by $(NH_4)_2SO_4$ (Pankhurst, 1971). However, the pH of continuous cultures of *Desulfovibrio desulfuricans* growing in the above medium, at a dilution rate of 0.05 h^{-1}, tended to remain stable at around 8.0 (Battersby and Stewart, unpublished observations).

11.5.4 Hydrogen-oxidizing (mixotrophic) *Desulfovibrio* spp.

The isolation of mixotrophic SRB, which use H_2 as an electron donor, and acetate and CO_2 for cell carbon (70% and 30% respectively) has been described by Badziong *et al.* (1978). The original medium has been modified slightly for marine strains:

Basal medium

KH$_2$PO$_4$	0.5 g
(NH$_4$)$_2$SO$_4$	5.3 g
MgSO$_4$·7H$_2$O	0.2 g
CaCl$_2$·2H$_2$O	0.1 g
NaCl	23.4 g
CH$_3$COONa	2 g
Trace element solution	10 ml
Resazurin (1 mg ml^{-1} soln.)	1 ml
Distilled water	1 l

(Autoclave and cool under oxygen-free nitrogen.)

Trace element solution

Nitriloacetic acid	12.8 g (adjust to pH 6.5 with NaOH)
FeCl$_2$·4H$_2$O	0.3 g
CuCl$_2$	20 mg
MnCl$_2$·4H$_2$O	0.1 g
CoCl$_2$·6H$_2$O	0.2 g
ZnCl$_2$	0.1 g
H$_3$BO$_3$	10 mg
Na$_2$MoO$_4$·2H$_2$O	10 mg
Distilled water	1 l

The following sterile additions are then made: 50 ml 8% (w/v) Na_2CO_3, 5.5 ml 25% (v/v) HCl and 1 ml 0.5 M $Na_2S_2O_4$. The pH of the medium should now be 7.2. Enrichment of SRB (1 ml inocula) is carried out in butyl-rubber stoppered tubes of 23 ml capacity, containing 10 ml of complete medium under an atmosphere of 80% H_2/20% CO_2. The tubes are incubated at 30 °C, with reciprocal shaking to equilibrate the liquid and gas phases. Pure cultures of hydrogen-oxidizing SRB are obtained using roll-tubes of the above medium ($10 \, g \, l^{-1}$ Agar no. 1), with a 80% H_2/20% CO_2 gas phase. Black SRB colonies appear with this medium.

11.6 ENUMERATION OF SRB

As mentioned above, the choice of medium for SRB enumeration will determine the types of SRB recovered. Postgate's medium E, for example, will yield predominantly *Desulfovibrio* spp. when inoculated with marine samples. The importance of acetate as a substrate for marine sulphate reduction (see above) and the inability of many acetate-oxidizing SRB to grow in simple lactate-based media (Pfennig *et al.*, 1981) suggest that SRB enumeration should involve the medium in Section 11.5.1 with acetate as the substrate. Parallel counts with propionate and lactate should also be performed to yield a fuller picture of the resident SRB population (Laanbroek and Pfennig, 1981; Battersby *et al.*, 1985b). This is because marine SRB genera such as *Desulfovibrio* (Postgate, 1984) and *Desulfobulbus* (Widdel and Pfennig, 1982) cannot use acetate as an electron donor for sulphate reduction. Similarly, although most marine SRB can utilize lactate as a growth substrate (Widdel, 1980), it is not used by *Desulfonema magnum* (Widdel *et al.*, 1983), *Desulfobacter postgatei* (Widdel and Pfennig, 1981) or *Desulfovibrio baarsii* (Widdel, 1980).

Care must be taken when interpreting SRB counts obtained with lactate as it can be rapidly fermented to propionate and acetate by other marine bacteria (Laanbroek and Pfennig, 1981; Taylor and Parkes, 1985). Hence, counts of SRB in lactate-containing media may not give an accurate estimation of the number of lactate-utilizers *in situ*. This is a particular problem in liquid culture. Representative SRB from high dilutions in lactate-containing media should be examined further.

11.6.1 Agar-shake and roll-tubes

Agar shake tubes, as described in Section 11.4.2, have been used frequently to enumerate SRB (Jørgensen 1977a,b; Laanbroek and Pfennig, 1981; Postgate, 1984; Battersby *et al.*, 1985b). Decimal dilutions of the sample are best performed in the enumeration medium (minus substrate), although reduced diluents can be used (see Winfrey *et al.*, 1981; Hines and

Buck 1982). Tubes should be set up in at least duplicate. The use of media containing $FeSO_4 \cdot 7H_2O$ as an indicator of SRB growth is recommended. Counts of black SRB colonies should be performed twice daily, as sulphide production from one active colony can blacken a whole tube, making accurate counting difficult. Ideally, between 50–200 SRB colonies should be counted in a tube (Postgate, 1984). However, this number of SRB can easily blacken the whole of the agar. Counting can be made easier by the use of roll-tubes (Section 11.4.1) and a dissection microscope. Excess medium blackening in roll-tubes can be reduced, before a final count, by removing the stopper and exposing the inside of the tube to the atmosphere.

11.6.2 Most probable number (MPN) estimations

Despite the requirement for a large number of tubes and the inherent low precision of the technique, MPN counts of SRB are reported widely in the literature (Nedwell and Floodgate, 1972; Tezuka, 1979; Schroder and Van Es, 1980; Hardy, 1981; Battersby *et al.*, 1985a). MPN methods have the advantage that small or slow growing SRB colonies are not masked by blackening caused by more rapidly growing cells (see above).

Ten-fold dilutions of sample are added to replicate bottles of medium containing $FeSO_4 \cdot 7 H_2O$ as an indicator of SRB growth. Inoculation can be as 1 ml aliquots to five bottles/dilution (see Alexander, 1965); or as 10 ml, 1 ml and 0.1 ml aliquots to 3, 4 or 5 bottles/dilution (see Taylor, 1962). Bottles showing SRB growth (blackening) after 2–4 weeks incubation are recorded as positives. The most probable number of SRB present in the original sample is calculated from the pattern of positive and negative bottles using standard MPN tables (or see above two references). The availability of programmable calculators and inexpensive, personal computers has made it possible to use arrangements of bottles and dilutions outwith those listed in standard MPN tables (see Koch, 1981; Deschacht, 1983).

MPN methods allow SRB to be enumerated even when other bacteria overgrow the culture. This allows the use of heterotrophic bacteria as biological reductants. Such a system removes the need for potentially toxic reducing agents and enables the medium to be re-reduced if allowed to oxidize. An MPN method, based on a simple medium reduced by respiring *Pseudomonas putida* cells and ascorbate, has been described by Battersby *et al.* (1985a):

$FeSO_4 \cdot 7 H_2O$	0.3 g
NH_4Cl	0.3 g
KH_2PO_4	0.2 g

Bacto-casamino acids (Difco)	1.0 g
Na ascorbate	1.0 g
$CH_3CH(OH)COONa$ (70% w/v soln.)	3.6 ml
Resazurin (BDH Chemicals, Poole)	1 tablet
Seawater	1 l

The pH of the medium is adjusted to 7.5 with 2 M NaOH. The medium is dispensed into 25 ml bottles before sterilization at 121 °C for 15 min. The tops of the bottles are sealed tightly as soon as the autoclave pressure is down to atmospheric. Decimal dilutions of the sample are made in the medium and 10, 1 and 0.1 ml aliquots added to five bottles/volume (remove 10 ml of medium from bottles to receive 10 ml inocula). Aliquots of 0.1 ml of a concentrated suspension of *Ps. putida* are then added to each bottle. The suspension is prepared by suspending the growth from three overnight plate cultures of *Ps. putida* (Tryptone Soya Agar, 30 °C) in 3 ml of sterile 0.9% (w/v) NaCl. The bottles are filled with sterile medium and sealed tightly. Incubation is at 30 °C. Bottles showing blackening are scored as positive SRB growth and the MPN of SRB in the original sample calculated as described above. Recoveries of marine SRB from estuarine sediments and from mixed and pure cultures, using this method, were comparable to other published techniques, even when the MPN medium was deliberately oxidized prior to inoculation (Battersby *et al.*, 1985a).

11.6.3 Plate counts

As has been mentioned in Section 11.4.3, the successful cultivation of SRB on the surface of agar plates is dependent on strict anaerobiosis. This is most easily attained within an anaerobic cabinet, although residual H_2 in the atmosphere of such cabinets means that nutritionally selective media are of little use. This problem can be overcome by incubating inoculated plates in an anaerobic jar, which has been flushed out with oxygen-free nitrogen. As the atmosphere within the jar now contains no H_2, the palladium catalyst within the jar cannot scavenge any O_2 that may enter. Hence, the jar should be leak-tight and maintained under a positive pressure of N_2. H_2 can be prevented from reaching the surface of spread plates by overlaying with agar. Twenty millilitres of semi-solid medium (0.5% agar) per plate has been used in the enumeration of SRB from saltmarsh sediments (Abdollahi and Nedwell, 1979). Tsuneishi and Goetz (1958) utilized an oxygen-scavenging overlayer of agar, inoculated with *Serratia marcescens*, during the cultivation of *Desulfovibrio aestuarii* on membrane filters incubated on agar plates.

Plates are inoculated usually with 0.1 ml of sample dilution, which is then allowed to dry into the agar (if the plate is to be overlayered).

Alternatively Pankhurst (1971) described a surface-drop method in which 8–10 drops of 0.02 ml volume are placed onto the surface of the agar. One plate/dilution is used and the plates incubated at 30 °C for between 1 and 2 weeks. Around 100 SRB colonies/plate are counted and the number of SRB in the original sample calculated from the average number of SRB/drop and the relevant dilution. This method gave recoveries of *Desulfovibrio vulgaris*, which were comparable to an MPN method (Pankhurst, 1971). Counting of blackened plates can be made easier by exposing the plates to the air for 2 h prior to counting (Iverson and Olson, 1984).

11.7 CHARACTERIZATION OF SRB

SRB were first isolated in pure culture by van Delden in 1903. The use of lactate-based media in the subsequent 70 years led to the isolation of only two, nutritionally limited genera: *Desulfovibrio* and *Desulfotomaculum*. The genus *Desulfotomaculum* contains anaerobic, Gram-negative rods, which are motile by means of peritrichous flagella (Campbell and Postgate, 1965). They are distinguished from *Desulfovibrio* spp. by sporulation, the possession of b-type cytochromes, the absence of desulfoviridin (a bisulphite reductase) and generally lower DNA base ratios (37.5 to 46.6 mol%). *Desulfotomaculum* spp. are differentiated by DNA base ratios, electron donors for sulphate reduction and growth by the fermentation of pyruvate in sulphate-free media. Members of the genus *Desulfovibrio* are anaerobic, Gram-negative vibrios (some spirilloid forms), which do not form endospores and are motile by polar flagellation (Postgate and Campbell, 1966). The majority of species possess desulfo-viridin and c-type cytochromes, and have higher DNA base ratios than *Desulfotomaculum* spp. (46.1 to 61.2 mol%). *Desulfovibrio* spp. are again differentiated by their growth substrates, DNA base ratios and NaCl requirement.

However, this taxonomy is confused due to the use of impure cultures in the past (which broadens the range of substrates) and to the limited number of diagnostic characters useful for differentiation. In a survey of 92 SRB isolates from freshwater, marine and highly saline environments, Skyring *et al.* (1977) found that only 25 out of 116 biochemical and physiological characteristics used were of any taxonomic use. The description of a range of new morphological and physiological types of SRB by Widdel *et al.* has complicated the taxonomy of the group, especially as few of these novel SRB have been studied taxonomically. The confusion, which exists in SRB taxonomy, can be illustrated by reference to two of the taxonomically significant characters proposed by Postgate and Campbell (1966) for classification of the genus *Desulfovibrio*, i.e. presence of desulfoviridin and c-type cytochromes. The bisulphite reductase, desulfo-viridin, is not present in *Desulfovibrio baculatus* (Rozanova and Nazina, 1976), *Desulfovibrio desulfuricans* strain Norway 4 (Miller and Saleh, 1964) or the *Desulfovibrio*-like 'sapovorans' group (Widdel, 1980). However, desul-foviridin has been detected in the newly described species *Desulfococcus multivorans* and *Desulfonema limicola*, and it now appears that the various types of bisulphite reductase are distributed at random throughout the SRB (Peck, 1984). Likewise, c-type cytochromes are found not only in *Desulfovibrio* spp. but also in *Desulfobacter*, *Desulfobulbus*, *Desulfococcus* and *Desulfonema* spp. (Widdel, 1980).

Pfennig *et al.* (1981) proposed a new working classification of the SRB as a physiological-ecological group. The group is differentiated by substrates utilized, the extent to which these substrates are oxidized, morphology and DNA base composition. Group 1 contains all the 'traditional' species of SRB which can utilize lactate for sulphate reduction but not acetate butyrate or propionate. The group contains the genera *Desulfovibrio* and *Desulfotomaculum* (including the lactate-negative *Dm. acetoxidans*). The individual species are distinguished as before. Group 2 contains SRB which can use acetate, butyrate, propionate or benzoate as electron donors for sulphate reduction and SRB which can grow chemoautotrophically on $H_2 + CO_2$. Group 2 is subdivided into SRB which perform an incomplete oxidation of at least one C3 to C18 fatty acid and SRB which oxidize their substrates completely to CO_2.

Characters useful in the differentiation of SRB are noted below. Further details on the various types of SRB can be found in the references given.

11.7.1 Group 1: *Desulfovibrio* and *Desulfotomaculum*

Desulfovibrio (Postgate and Campbell, 1966; Postgate, 1984)
Gram negative
Non-sporulating
Vibrio-shaped (some spirilloid forms) length 3–5 μm × width *ca.* 0.5–1 μm
Motile (polar flagella)
Cytochrome c_3 as principal cytochrome (majority of spp.)
Contain desulfoviridin (some exceptions)
Lactate incompletely oxidized to acetate
Substrates incompletely oxidized
Acetate, butyrate or propionate not utilized as electron donors

Desulfovibrio desulfuricans
G + C 55.3 mol%
Growth on malate plus sulphate
Growth by fermentation of pyruvate or choline in absence of sulphate

Desulfovibrio vulgaris
G + C 61.2 mol%
No growth on malate plus sulphate
No growth on pyruvate or choline in absence of sulphate

Desulfovibrio salexigens
G + C 46.1 mol%
Growth on malate plus sulphate
No growth on pyruvate or choline in absence of sulphate
Media containing 2.5–5% w/v NaCl required for growth
Generally exhibit a high resistance (*ca.* 0.1–1.0 g/l) to bis-(*p*-chlorophenyl-
    diguanide)-hexane diacetate (Hibitane; ICI PLC, Manchester).

Desulfovibrio gigas
G + C 60.2 mol%
Length 5–10 μm × width 1.2–1.5 μm, often in chains as spirilla
No growth on malate plus sulphate
No growth on pyruvate or choline in absence of sulphate

Desulfovibrio baculatus
G + C 56.8 mol%
Rod shaped
Desulfoviridin absent
Growth on malate plus sulphate
No growth on pyruvate or choline in absence of sulphate

Desulfovibrio africanus
G + C 61.2 mol%
Growth on malate plus sulphate
No growth on pyruvate or choline in absence of sulphate

Desulfotomaculum (Campbell and Postgate 1965; Postgate, 1984):
Gram-negative
Terminal or subterminal sporulation
Rods
Motile (generally peritrichous flagella)
Cytochrome b as principal cytochrome
Do not contain desulfoviridin
Optimum growth temperatures *ca.* 30–40 °C
Lactate incompletely oxidized to acetate ⎫
Substrates incompletely oxidized   ⎬   except *Dm. acetoxidans*
Acetate, butyrate or propionate not utilized ⎪
    as electron donors   ⎭

Recovered only rarely from marine waters (ZoBell and Rittenberg, 1948; Ochynski and Postgate, 1963), generally found in freshwaters, rumen, faeces, thermal regions and some spoiled foods.

Desulfotomaculum acetoxidans (Widdel and Pfennig, 1977):
G + C 37.5 mol%
Oxidizes acetate and butyrate to CO_2
No growth on lactate plus sulphate
Length 3.5–9 μm × width 1–1.5 μm
Normally of intestinal origin, introduced into sea by faecal pollution (Widdel and Pfennig, 1981)

Desulfotomaculum orientis
G + C 41.7 mol%
Curved rod length 5 μm × width 1.5 μm
No growth on pyruvate in absence of sulphate

11.7.2 Group 2a: incomplete oxidation of substrate

Desulfobulbus propionicus (Widdel and Pfennig, 1982):
Gram-negative
Lemon/onion shaped cells with pointed ends (length 1.8–2 μm × width 1–1.3 μm)
Propionate (some strains also use butyrate) incompletely oxidized to acetate
No growth on acetate plus sulphate
G + C 59.9 mol%
Growth by fermentation of pyruvate or lactate in the absence of sulphate
Motile (single, polar flagellum) and non-motile forms
Non-sporulating

Vibrioid 'sapovorans group' (Pfennig *et al.*, 1981):
Gram-negative
Vibrios or spirilloid forms (length 3–6 μm × width 0.5–1.5 μm)
Motile (single or lophotrichous, polar flagella)
Oxidize C4 to C14 fatty acids
Even-numbered fatty acids oxidized to acetate
Odd-numbered fatty acids oxidized to acetate + propionate
G + C 52.7 mol% (one strain)
Motile (polar flagella, either singly or lophotrichous)
Growth by fermentation of pyruvate (but not lactate) in the absence of sulphate

11.7.3 Group 2b: complete oxidation of substrate

Desulfobacter postgatei (Widdel and Pfennig, 1981)
Gram-negative
Rods or ellipsoidal cells (length $1.7-3.5\,\mu m \times 1-2\,\mu m$)
Acetate (often only electron donor) completely oxidized to CO_2
Some strains may use ethanol or lactate (slow growth) as electron donors
Non-sporulating
$G + C$ 45.9 mol %
Motile (single, polar flagella) or non-motile
Desulfoviridin absent

Desulfonema limicola (Widdel *et al.*, 1983)
Gram-positive (cell walls characteristic of Gram-negative bacteria)
Gliding filaments (length $0.05-1\,mm \times 3\,\mu m$ diameter)
Autotrophic growth on $H_2 + CO_2$
Growth on acetate, propionate and higher fatty acids (up to C14)
Benzoate not utilized
$G + C$ 34.5 mol %

Desulfonema magnum (Widdel *et al.*, 1983)
Gram-positive (cell walls characteristic of Gram-negative bacteria)
Gliding filaments (length $0.1-2\,mm \times 6-8\,\mu m$ diameter)
No autotrophic growth
Growth on higher fatty acids (up to C10)
Growth on benzoate
$G + C$ 41.6 mol%

Desulfosarcina variabilis (Widdel, 1980)
Gram-negative
Sarcina packets or free-living rods/ovoid cells (length $1.5-2.5\,\mu m \times$ width
    $1-1.5\,\mu m$)
Non-motile
Autotrophic growth on $H_2 + CO_2$
Growth on higher fatty acids (up to C14)
Growth on benzoate
Growth by fermentation of pyruvate or lactate in the absence of sulphate
Cells sensitive to light

Desulfococcus multivorans (Widdel, 1980)
Gram-negative
Spherical cells (diameter $1.5-2.2\,\mu m$)
Non-motile
Growth on formate does not require further carbon sources

Growth on higher fatty acids (up to C14)
Growth on benzoate
Growth by fermentation of pyruvate or lactate in the absence of
  sulphate
$G + C$ 57.4 mol%

11.8 CHARACTERIZATION TESTS FOR SRB

Morphology. Young cultures of SRB in a suitable medium must be examined
as pleomorphism caused by age or environment is common in the group
(Postgate, 1984).

Substrate utilization. SRB are inoculated into a suitable medium containing
the substrate, with growth being compared to basal medium only blanks.
Substrates which support growth over at least three subcultures are
recorded as positive. Yeast extract is normally replaced by a trace element
solution, and vitamins if required (e.g. see Section 11.5.1). As yeast extract
reportedly assists the metabolic adjustment of *Desulfovibrio* spp. to
unfamiliar carbon sources (Postgate, 1984), it may be retained at a reduced
concentration of 0.001% (w/v). However, as yeast extract at this concentration can support the growth of approximately 10^7 SRB ml^{-1}, comparison
with blank bottles is essential. Utilization of H_2 is tested in tubes or bottles
filled to one-third their volume with medium containing a source of cell
carbon (e.g. acetate, yeast extract or CO_2 in the case of autotrophic
growth). Incubation is under an atmosphere of 80% H_2/20% CO_2, with
shaking (Widdel and Pfennig, 1981).
  Many *Desulfovibrio* spp. may require an additional source of organic
carbon when growing on substrates such as butanol, iso-butanol, ethanol,
formate, propanol or oxamate (Wake *et al.*, 1977; Postgate, 1984). These
'incomplete' or mixotrophic substrates act as electron donors only. Such
growth can be detected by the fact that an additional carbon source (e.g.
acetate, yeast extract) is always required for growth—which is greater than
that on the carbon source alone (Postgate, 1984).
  Growth by fermentation of lactate, pyruvate or choline is tested in media
containing no sulphate. Growth in basal medium and test bottles is often
difficult to distinguish (Hardy, 1981), although fermentation of choline can
be detected by the smell of the trimethylamine end-product (Stams *et al.*,
1985).

Sporulation. A presumptive test for sporulation is the viability of cells after
exposure to 80 °C for 5 min (Postgate, 1984). The presence of spores is

confirmed by phase contrast microscopy (spores appear phase-bright). The overnight exposure of *Desulfotomaculum* cultures to air can often promote sporulation.

Desulfoviridin. A simple test for the detection of this bisulphite reductase pigment has been described by Postgate (1984). Cells from *ca.* 15 ml of culture are harvested by centrifugation and resuspended in a few drops of supernatant. One drop of 2 M NaOH is added under ultraviolet light (365 nm). A red fluorescence indicates the presence of desulfoviridin. Alternatively, the pigment can be detected by a peak at 630 nm when the spectrum of the cytoplasmic fraction of centrifuged cells is examined (Widdel and Pfennig, 1981).

Cytochromes. The extraction of SRB pigments and the determination of their difference spectra are described by Badziong *et al.* (1978) and Widdel and Pfennig (1981). Cytochrome b shows peaks at 420, 525 and 556 nm in redox-difference spectra, whilst c-type cytochromes have maxima at 415, 521–523 and 550–554 nm (Badziong *et al.*, 1978; Hardy, 1981, Widdel and Pfennig, 1981, 1982; Widdel *et al.*, 1983; Postgate, 1984).

DNA base composition. DNA is usually extracted from lysozyme-treated SRB cells by the method of Marmur (1961). DNA base compositions are determined by thermal denaturation in standard saline buffer (De Ley, 1970).

11.9 OTHER METHODS FOR STUDYING SRB IN THE MARINE ENVIRONMENT

11.9.1 Fatty acid biomarkers

Analysis of the lipid fatty acids of SRB has been postulated as a means of estimating the different types of SRB present in marine sediments (Taylor and Parkes, 1985). Earlier work had shown that certain fatty acids could act as biomarkers for *Desulfobacter*, *Desulfobulbus* and *Desulfovibrio* spp. (Taylor and Parkes, 1983). The fatty acid composition of *Desulfobulbus* varied with the growth substrate but always contained large amounts of $C_{15:0}$ and $C_{17:\omega 8}$ fatty acids (5–23% and 11–52%, respectively). *Desulfovibrio desulfuricans* when grown on lactate or H_2 as electron donor contained large amounts of iso-$C_{15:0}$ and iso-$C_{17:1\omega 7}$ fatty acids. Similar fatty acid profiles for *Desulfovibrio desulfuricans* have been described by Boon *et al.* (1977) and Parkes and Taylor (1983). The fatty acid composition of *Desulfobacter postgatei* growing on acetate was dominated by even numbered, straight chain acids such as $C_{14:0}$ and $C_{16:0}$ and contained *ca.* 11% of 10 Me-$C_{16:0}$ which was not detected in the two other SRB investigated. However, even chain fatty acids are very common in bacteria and such limitations in the use of SRB fatty acid biomarkers when investigating complex marine environments are discussed in detail by Taylor and Parkes (1985). Nevertheless, the similar fatty acid profiles of *Desulfobacter* growing on acetate and the zone of sulphate reduction in a Scottish sea-loch (Parkes and Taylor, 1983) support the currently held view that acetate is the main substrate for sulphate reduction in marine sediments (see above). The use of fatty acid biomarkers for the study of mixed populations of SRB holds much promise at the time of writing.

11.9.2 Immunofluorescence

The enumeration of SRB by conventional means can take from 2 to 4 weeks. Immunofluorescence is one possible technique for the rapid detection and enumeration of SRB in marine samples. However, serological studies of *Desulfovibrio* and *Desulfotomaculum* spp. have yielded conflicting results, with Postgate (1984) reporting cross-reactions between strains of the same species and with other members of the same genus. Strain-specific and genus-specific antigens have been observed in *Desulfovibrio* spp. (Abdollahi and Nedwell, 1980; Norqvist and Roffey, 1985). An indirect fluorescent antibody technique for the detection of SRB has been described by Smith (1982). Antisera to *Desulfovibrio desulfuricans, D. salexigens, D. vulgaris* and *Desulfotomaculum nigrificans* tended to be strain specific. A polyvalent cocktail containing antisera to the above bacteria gave positive fluorescence against homologous bacteria, five *Desulfovibrio desulfuricans* strains, two *D. africanus* strains, *D. salexigens* and *D. vulgaris.* There was no interference from *Streptococcus faecalis, Escherichia coli, Flavobacterium aquatile* or *Pseudomonas fluorescens.*

11.9.3 Radiotracer techniques

The activity of SRB in the marine environment can be measured directly by radiotracer incubations using $^{35}SO_4^{2-}$ (Jørgensen, 1978). The inhibition of $^{35}SO_4^{2-}$ reduction with 20 mM sodium molybdate and the measurement of accumulating fatty acids (e.g. acetate, propionate) by gas chromatography has led to a better understanding of the substrates utilized by SRB *in situ* (Sørensen *et al.*, 1981; Christensen, 1984). However, these techniques yield little information on the numbers of bacteria present in the area under study.

Sand *et al.* (1975) described a radioisotope enrichment culture technique to evaluate the active component of SRB in marine samples. Samples are incubated in API broth containing $Na_2^{35}SO_4$ and the residual $^{35}SO_4^{2-}$ determined by liquid scintillation counting. The loss of sulphate is converted to the number of *Desulfovibrio desulfuricans* cells that would reduce an equivalent amount of sulphate in the same time by means of a series of growth curves covering a range of inocula. Counts obtained by this method, of SRB from an anoxic stream, were comparable to anaerobic plate counts. However, other natural samples may not behave in an analogous way to *D. desulfuricans*.

11.9.4 Impedance measurements

Oremland and Silverman (1979) reported on the change in electrical impedance ratios (R_i) of sediment slurries and pure cultures of SRB during SRB growth. R_i was defined as:

$$R_i = \frac{I_r}{I_r + I_s}$$

where I_s is the impedance of the sample and I_r is the impedance of the reference.

Increase in R_i in pure cultures of *Desulfovibrio aestuarii* correlated with growth and sulphide production. Both R_i and $^{35}S^{2-}$ production in surface marine sediments were stimulated by the addition of lactate and inhibited by the presence of 5–20 mM molybdate. The authors suggest that the continuous recording of R_i by their machine (a Bactomatic model 32, Bactomatic Inc., California) could serve as a useful tool in the evaluation of substrates utilized for sulphate reduction in marine systems. However, care must be taken when interpreting the results of stimulation experiments using relatively high concentrations (75 mM) of a fermentable substrate such as lactate (see Section 11.6).

11.10 REFERENCES

Abdollahi, H. and Nedwell, D. B. (1979). Seasonal temperature as a factor influencing bacterial sulfate reduction in a saltmarsh sediment. *Microbial Ecology*, **5**, 73–79.

Abdollahi, H. and Nedwell, D. B. (1980). Serological characteristics within the genus *Desulfovibrio*. *Antonie van Leeuwenhoek, Journal of Microbiology and Serology*, **56**, 73–83.

Alexander, M. (1965). Most-probable-number method for microbial populations. In *Methods of Soil Analysis*, Part 2, *Chemical and Microbiological Properties*, C. A. Black (Ed.), American Society of Agronomy, Wisconsin, pp. 1467–1472.

Anon. (1975). *Recommended Practice for Biological Analysis of Subsurface Injection Waters*, 3rd edn, API RP 38, American Petroleum Institute, Dallas.

Badziong, W., Thauer, R. K. and Zeikus, J. G. (1978). Isolation and characterization of *Desulfovibrio* growing on hydrogen plus sulphate as the sole energy source. *Archives of Microbiology*, **116**, 41–49.

Barghoorn, E. S. and Nichols, R. L. (1961). Sulphate-reducing bacteria and pyritic sediments in Antarctica. *Science*, **134**, 190.

Battersby, N. S. (1983). *Dissimilatory Nitrate and Sulphate Reduction in Marine Sediments*. Ph.D. thesis, Heriot-Watt University.

Battersby, N. S., Stewart, D. J. and Sharma, A. P. (1985a). A simple most probable number method for the enumeration of sulphate-reducing bacteria in biocide containing waters. *Journal of Applied Bacteriology*, **58**, 425–429.

Battersby, N. S., Malcolm, S. J., Brown, C. M. and Stanley, S. O. (1985b). Sulphate reduction in oxic and sub-oxic North-East Atlantic sediments. *FEMS Microbiology Ecology*, **31**, 225–228.

Battersby, N. S., Stewart, D. J. and Sharma, A. P. (1985c). Microbiological problems in the offshore oil and gas industries. *Journal of Applied Bacteriology Symposium Supplement*, **1985**, 227S–235S.

Boon, J. J., De Leeuw, J. W., Hoek, G. J. and Vosjan, J. H. (1977). Significance and taxonomic value of iso and anteiso monoenoic fatty acids and branched β-hydoxy acids in *Desulfovibrio desulfuricans*. *Journal of Bacteriology*, **129**, 1183–1191.

Bridson, E. Y. and Brecker, A. (1970). Design and formulation of microbial culture media. In *Methods in Microbiology*, vol. 3A, J. R. Norris and D. W. Ribbons (Eds), Academic Press, London, pp. 229–295.

Brown, D. E. Groves, G. R. and Miller, J. D. A. (1973) pH and Eh control of cultures of sulphate-reducing bacteria. *Journal of Applied Chemistry and Biotechnology*, **23**, 141–149.

Bühler, H. and Baumann, R. (1986). Improved pH control by new pH sensor and an automatic calibration. *International Labmate*, **11**, 34–37.

Bull, A. T. and Brown, C. M. (1979). Continuous culture applications to microbial biochemistry. In *International Review of Biochemistry, Microbial Biochemistry*, vol. 21, J. R. Quayle (Ed.), University Park Press, Baltimore, pp. 177–226.

Campbell, L. L. and Postgate, J. R. (1965). Classification of the spore-forming sulfate-reducing bacteria. *Bacteriological Reviews*, **29**, 359–363.

Christensen, D. (1984). Determination of substrates oxidized by sulfate reduction in intact cores of marine sediments. *Limnology and Oceanography*, **29**, 189–192.

Compeau, G. C. and Bartha, R. (1985). Sulfate-reducing bacteria: principal methylators of mercury in anoxic estuarine sediment. *Applied and Environmental Microbiology*, **50**, 498-502.

Costilow, R. N. (1981). Biophysical factors in growth. In *Manual of Methods for General Bacteriology*, P. Gerhardt (Ed.), American Society for Microbiology, Washington, pp. 66–78.

Cypionka, H., Widdel, F. and Pfennig, N. (1985). Survival of sulfate-reducing bacteria after oxygen stress, and growth in sulfate-free oxygen-sulfide gradients. *FEMS Microbiology Ecology*, **31**, 39–45.

De Ley, J. (1970). Reexamination of the association between melting point, buoyant density and the chemical base composition of deoxyribonucleic acid. *Journal of Bacteriology*, **101**, 738–754.

Deschacht, W. (1983). A note on the use of a pocket-calculator to work out MPN-values. *Journal of Applied Bacteriology*, **55**, 499–500.

Fonselius, S. H. (1976). Determination of hydrogen sulphide. In *Methods of Seawater Analysis*, K. Grasshoff (Ed.), Verlag Chemie, Weinheim, pp. 71–78.

Goldhaber, M. B. and Kaplan, I. R. (1974). The sulfur cycle. In *The Sea*, E. D. Goldberg (Ed.), Wiley-Interscience, New York. pp. 569–655.

Hardy, J. A. (1981). The enumeration, isolation and characterization of sulphate-reducing bacteria from North Sea waters. *Journal of Applied Bacteriology*, **51**, 505–516.

Hardy, J. A. and Hamilton, W. A. (1981). The oxygen tolerance of sulfate-reducing bacteria isolated from North Sea waters. *Current Microbiology*, **6**, 259–262.

Herbert, B. N. and Gilbert, P. D. (1984). Isolation and growth of sulphate-reducing bacteria. In *Microbiological Methods for Environmental Biotechnology*, J. M. Grainger and J. M. Lynch (Eds), Academic Press, London, pp. 235–257.

Hines, M. E. and Buck, J. D. (1982). Distribution of methanogenic and sulfate-reducing bacteria in near-shore marine sediments. *Applied and Environmental Microbiology*, **43**, 447–453.

Hungate, R. E. (1969). A roll tube method for the cultivation of strict anaerobes In *Methods in Microbiology*, vol. 3B, J. R. Norris and D. W. Ribbons (Eds), Academic Press, London, pp. 117–132.

Ingvorsen, K., Zehnder, A. J. B. and Jørgensen, B. B. (1984). Kinetics of sulfate and acetate uptake by *Desulfobacter postgatei*. *Applied and Environmental Microbiology*, **47**, 403–408.

Iverson, W. P. and Olson, G. J. (1984). Problems related to sulfate-reducing bacteria in the petroleum industry. In *Petroleum Microbiology*, R. M. Atlas (Ed.), Macmillan, New York, pp. 619–641.

Jørgensen, B. B. (1977a). Bacterial sulfate reduction within reduced microniches of oxidized marine sediments. *Marine Biology*, **41**, 7–17.

Jørgensen, B. B. (1977b). The sulfur cycle of a coastal marine sediment (Limfjorden, Denmark). *Limnology and Oceanography*, **22**, 814–832.

Jørgensen, B. B. (1978). A comparison of methods for the quantification of bacterial sulfate reduction in coastal marine sediments. I. Measurement with radiotracer techniques. *Geomicrobiology Journal*, **1**, 11–27.

Jørgensen, B. B. (1982). Mineralization of organic matter in the sea bed–the role of sulphate reduction. *Nature*, **296**, 643–645.

Keith, S. M., Herbert, R. A. and Harfoot, C. G. (1982). Isolation of new types of sulphate-reducing bacteria from estuarine and marine sediments using chemostat enrichments. *Journal of Applied Bacteriology*, **53**, 29–33.

Khosrovi, B. and Miller, J. D. A. (1985). A comparison of the growth of *Desulfovibrio vulgaris* under a hydrogen and under an inert atmosphere. *Plant and Soil*, **43**, 171–187.

Kimata, M., Kadota, H., Hata, Y. and Tajima, T. (1955). Studies on the marine sulfate-reducing bacteria. I. Distribution of marine sulfate-reducing bacteria in the coastal waters receiving a considerable amount of pulp mill drainage. *Bulletin of the Japanese Society of Scientific Fisheries*, **21**, 102–108.

Koch, A. L. (1981). Growth measurement. In *Manual of Methods for General Bacteriology*, P. Gerhardt (Ed.), American Society for Microbiology, Washington, pp. 179–207.

Laanbroek, H. J. and Pfennig, N. (1981). Oxidation of short-chain fatty acids by sulfate-reducing bacteria in freshwater and marine sediments. *Archives of Microbiology*, **128**, 330–335.

Leban, M., Edwards, V. H. and Wilke, C. R. (1966). Sulfate reduction by bacteria. *Journal of Fermentation Technology*, **44**, 334–343.

Marmur, J. (1961). A procedure for the isolation of deoxyribonucleic acid from micro-organisms. *Journal of Molecular Biology*, **3**, 208–218.

Miller, J. D. A. and Saleh, A. M. (1964). A sulphate-reducing bacterium containing cytochrome c_3 but lacking desulfoviridin. *Journal of General Microbiology*, **37**, 419–423.

Nedwell, D. B. (1982). Sulphur cycling. In *Sediment Microbiology*, D. B. Nedwell and C. M. Brown (Eds), Academic Press, London, pp. 73–106.

Nedwell, D. B. and Floodgate, G. D. (1972). The effect of microbial activity upon the sedimentary sulphur cycle. *Marine Biology*, **16**, 192–200.

Norqvist, A. and Roffey, R. (1985). Biochemical and immunological study of cell envelope proteins in sulfate-reducing bacteria. *Applied and Environmental Microbiology*, **50**, 31–37.

Ochynski, F. W. and Postgate, J. R. (1963). Some biochemical differences between fresh water and salt water strains of sulfate-reducing bacteria. In *Symposium on Marine Microbiology*, C. H. Oppenheimer (Ed.), Charles C. Thomas, Illinois, pp. 426–441.

Oremland, R. S. and Silverman, M. P. (1979). Microbial sulfate reduction measured by an automated electrical impedance technique. *Geomicrobiology Journal*, **1**, 355–372.

Pankhurst, E. A. (1971). The isolation and enumeration of sulphate-reducing bacteria. In *Isolation of Anaerobes*, D. A. Shapton and R. G. Board (Ed.), Academic Press, London, pp. 223–240.

Parkes, R. J. and Taylor, J. (1983). The relationship between fatty acid distributions and bacterial respiratory types in contemporary marine sediments. *Estuarine, Coastal and Shelf Science*, **16**, 173–189.

Peck, H. D. (1984). Physiological diversity of the sulfate-reducing bacteria. In *Microbial Chemoautotrophy*, W. R. Strohl and O. H. Tuovinen (Eds), Ohio State University Press, Columbus, pp. 309–335.

Peck, H. D. and Odom, J. M. (1984). Hydrogen cycling in *Desulfovibrio*: A new mechanism for energy coupling in anaerobic microorganisms. In *Microbial Mats; Stromatolites*, Y. Cohen, R. W. Castenholz and H. O. Halvorson (Eds) Alan R. Liss, New York, pp. 215–243.

Pfennig, N., Widdel, F. and Trüper, H. G. (1981). The dissimilatory sulfate-reducing bacteria. In *The Prokaryotes*, M. P. Starr, H. Stolp, H. G. Trüper, A. Balows and H. G. Schlegel (Eds), Springer-Verlag, New York, pp. 926–940.

Postgate, J. R. (1984). *The Sulphate-Reducing Bacteria*, 2nd edn, Cambridge University Press, Cambridge.

Postgate, J. R. and Campbell, L. L. (1966). Classification of *Desulfovibrio* species, the nonsporulating sulfate-reducing bacteria. *Bacteriological Reviews*, **30**, 732–738.

Postgate, J. R., Kent, H. M., Robson, R. L. and Chesshyre, J. A. (1984). The genomes of *Desulfovibrio gigas* and *D. vulgaris*. *Journal of General Microbiology*, **130**, 1597–1601.

Rozanova, E. P. and Nazina, T. N. (1976). A mesophilic, sulfate-reducing rod-shaped, nonspore-forming bacterium. *Microbiologiya*, **45**, 825–830 (English translation pp. 711–716).

Sand, M. D., LaRock, P. A. and Hodson, R. E. (1975). Radioisotope assay for the quantification of sulfate-reducing bacteria in sediment and water. *Applied Microbiology*, **29**, 626–634.

Schneider, J. (1977). Desulfurication and sulfur oxidation. In *Microbial Ecology of a Brackish Water Environment*, G. Rheinheimer (Ed.), Springer-Verlag, Berlin, pp. 244–248.

Schroder, H. G. J. and Van Es, F. B. (1980). Distribution of bacteria in intertidal sediments of the Ems-Dollard Estuary. *Netherlands Journal of Sea Research*, **14**, 268–287.

Skyring, G. W., Jones, H. E. and Goodchild, D. (1977). The taxonomy of some new isolates of dissimilatory sulfate-reducing bacteria. *Canadian Journal of Microbiology*, **23**, 1415–1425.

Smith, A. D. (1982). Immunofluorescence of sulphate-reducing bacteria. *Archives of Microbiology*, **133**, 118–121.

Sørensen, J., Christensen, D. and Jørgensen, B. B. (1981). Volatile fatty acids and hydrogen as substrates for sulfate-reducing bacteria in anaerobic marine sediment. *Applied and Environmental Microbiology*, **42**, 5–11.

Spencer, C. P. (1957). Utilization of trace elements by marine unicellular algae. *Journal of General Microbiology*, **16**, 282–285.

Stams, A. J. M., Hansen, T. A. and Skyring, G. W. (1985). Utilization of amino acids as energy substrates by two marine *Desulfovibrio* strains. *FEMS Microbiology Ecology*, **31**, 11–15.

Taylor, J. (1962). The estimation of numbers of bacteria by tenfold dilution series. *Journal of Applied Bacteriology*, **25**, 54–61.

Taylor, J. and Parkes, R. J. (1983). The cellular fatty acids of the sulphate-reducing bacteria, *Desulfobacter* sp., *Desulfobulbus* sp. and *Desulfovibrio desulfuricans*. *Journal of General Microbiology*, **129**, 3303–3309.

Taylor, J. and Parkes R. J. (1985). Identifying different populations of sulphate-reducing bacteria within marine sediment systems, using fatty acid biomarkers. *Journal of General Microbiology*, **131**, 631–642.

Tezuka, Y. (1979). Distribution of sulfate-reducing bacteria and sulfides in aquatic sediments. *Japanese Journal of Ecology*, **29**, 95–102.

Trüper, H. G. Kelleher, J. J. and Jannasch, H. W. (1969). Isolation and characterization of sulfate-reducing bacteria from various marine environments. *Archiv für Mikrobiologie*, **65**, 208–217.

Tsuneishi, N. and Goetz, A. (1958). A simple method for the rapid cultivation of *Desulfovibrio aestuarii* on filter membranes. *Applied Microbiology*, **6**, 42–44.

Veldkamp, H. (1976). *Continuous Culture in Microbial Physiology and Ecology*, Meadowfield Press Ltd. Durham.

Vosjan, J. H. (1975). Respiration and fermentation of the sulphate-reducing bacterium *Desulfovibrio desulfuricans* in a continuous culture. *Plant and Soil*, **43**, 141–152.

Wake, L. V., Christopher, R. K., Rickard, A. D., Andersen, J. E. and Ralph, B. J. (1977). A thermodynamic assessment of possible substrates for sulphate-reducing bacteria. *Australian Journal of Biological Sciences*, **30**, 155–172.

Wardell, J. N., Battersby, N. S. and Stewart, D. J. (1986). A note on the control of sulphate-reducing bacteria in seawater by u. v. irradiation. *Journal of Applied Bacteriology*, **60**, 73–76.

Ware, D. A. and Postgate, J. R. (1971). Physiological and chemical properties of a reductant-activated inorganic pyrophosphatase from *Desulfovibrio desulfuricans*. *Journal of General Microbiology*, **67**, 145–160.

Widdel, F. (1980). *Anaerober Abbau von Fettsäuren und Benzoesäure durch neu isolierte Arten Sulfat-reduzierender Bakterien*. Doctoral thesis, University of Gottingen, F. R. G.

Widdel, F. (1983). Methods for enrichment and pure culture isolation of filamentous gliding sulfate-reducing bacteria. *Archives of Microbiology*, **134**, 282–285.

Widdel, F. and Pfennig, N. (1977). A new anaerobic, sporing, acetate-oxidizing sulfate-reducing bacterium, *Desulfotomaculum* (emend.) *acetoxidans*. *Archives of Microbiology*, **112**, 119–122.

Widdel, F. and Pfennig, N. (1981). Studies on dissimilatory sulfate-reducing bacteria that decompose fatty acids. I. Isolation of new sulfate-reducing bacteria enriched with acetate from saline environments. Description of *Desulfobacter postgatei* gen. nov., sp. nov. *Archives of Microbiology*, **129**, 365–400.

Widdell, F. and Pfennig, N. (1982). Studies on dissimilatory sulfate-reducing bacteria that decompose fatty acids. II. Incomplete oxidation of propionate by *Desulfobulbus propionicus* gen. nov., sp. nov. *Archives of Microbiology*, **131**, 360–365.

Widdel, F., Kohring, G-W and Mayer, F. (1983). Studies on sulfate-reducing bacteria that decompose fatty acids. III. Characterization of the filamentous gliding *Desulfonema limicola* gen. nov. sp. nov., and *Desulfonema magnum* sp. nov. *Archives of Microbiology*, **134**, 286–294.

Wilson, T. R. S. (1975). Salinity and the major elements of sea water. In *Chemical Oceanography* vol. 1, J. P. Riley and G. Skirrow (Eds), Academic Press, London, pp. 365–413.

Winfrey, M. R., Marty, D. G., Bianchi, A. J. M. and Ward, D. M. (1981). Vertical distribution of sulfate reduction, methane production, and bacteria in marine sediments. *Geomicrobiology Journal*, **2**, 341–362.

ZoBell, C. E. (1958). Ecology of sulphate-reducing bacteria. *Producers Monthly*, **22**, 12–29.

ZoBell, C. E. and Rittenberg, S. C. (1948). Sulfate-reducing bacteria in marine sediments. *Journal of Marine Research*, **7**, 602–617.

Methods in Aquatic Bacteriology
Edited by B. Austin
© 1988 John Wiley & Sons Ltd.

12

Methods of Studying Methanogenic Bacteria and Methanogenic Activities in Aquatic Environments

Ralf Conrad* and Helmut Schütz[†]

* Universität Konstanz, Fakultät Für Biologie, Postfach 5560, D-7750 Konstanz 1, FRG

[†] Fraunhofer Institut Für Atmosphärische Umweltforschung, Kreuzeckbahnstrasse 19, D-8100 Garmisch-Partenkirchen, FRG

12.1 INTRODUCTION

The interest in methanogenesis and methanogenic bacteria has recently increased considerably, notably when chemists discovered that the methane (CH_4) concentration of the atmosphere is increasing by ca. 2% per annum (Rasmussen and Khalil, 1981; Seiler, 1985; Blake and Rowland, 1986). The increase is believed to be partially due to changes in biogenic CH_4 production and is considered to affect tropospheric and stratospheric chemistry, as well as climate. Without doubt, methanogenesis is one of the most important biogeochemical processes. On a global scale, CH_4 production must have been the most important process for mineralization of organic matter in the anoxic conditions prevailing during the first 80% of earth's history (Schidlowski, 1980; Holland et al., 1986). In the oxic conditions of the present era, methanogenesis still accounts for 1–2% of mineralization of the global primary productivity (Ehhalt, 1979). In the absence of oxygen, nitrate, ferric iron or sulphate, methanogenic bacteria are essential for the complete mineralization of organic materials to gaseous products, mainly because of their hydrogen-consuming activity (Zehnder, 1978). In lake ecosystems, the emerging CH_4 is an important substrate for oxygen-consuming methanotrophic bacteria, and often is determinative for the oxygen deficiency in lake water (Rudd and Taylor, 1980). A large portion of the released CH_4 is recycled within the water body and, thus, may stimulate productivity and eutrophication of a lake (Fallon et al., 1980).

Biogenic methane production is exclusively due to the activity of the strictly anaerobic methanogenic bacteria. Methane production by aerobic bacteria, such as during degradation of alkylphosphonates (Daughton et al., 1979), is most probably a rare and exotic event in nature. Because of the requirement of anoxic conditions, CH_4 production is very common in aquatic environments where the diffusion of oxygen is restricted. Besides the intestinal system (Hungate, 1966), the biogas digestors (Zehnder et al., 1982), the landfills (Farquhar and Rovers, 1973), and the anoxic wetwood trees (Zeikus and Ward, 1974), biogenic CH_4 production occurs only in aquatic ecosystems.

Interestingly, CH_4 apparently is produced not only in anoxic aquatic environments, but also in oxygenated surface and subsurface water. Ocean water often exhibits a substantial supersaturation with respect to atmospheric CH_4 that can only be explained by in situ CH_4 production (Seiler and Schmidt, 1974). Supersaturation of surface water with CH_4 has subsequently been reported by many researchers including the authors' laboratory (Seiler and Conrad, manuscript in preparation), but so far nobody has provided a satisfactory explanation of how anaerobic methanogenic bacteria can thrive in such an apparently hostile environment. It is suggested that CH_4 is produced within anoxic micro-environments provided that marine particles or zooplankton are present (Lilley et al., 1982; Traganza et al., 1979; Oremland, 1979; Burke et al., 1983). Recent measurements with oxygen microelectrodes by Joergensen and Revsbech (1985) indicate that flocs may provide anoxic micro-niches within oxic water, providing that the respiration rate is sufficient to result in a high transfer resistance of oxygen diffusion across the water–solid interface. However, Rudd and Taylor (1980) calculated that even in completely anoxic marine flocs, the numbers and activities of methanogenic bacteria would not be sufficient to explain the in situ CH_4 production.

Methanogenic bacteria have so far only been isolated from anoxic environments. They are most abundant below Eh values of $-200\,mV$ (Mah et al., 1977), being inactivated by the presence of oxygen, although not every species is rapidly killed by oxygen (Kiener and Leisinger, 1983). The latter finding may explain how methanogenic bacteira can survive in an oxic environment, e.g. dry soil, and develop rapidly as soon as conditions become anoxic again, e.g. flooded paddy soil. However, to date, no attempt has been made to study the occurrence and abundance of methanogenic bacteria in oxic environments.

Studies of CH_4 production in aquatic environments have been done almost exclusively in sediments; very little has been done in anoxic water bodies such as hypolimnetic water or groundwater (Winfrey and Zeikus,

1979a; Oremland and DesMarais, 1983; Matthess, 1985). New isolates of methanogenic bacteria have been almost exclusively obtained from sediments except those from hot hydrothermal vent fluid (Jones *et al.*, 1983b) or from hot spring water (Stetter *et al.*, 1981). Another interesting habitat for methanogenic bacteria is the sapropelic protozoa, which may be abundant in freshwater sediments. The protozoa apparently contain symbiotic methanogenic bacteria and, thus, are able to produce CH_4 (Van Bruggen *et al.*, 1983, 1985).

The methanogens which have so far been isolated utilize only a very limited number of simple substrates. With the exception of acetate, the substrates contain no C–C bonds. Most recently, however, two new isolates were described which are the first reported to utilize alcohols with one to three C–C bonds (Widdel, 1986). In aquatic ecosystems, CH_4 is usually produced from a combination of several methanogenic substrates. Acetate and H_2/CO_2 are generally the only significant CH_4 precursors in freshwater environments, with acetate usually being the predominant substrate (Takai, 1970; Cappenberg and Prins, 1974; Winfrey and Zeikus, 1979b; Lovely and Klug, 1982; Phelps and Zeikus, 1984). In some environments, however, H_2/CO_2 was the more important substrate (Belyaev *et al.*, 1975; Zaiss, 1981; Jones *et al.*, 1982). Methane production from methanol or methylamines seems to be insignificant in freshwater ecosystems (Lovley and Klug, 1983a). Conversely, in marine sediments, where mineralization of organic matter is dominated by sulphate reduction (Joergensen, 1977; Nedwell, 1984), the little CH_4 being formed appears to originate mainly from methylamines for which sulphate reducers apparently are not competitive (Oremland and Polcin, 1982; Winfrey and Ward, 1983; King *et al.*, 1983; King, 1984a, b). There are also a few reports, however, showing that either H_2/CO_2 (Warford *et al.*, 1979; Lein *et al.*, 1981) or acetate (Sansone and Martens, 1981) were the dominant CH_4 precursors in marine sediments.

There have been many excellent reviews dealing with methanogenic bacteria and/or methanogenesis (Mah *et al.*, 1977; Mah and Smith, 1981; Zeikus, 1977, 1983; Wolfe, 1979, 1980; Balch *et al.*, 1979; Zehnder, 1978; Daniels *et al.*, 1984; Fuchs and Stupperich, 1985; Rudd and Taylor, 1980; Reeburgh, 1983; Nedwell, 1984; Dubach and Bachofen, 1985; Whitman, 1985). Some of these reviews contain detailed descriptions for growth media and cultivation techniques (Balch *et al.*, 1979; Mah and Smith, 1981) which are useful for detection, isolation and description of methanogenic bacteria in aquatic environments. Rudd and Taylor (1980) described techniques for quantification of rate of CH_4 production. However, a general overview of methods is lacking. In the following sections techniques will be presented which are especially useful for evaluation of the role of methanogenic bacteria in the biogeochemistry of aquatic ecosystems.

12.2 METHANOGENIC POPULATIONS

In order to understand methane production in aquatic environments, it is necessary to know the microbial community which is involved in methanogenesis. The most important members of this community are those bacteria which actively produce methane in the final step of the mineralization of organic matter, i.e. the methanogens. To learn about these bacteria, they must be isolated and characterized. For quantitative aspects, bacterial numbers and biomass have to be determined among the various methanogenic communities.

12.2.1 Strategies for isolation

Enrichment of a bacterial population means to give a selective advantage to its development by providing optimal growth conditions. Thus, even a population which represents only a fraction of the total microbial community in the natural environment, can overgrow others. Basically, enrichment of methanogens is achieved in mineral media supplemented for growth with an energy source (substrate), a nitrogen source, a sulphur source, and in some cases an additional carbon source (acetate) and vitamins. Depending on the origin of the methanogenic sample, the ecophysiological characteristics, notably temperature, pH, salinity, and substrate concentrations, may be adjusted. Starting enrichment cultures with serial dilution of the sample (MPN-technique), the size of the dominant methanogenic population may be estimated simultaneously. Thus, the highest dilution revealing CH_4-production may already be a highly enriched culture of the dominant methanogen, and may be used for further isolation. However, such enrichment cultures will miss those methanogenic species that constitute only a minor fraction of the total methanogenic community. Moreover, depending on growth conditions, the resulting isolate will only be the organism that grew best and will not necessarily be the organism that is mainly responsible for methanogenesis in the original habitat. Therefore, other strategies are necessary if the whole spectrum of methanogens is to be described.

For this purpose, a direct isolation technique may be applied in which the original sample is distributed in solid growth medium (roll-tubes, agar-deeps) so that the individual bacteria are fixed before growth starts. Methanogenic colonies are identified by their unique fluorescence properties, and picked for further enrichment and isolation in pure culture. The direct isolation procedure may give information not only on methanogens but also on methanogenic microcosms, e.g. synthrophic consortia. The best known example is *Methanobacterium omelianskii* which, for a long time, was believed to be a pure culture of an ethanol-degrading methanogen, and which turned out to be a synthrophic association of an ethanol-fermenting, hydrogen-producing organism, and hydrogen-consuming *Methanobacterium bryantii*. This observation led to the discovery of interspecies hydrogen transfer (Bryant *et al.*, 1967). In the same way, synthrophic consortia with methanogens have been found to degrade fatty acids like *Syntrophobacter wolinii* (Boone and Bryant, 1980), *Synthrophomonas wolfei* (McInerney *et al.*, 1981) and other species (Schink, 1988). The presence of syntrophic consortia, on the basis of other substrates (e.g. formate, acetate) as the transferred metabolite can not be excluded, as yet.

Other strategies of enrichment refer to the nature of the substrate. Either methanogens may be enriched and isolated on methanogenic substrates that are well-known, e.g. H_2/CO_2, formate, acetate, methanol, or methylamines; or methanogens may be enriched and isolated on potential, as yet unknown methanogenic substrates. Thus just recently, two new isolates producing methane and growing on alcohols, other than methanol have been described by Widdel (1986). Application of antibiotics, specific to eubacteria but not to archaebacteria (Böck and Kandler, 1985), may give the methanogens a selective advantage against fermenting bacteria (Huser *et al.*, 1982). Finally, new methanogens may be isolated with respect to substrate concentration. *In situ* concentrations of methanogenic substrates may be much lower than those normally used in growth media. Thus, K-strategists, having a high affinity for substrate (low K_s) but a low maximum growth rate, can only be enriched against r-strategists having a high growth rate but a low substrate affinity, providing that substrate concentrations are kept low, as in substrate-limited continuous culture (Kuenen and Harder, 1982).

12.2.2 Culture requirements

Media for methanogens must be strictly oxygen-free and reduced ($E_h < -200$ mV). This has been achieved after development of the so-called Hungate technique (see review by Hungate, 1969) and was further improved by introducing serum bottles (Miller and Wolin, 1974), gassing manifolds (Balch and Wolfe, 1976) and anaerobic glove boxes (Edwards and McBride, 1975). Dissolved oxygen is removed from the media by boiling and cooling under oxygen-free atmospheres. The medium is reduced by the addition of sulphide, thioglycollate, dithionite or other reducing agents. Resazurin may be added as redox indicator, turning red at Eh-values higher than -42 mV.

In the simplest case, autotrophic methanogens may be enriched using hydrogen as the electron donor and energy source and CO_2 as the sole carbon source. Carbon dioxide is supplied as bicarbonate solution and as gas (H_2/CO_2) in the headspace. Because of hydrogen depletion by

methanogens, the headspace has to be refilled regularly. Some methanogens need acetate in addition to CO_2 for cell carbon synthesis (e.g. Balch
et al., 1979; Corder et al., 1983; Miller and Wolin, 1985). Heterotrophically
growing methanogens may be enriched on formate, methylamines,
methanol or acetate as energy, electron and carbon sources. However, CO_2
may be required for cell carbon synthesis (Miller and Wolin, 1983).

Enrichment on complex organic substrates, for example carbohydrates,
may result in overgrowth of methanogens by fermentative bacteria.
Selection of methanogens may be favoured, however, if the culture
medium is supplemented with antibiotics which selectively inhibit the
fermentative eubacteria but not the methanogenic archaebacteria (Böck
and Kandler, 1985). Enrichment on complex organic substrates such as
long chain fatty acids or alcohols may result in the enrichment of
syntrophic associations with methanogens (e.g., McInernay et al., 1979).

Ammonium is the preferred nitrogen source for methanogenic bacteria
(Mah and Smith, 1981), although in one species of Methanobacterium it
could be replaced by glutamine (Bhatnagar et al., 1984). It should further be
mentioned that even molecular nitrogen fixation was found in methanogens (Belay et al., 1984; Murray and Zinder, 1984; Bomar et al., 1985).
Nitrate is not a suitable nitrogen source for methanogens. In addition, it
increases the redox potential of media and may result in overgrowth of
enrichment cultures by nitrate-reducing bacteria.

Sulphide, commonly added as the reducing agent, also serves as sulphur
source for methanogens, but in high concentrations it may inhibit growth
(Scherer and Sahm 1981; Rönnow and Gunnarson, 1981, 1982). Organic
sulphur sources (cysteine, methionine) are utilized by Methanobacterium
(Bhatnagar et al., 1984) and by Methanosarcina (Scherer and Sahm, 1981).
Methanogenic bacteria are also able to reduce molecular sulphur to H_2S
(Stetter and Gaag, 1983). Assimilatory sulphate reduction was described
for Methanococcus (Daniels et al., 1986). Enrichment media, however, must
usually be free of sulphate, because methanogens may otherwise be
outcompeted by sulphate-reducing bacteria at least for the most common
substrates H_2 and acetate.

Special growth factors include nickel (Thauer et al., 1983) and vitamin
B12 (Krzycki and Zeikus, 1980). For enrichment, other trace elements or
vitamins may be added (Whitman, 1985), or as yet unknown growth
factors which are present in rumen fluid, mud water, yeast extract or
trypticase.

The ecophysiological characteristics of methanogens cover a broad range
of temperature, salinity and pH tolerance (Dubach and Bachofen, 1985),
including alkaliphilic (Blotevogel et al., 1985; Boone et al., 1986), acid-
tolerant (Williams and Crawford, 1985), halophilic (Zhilina, 1983; 1986) and
extremely thermophilic (Stetter et al., 1981) species. In their natural habitats,

methanogenic bacteria can exhibit an even broader range of tolerances than in pure culture. Hence, methanogenic bacteria have been found in hot (90 °C) spring water (Zeikus *et al.*, 1980), in hypersaline lake sediments (Giani *et al.*, 1984; Paterek and Smith, 1985), in alkaline (pH 9.7) lake sediments (Oremland *et al.*, 1982) and in acidic (pH 3.6) peat (Svensson and

TABLE 12.1 Carbonate-buffered mineral medium for cultivation of freshwater methanogens as described by Widdel (1986)

Compound	Concentration (gl^{-1})	
(A) Basal medium		
KH_2PO_4	0.2	
NH_4Cl	0.3	
NaCl	1.0	
$MgCl_2 \cdot 2H_2O$	0.4	
KCl	0.5	
$CaCl_2$	0.15	
(B) Bicarbonate solution		
$NaHCO_3$	84	$30\,ml\,l^{-1}$ medium
(C) Vitamins		
4-Aminobenzoic acid	0.01	
D (+)-Biotin	0.01	
Nicotinic acid	0.02	
Ca-D-pantothenate	0.01	
Pyridoxamine 2HCl	0.02	
Thiamine HCl	0.02	$5\,ml\,l^{-1}$ medium
(D) Vitamin B_{12}		
Cyanocobalamine	0.05	$5\,ml\,l^{-1}$ medium
(E) Trace elements		
HCl, 25%	10 ml	
$FeCl_2 \cdot 4H_2O$	1.0	
$CoCl_2 \cdot 6H_2O$	0.2	
$MnCl_2 \cdot 4H_2O$	0.1	
$ZnCl_2$	0.07	
$Na_2MoO_4 \cdot 2H_2O$	0.036	
$NiCl_2 \cdot 6H_2O$	0.12	
H_3BO_3	0.062	
$CuCl_2 \cdot 2H_2O$	0.017	$1\,ml\,l^{-1}$ medium
(F) Additional trace elements		
$Na_2SeO_3 \cdot 5H_2O$	0.006	
$NaMoO_4 \cdot 2H_O$	0.024	
$Na_2WO_4 \cdot 2H_2O$	0.033	$1\,ml\,l^{-1}$ medium
(G) Reducing agent		
$Na_2S \cdot 9H_2O$	120	$4\,ml\,l^{-1}$ medium

Rosswall, 1984; Williams and Crawford, 1984; Harriss et al., 1985). The conditions of the enrichment cultures have to be adjusted so that methanogens of a particular ecotype can develop.

Several media compositions are presently used by different research groups for enrichment and cultivation of methanogens (e.g., Balch et al., 1979; Huser et al., 1982; Lynd et al., 1982; and the review by Mah and Smith, 1981). They are either phosphate-buffered or carbonate-buffered media. In aquatic environments, phosphate is certainly not an important buffering substance and, actually, many bacteria are sensitive to phosphate (Pfennig, 1984). Therefore, carbonate-buffered media may be preferred. Thus, a carbonate-buffered medium has successfully been used for the isolation of novel sulphate-reducing bacteria (Widdel and Pfennig, 1981, 1984). The same mineral medium has also been used to isolate new species of fermentative bacteria (e.g. Schink and Pfennig, 1982a,b), homoace-togenic bacteria (e.g. Eichler and Schink, 1984) and obligately syntrophic bacteria (e.g. McInerney et al., 1979; Stieb and Schink, 1985). Recently, in a slightly modified composition it has been used to isolate novel metha-nogenic bacteria which are able to use 2-propanol or other alcohols as the sole substrate for growth (Widdel, 1986). The composition of the medium is given in Table 12.1. The individual solutions are autoclaved or filter sterilized and mixed under anoxic, aseptic conditions. The pH is adjusted with sterile HCl or Na_2CO_3 solutions and then the medium is distributed to sterile culture vessels. The medium is kept under oxygen free atmosphere containing 20% carbon dioxide.

12.2.3 Isolation

Pure cultures of methanogens are usually isolated from enrichment cultures by cloning techniques. For this purpose, the roll-tube technique has been developed by Hungate (1969). In addition, colony-forming units may be obtained in deep-agar (Agar-shakes), by plating on agar plates inside anaerobic chambers or by streaking on agar surfaces in anaerobic bottles.

Agar deep cultures are prepared by serial dilution of the inoculum with culture medium mixed with liquid agar (e.g. Pfennig, 1978). Colonies are isolated, suspended and serial dilutions in agar deeps are repeated until pure cultures are obtained. The agar shake technique allows the observa-tion of gas-producing colonies by small gas bubbles adjacent to the colony. Using shakes, roll-tubes, etc., it is sometimes possible to recognize synthrophic associations. These may appear as a central colony (e.g. a fatty acid-utilizing H_2-producer) surrounded by satelite colonies (e.g. H_2-utilizing methanogens). However, syntrophic methanogens growing as satelites can only be observed if they are more numerous in the enrichment culture than the fermentative bacteria. Hence, this technique is usually applied to isolate obligately syntrophic fermentative bacteria rather than methanogens. Pure cultures of methanogens, sulphate reducers or homoacetogens are even added to obtain the syntrophs (e.g., McInerney *et al.*, 1979; Mountford and Bryant, 1982).

Plating of methanogenic samples on agar petri dishes can be done in anaerobic chambers (Edwards and McBride, 1975; Balch and Wolfe, 1976; Jones *et al.*, 1983c) which needs considerable investment in equipment. A plating method using flat bottles (Braun *et al.*, 1979) has recently been improved for the isolation of methanogens (Hermann *et al.*, 1986). This technique combines the advantages of plating with the easy achievement of strictly anoxic conditions. Methanogenic samples are streaked on the agar inside the culture bottle. Gas atmospheres can easily be changed and pressurized inside the bottles.

Problems may arise during purification of bacteria, for instance in the case of slowly growing strains or filamentous bacteria, e.g. the filamentous methanogenic bacterium *Methanothrix soehngenii* (Huser *et al.*, 1982). In principle, filamentous bacteria may be purified by a washing procedure as described for the isolation of the filamentous sulphate reducer *Desulfonema* (Widdel, 1983).

12.2.4 Characterization

Taxonomic, and especially ecophysiological characterization, of isolated methanogens, or of environmental methanogenic populations, is desirable in order to evaluate their potential role in methane formation under *in situ* conditions. Methanogens are members of the kingdom Archaebacteria and are classified in three orders, namely the *Methanomicrobiales*, *Methanococcales* and *Methanobacteriales* (Whitman, 1985). Taxonomic characterization requires the study of cell morphology and especially of cell molecular properties such as cell wall composition, lipids, rRNA sequence, immunological properties and DNA homologies (Balch *et al.*, 1979; Woese and Olsen, 1986; Kandler, 1979; for recent overviews see Kandler and Zillig, 1985; Woese and Wolfe, 1985).

Methods are now being developed to allow the characterization of methanogenic populations in the natural ecosystem, e.g. the analysis of so-called biomarkers such as characteristic fatty acids. These have been used to characterize communities of sulphate-reducing bacteria in sediments (Taylor and Parkes, 1985) and to determine the biomass of methanogenic bacteria (Martz *et al.*, 1983). The method, however, is based on the assumption that the distribution of the biomarker in the organism grown under laboratory conditions is identical to that under *in situ* conditions and that no other organisms share the same biomarker.

Another approach is to analyse 5s rRNA sequences isolated from the natural environment, and compare them to reference rRNA sequences (Pace *et al.*, 1986). In a similar way, Sayler *et al.* (1985) used DNA-probes of a particular micro-organism to detect, by DNA–DNA colony hybridization, this catabolic genotype in the environment. The latter technique is very promising for evaluating the existence of specific genotypes in the various environments. All these methods are limited, however, by the availability of reference organisms. In other words, these techniques can only be used to detect micro-organisms that have previously been isolated, but cannot be used when as yet unknown species prevail.

To account for ecophysiological characteristics, temperature range, pH range, salinity range, substrate utilization and kinetics of growth and methanogenic substrate utilization may be of importance.

Most isolates of methanogens have temperature optima of 30 to 40 °C, except for *Methanobacterium thermoautotrophicum* (65–70 °C; Mah and Smith, 1981). In recent years, more thermophilic methanogens have been described (Harris *et al.*, 1984; Winter *et al.*, 1984; Ahring and Westermann, 1985; Blotevogel *et al.*, 1985; Zinder *et al.*, 1985) and extremely thermophilic (up to 97 °C) methanogens have been isolated from hot environments, for example *Methanothermus fervidus* (Stetter *et al.*, 1981) and *Methanothermus sociabilis* (Lauerer *et al.*, 1986) from an Icelandic hot spring, and *Methanococcus jannaschii* (Jones *et al.*, 1983b) from a submarine hydrothermal vent. Although CH_4-production has been detected at 5 °C in enrichments from low-temperature adapted methanogenic peatland flora, psychrophilic methanogens have not yet been isolated. However, two different methanogenic populations were distinguished in acidic peat by studies of the temperature optimum for CH_4-formation from CO_2 and acetate, respectively (Svensson, 1984).

Pure cultures of methanogens usually show optimal growth in the pH-range of 6.5 to 7.5 (Mah and Smith, 1981). However, pH-values in the natural environment may differ significantly. From a hydrothermal vent, *Methanococcus jannaschii* was selectively isolated at pH 6.0 but not at pH 4.0 and 7.0 (Jones *et al.*, 1983b). The lower limit for CH_4 formation in acid peat was found to be pH 3.1 but optimal activity was recorded at pH 6.0 to 7.0 and growth did not occur at pH less than 5.3 (Williams and Crawford, 1985). Thus, CH_4-production in acidic environments (Svensson, 1984; Williams and Crawford, 1984; Harriss *et al.*, 1985) is attributed to acid-tolerant rather than acidophilic methanogens. By contrast, alkaliphilic methanogenic enrichments from Big Soda Lake revealed highest methane production at the *in situ* pH, which is extremely alkaline (pH 9.7) (Oremland *et al.*, 1982).

Marine taxa, e.g. *Methanogenium*, require 1.5 to 3.0% (w/v) NaCl for optimal growth (Mah and Smith, 1981). Obligate, halophilic methanogens, e.g. *Methanococcus halophilus*, showed optimum growth at 7% (w/v) NaCl with a range of 1.5 to 15% (w/v) (Zhilina, 1983, 1986). The optimal NaCl concentration for a methanogen from Great Salt Lake was 11.6% (w/v), and growth occurred between 5.8 and 14.4% (w/v) (Paterek and Smith, 1985).

Most of the methanogenic species utilize H_2/CO_2 or formate as energy substrate (Balch *et al.*, 1979; Dubach and Bachofen, 1985). Only two genera are able to utilize acetate, i.e. *Methanosarcina* (Smith and Mah, 1978; Weimer and Zeikus, 1978a; Mah, 1980; Zinder and Mah, 1979; Sowers *et al.*, 1984) and *Methanothrix* (Huser *et al.*, 1982; Fathepure, 1983; Ahring and Westermann, 1984; Patel, 1984). *Methanosarcina* is also able to utilize methanol and methylamines (Smith and Mah, 1978; Weimer and Zeikus, 1978b; Hippe *et al.*, 1979). Other more recently isolated species which utilize methanol and methylamines include *Methanosphaera stadtmaniae* (Miller and Wolin,1985), *Methanolobus tindarius* (König and Stetter, 1982), *Methanococcoides methylutens* (Sowers and Ferry, 1983) and *Methanococcus frisius* (Blotevogel *et al.*, 1986). Many of the methanogenic bacteria are able to utilize CO (Balch *et al.*, 1979; Dubach and Bachofen, 1985). Among the hydrogen-utilizing methanogens, this ability seems to be restricted to those species which grow autotrophically and do not require acetate as a carbon source (Bott *et al.*, 1985). Besides utilization of single substrates, diauxie and biphasic CH_4-formation was observed on H_2/CO_2 and acetate, or on methanol and acetate with *Methanosarcina* species (Ferguson and Mah, 1983), but an acetate-adapted strain of *Methanosarcina barkeri* catabolized methanol and acetate simultaneously (Krzycki *et al.*, 1982).

12.2.5 Quantitative determination of methanogenic populations

In order to evaluate the population size of methanogens in the environment, the bacterial cells must be counted or their biomass determined. This allows the recording of methanogenic activities relative to cellular or biomass basis. Counting may be conducted in a way to obtain figures for all methanogenic populations or only for populations of specific methanogenic ecotypes (e.g. acetate-utilizing). Furthermore, total cells or viable cells of methanogens may be distinguished.

The total number of methanogens may be determined by direct counting with an epifluorescence microscope (Mink and Dugan, 1977; Doddema and Vogels, 1978; Miller and Wolin, 1982; Smith, 1966). The technique is based on the fluorescence properties of factor F_{420}, which is a coenzyme typical of methanogens (Cheeseman *et al.*, 1972; Eirich *et al.*, 1978; Wolfe, 1980). Doddema and Vogels (1978) applied excitation and barrier filters to select for the fluorescent compounds F_{420} and F_{350}, respectively. Factor F_{350} is the so-called yellow fluorescent compound or methanofuran (Leigh *et al.*, 1985), which is also unique to methanogenic bacteria. An example for the

detection of different methanogenic species, by epifluorescence microscopy of F_{420} and F_{350} is given in Fig 12.1. Using this technique,
methanogenic bacteria have been detected and counted as endosymbionts

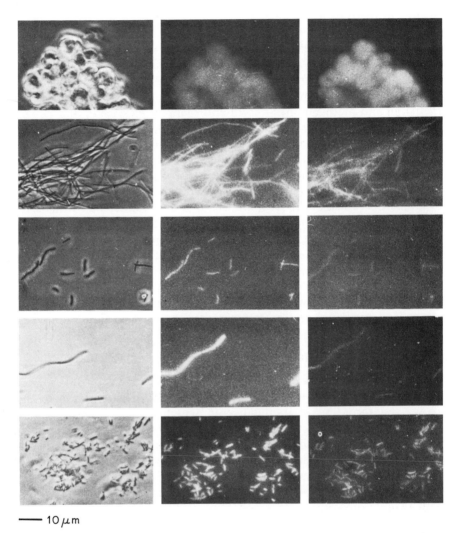

— 10 μm

Fig. 12.1 Photomicrographs of methanogenic bacteria. The columns show, from
top to bottom: *Methanosarcina barkeri. Methanobacterium formicicum, Methanobacterium bryantii* strain M.O.H., *Methanospirillum hungatii,* and *Methanobacterium* strain
TH. The columns from left to right show identical fields of phase-contrast
illumination, fluorescence when illuminated at 420 nm, and fluorescence at
355 nm (from Doddema and Vogels, (1978) and reproduced by premission of the
American Society for Microbiology, Washington DC).

in sapropelic protozoa which possibly constitute important methanogenic communities within aquatic sediments (Van Bruggen *et al.*, 1983, 1985). However, microscopic counting techniques need experience to distinguish fluorescent methanogens from autofluorescent particles and from micro-organisms containing compounds with similar fluorescense properties. A fluorescent-antibody technique has been proposed by Strayer and Tiedje (1978a) to enumerate methanogens in the sediment matrix. However, this technique would be specific to only those methanogens which possess the particular antigen.

Viable cells that are able to grow in a particular medium or under specific conditions (e.g. temperature), may be enumerated by plating. This requires an anaerobic glove box in which transfers can be made (Edwards and McBride, 1975; Balch and Wolfe, 1976). Jones *et al.* (1983c) optimized this method for the counting of *Methanococcus* on agar plates, with plating efficiencies of 90%. As an alternative to plating, methanogens may also be counted by using roll-tubes, agar-shakes or flat agar bottles (Braun *et al.*, 1979; Hermann *et al.*, 1986). The methanogenic colonies may easily be detected by their fluorescence.

The most common way to enumerate specific ecotypes of methanogens is by the MPN-technique (e.g. Heukelikian and Heinemann, 1939; Siebert and Hattingh, 1967; Zeikus and Winfrey, 1976, Williams and Crawford, 1983), which is done by serial dilution of the sample in liquid growth medium suitable for the ecotype in question. Counting is done usually by testing for methane production rather than by evaluation of growth, which may be too faint to detect.

Instead of bacterial counts, the total biomass of the methanogenic population may be determined. This can be achieved by extraction and quantification of cell compounds that are unique for methanogens. Extraction and quantification of typical lipids was used to estimate the biomass of methanogens in sediments (Martz *et al.*, 1983). Coenzyme F_{420}, which is also unique for methanogens, but also for one strain of *Streptomyces griseus* (Eker *et al.*, 1980) was extracted and quantified in sludge as a measure to characterize the methanogenic activity (Dolfing and Mulder, 1985; Whitmore *et al.*, 1986). However, the cellular F_{420} content of methanogens, as well as the ratio of methanogenic activity to F_{420} content, was found to vary by a factor of about 100.

12.3 *IN SITU* METHANOGENESIS

Methane is produced in strictly anoxic sites of sediments and other aquatic habitats. *In situ* rates of methanogenesis can only be obtained if the methanogenic food chain is not affected by the measurement procedure.

Ideally, turnover and pool sizes of all substrates involved should be as under *in situ* conditions when rates of methanogenesis are determined. There are different approaches to accomplish this task, i.e. determination of concentration gradients, measurement of fluxes and assay of methanogenic activity in samples.

12.3.1 Vertical gradients

CH_4 production rates in aquatic sediments may be quantified by measuring the vertical distribution of CH_4 in the sediment. For this purpose, dissolved CH_4 is extracted from sediment cores and analysed. Dissolved gases may also be analysed in pore water that is collected by special *in situ* pore water samplers (Hesslein, 1976; Winfrey and Zeikus, 1977; Howes *et al.*, 1985). Recently, probes were developed allowing the direct measurement of dissolved CH_4 by membrane-inlet mass spectroscopy (Lloyd *et al.*, 1986). The resulting vertical CH_4 concentration profiles may then be used to calculate the diffusional flux across the sediment–water interface (e.g. Martens and Val Klump, 1980; Kipphut and Martens, 1982). This flux is identical to the *in situ* CH_4 production rate, providing that both sediment and overlying water, are anoxic. Otherwise, the vertical gradient may be influenced by CH_4 oxidation reactions. In fact, the particular shape of CH_4 gradients observed in some aquatic environments was taken as evidence for the occurrence of anaerobic, CH_4 oxidation (Alperin and Reeburgh, 1984). In practice, the determination of flux rates from vertical gradients may be complicated by spatial and temporal variations in the actual diffusion coefficient which may drastically increase if open bubble tubes or benthic animals exist (Martens and Val Klump, 1980).

12.3.2 Collecting chambers

If aquatic environments are shallow enough to allow the installation of collecting chambers, CH_4 flux rates may also be directly determined. The box method may be used to measure fluxes across the sediment-water interface by collecting the methane which is diffusing out of the sediment into the overlaying water (Martens and Val Klump, 1980; Oremland, 1975). Usually, however, it is applied to determine CH_4 emission from shallow aquatic environments into the atmosphere, e.g. in rice paddies (Cicerone and Shetter, 1981; Seiler *et al.*, 1984; Holzapfel-Pschorn and Seiler, 1986), tundra regions (Svensson and Rosswall, 1984; Harriss *et al.*, 1985), swamps (Harriss and Sebacher, 1981; Harriss *et al.*, 1982), and coastal marshes (De Laune *et al.*, 1983; King and Wiebe, 1978). Boxes may be applied as static enclosures measuring the temporal increase of the CH_4 concentration within the enclosure (e.g. Seiler *et al.*, 1984), or as dynamic systems being flushed with air either in a flow-through way or by circulation (e.g., Sebacher and Harriss, 1982). Especially in the latter case, it is very important that the measurements are done under ambient atmosphere. Otherwise, e.g. by flushing with an inert gas, gradients will be changed from the beginning and, usually, fluxes will be overestimated.

Emission of CH_4 may be due to different pathways of CH_4 transport from the sediment into the atmosphere; namely diffusion, ebullition and transport through aquatic plants. Shallow, waterlogged areas are often

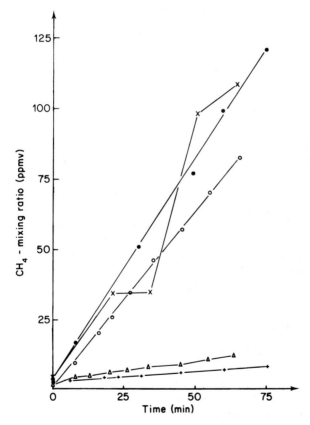

Fig. 12.2 Temporal increase of CH_4-mixing ratios in collector boxes placed over intact rice plants (●), over submerged soil without plants (×), over rice plants which were cut above the water surface (○), over rice plants which were cut below the water surface (△), and over soil which has been covered with a polypropylene screen (250 μm) (+). The measurements were conducted in rice paddies near Vercelli, Italy.

covered with plants, and methane may be released from the sediment via the intercellular gas space system of the aquatic plants (Seiler *et al.*, 1984; Dacey and Klug, 1979; Holzapfel-Pschorn *et al.*, 1986; Sebacher *et al.*, 1985). The importance of the plant transport can easily be demonstrated by cutting the plants above and below the water surface, which in the latter case results in a dramatic decrease of CH_4-emission if plants constitute the major transport pathway (Fig. 12.2). The residual emission is due to ebullition and diffusional flux. When the emission event is mainly due to ebullition, CH_4 increase in the collector box occurs in a stepwise pattern (Fig. 12.2). Funnels have been used as traps to collect gas bubbles and determine ebullition fluxes (e.g. Martens and Val Klump, 1980; Holzapfel-Pschorn *et al.*, 1986). The diffusional flux may also be determined with the box method by measuring the CH_4 emission rate after covering the sediment with a screen that impedes the ebullition flux (Fig. 12.2).

12.3.3 Sampling and incubation

CH_4-production rates may be measured in samples that are taken out of the natural environment, and are incubated under conditions close to those *in situ*. All transfers have to be done in a way that minimizes changes in pool sizes and activities. In addition, poisoned controls have to be used to distinguish CH_4 production from degassing of the CH_4 from sediment pore water (Kiene and Capone, 1985). CH_4-production may be measured in grab samples of sediment (e.g., Winfrey and Zeikus, 1979b), or in sediment cores from which subsamples may be taken to obtain a vertical profile of the CH_4 production activity (e.g., Mountfort and Asher, 1981; Jones *et al.*, 1982; Lovley and Klug, 1982).

Measurement of CH_4-production by *in situ* enclosure of the sediment has been reported by Holzapfel-Pschorn *et al.* (1985). Corers were pressed into the submerged soil of rice paddies, left for different time periods and retrieved to analyse the CH_4 concentration within the sediment core. However, high variations due to spatial inhomogeneities were found so that the precision of this method is very low.

12.4 PRECURSORS OF METHANOGENESIS

The determination of total methanogenesis is relatively straightforward. However, it tells little of its role in the mineralization process unless the pathway of carbon and electron flow is known. At present, we have only a very rough idea on carbon flow in the various aquatic ecosystems. In general, it is assumed that the breakdown of organic matter (e.g.

carbohydrates) is achieved by four major steps (Zehnder, 1978), namely: (1) depolymerization, (2) fermentation to hydrogen, CO_2, fatty acids and alcohols, (3) conversion to acetate and hydrogen and (4) methanogenesis or/and sulphate reduction. Although this flow scheme may not be valid for every case, it represents a good working hypothesis for carbon flow analysis. A first step in such an analysis is usually the question for the *immediate* precursors of methanogenesis, i.e. the substrates that the methanogenic microbial population is using for CH_4 production. More precisely, it has to be determined, to what extent the methanogens are utilizing hydrogen/CO_2, formate, CO, acetate, methanol or methylamines as substrates under *in-situ* conditions. To this end, three approaches have been used i.e. (1) application of radioactive substrates, (2) comparison of *in situ* rates of CH_4 production to the turnover rates of the pools of immediate CH_4 precursors and (3) stimulation of methanogenesis by increasing the substrate pool size.

12.4.1 Tracer experiments

To determine the percentage contribution of individual immediate CH_4 precursors, samples are taken from an aquatic environment and incubated in the presence of a ^{14}C-labelled substrate. The specific radioactivity (SR) of the substrate and of the produced CH_4 are determined. The fraction (f) of CH_4 produced from the labelled C in the substrate is calculated by

$$f = SR_{CH_4}/SR_{Substrate-C}$$

$$f = \frac{dpm\ CH_4}{mol\ CH_4} \bigg/ \frac{dpm\ Substrate-C}{mol\ Substrate-C}$$

e.g. in the case of $2\text{-}^{14}C$-acetate as substrate

$$f = SR_{CH_4}/SR_{acetate}$$

or, in the case of $NaH^{14}CO_3$

$$f = SR_{CH_4}/SR_{CO_2}$$

The tracer method is a rather straightforward approach, since the label is measured in both: the substrate and the product. The method has been used by Cappenberg and Prins (1974) in analysis of anaerobic mineralization process in Lake Vechten sediment and, subsequently, has been applied by many other researchers, e.g. Winfrey and Zeikus (1979b), Zaiss (1981), Jones *et al.* (1982), Phelps and Zeikus (1985).

A technical difficulty of this method is the necessity to measure the specific radioactivity of the produced CH_4. The most elegant and rapid method is the detection of radioactivity by a gas proportional detector, which is attached to the gas chromatograph by which the CH_4 is separated from other gases and is quantified by means of a thermal conductivity detector (TCD) or a flame ionization detector (FID). Analysis of a gas sample immediately gives the amount of the radioactivity of CH_4 in the sample (Nelson and Zeikus, 1974). However, the quenching effect of the gas sample must be standardized. Since CH_4 itself exhibits a strong quenching effect in the gas proportional counter, it is usually necessary to oxidize the CH_4 to CO_2 before counting. This is usually achieved by the use of commercial gas proportional detectors.

An alternative technique is to measure the specific radioactivity of CH_4 by trapping the $^{14}CO_2$ that is formed from $^{14}CH_4$ within the FID of the gas chromatograph. The CO_2 can be trapped in scintillation vials containing ethanolamine or phenethylamine scintillation mixtures, followed by routine scintillation counting (Zehnder and Brock, 1979; Iversen and Blackburn, 1981). Again technique requires the injection of only one gas sample, but has the disadvantage of almost no resolution of the radioactive signal of the gas peaks leaving the chromatograph.

Another technique requires one aliquot of gas sample for the chromatographic analysis of CH_4, and a second aliquot for determination of $^{14}CH_4$ by liquid scintillation counting (Zehnder et al., 1979). For this purpose, the gas sample is stripped off contaminating $^{14}CO_2$ by alkali and then injected into a scintillation vial, which is equipped with a septum, and filled with at least 20 ml of a scintillation cocktail with high dissolution capacity for CH_4 (e.g. toluene-based cocktails).

From the conceptual point of view, it must be emphasized that: (1) the radioactive substrate added does not significantly change the pool size of the substrate, (2) the fraction (f) of CH_4 produced from the labelled substrate remains constant with incubation time and, (3) the contribution of the individual immediate CH_4 precursors is balanced by the total CH_4 production. The first requirement may be achieved by using radioactive substrates with sufficiently high relative radioactivity.

The second requirement is not always met (e.g. Zaiss, 1981) and data then have to be analysed carefully. An example is shown in Fig. 12.3, where CH_4 production from labelled acetate and bicarbonate was measured in anoxic sediments of the River Saar (Zaiss, 1981). The specific radioactivity of CH_4 derived from $2\text{-}^{14}C$-acetate decreased dramatically during the incubation, since acetate presumably was turned over so rapidly in comparison to CH_4 production that the radioactive label was diluted by unlabelled substrate being produced during the incubation period. Therefore, when calculating the fraction (f) of CH_4 produced from the labelled precursor, the temporal change of $SR_{Substrate-C}$ has to be taken into account. This temporal change is technically difficult to measure in case of organic substrates, but is easily determined with bicarbonate, whose SR is identical to the SR of CO_2 in the gas phase of the incubated sample (Winfrey and Zeikus, 1979b; Zaiss, 1981).

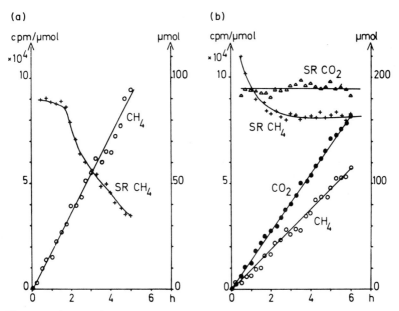

Fig. 12.3 (a) Methane production and specific radioactivity (SR) of methane during the tracer study with methyl-labelled acetate. (b) Methane and carbon dioxide production, specific radioactivity of methane and carbon dioxide during the tracer study with labelled bicarbonate (from Zaiss, (1981) and reproduced by permission of Gustav Fischer Verlag, Stuttgart).

Changes in specific radioactivity may also occur, when the potential CH_4 precursor is simultaneously utilized by micro-organisms others than methanogens at a comparable rate so that CH_4 is also produced from the metabolic products of the CH_4 precursor. For example, this may be the case when CO_2 is converted to acetate by homoacetogens (Phelps and Zeikus, 1984; Jones and Simon, 1985), or when acetate is oxidized to CO_2 by syntrophic micro-organisms (Zinder and Koch, 1984). During the incubation experiment, the relative contribution to CH_4 production by hydrogen/ CO_2 or acetate, respectively, would then decrease. The existence of such diverted pathways may be checked by adding unlabelled intermediates, e.g. acetate in the presence of $^{14}CO_2$ or hydrogen in the presence of 2-^{14}C-acetate (Winfrey and Zeikus, 1979b). Changes in specific radioactivity may also occur when the labelled substrate is exchanged to substrate pool which is less available to biological activity. The existence of different substrate pools with different biological availability has been shown for acetate turnover in marine sediments where up to 80% of the acetate may be unavailable (Ansbaek and Blackburn, 1980; Christensen and Blackburn, 1982; Parkes *et al.*, 1984).

The third requirement is usually met by doing parallel experiments using $^{14}CO_2$ and $2\text{-}^{14}C$-acetate as substrates. $^{14}CH_4$ is then formed from methanogens utilizing hydrogen/CO_2, and from methanogens utilizing acetate, respectively. If hydrogen/CO_2 and acetate are the only CH_4 precursors, their individual contribution should balance total CH_4 production. If the contribution is more, the diverted pathways may be operative. Conversely, if the contribution is less, additional CH_4 precursors may be present. However, the latter conclusion is only true, if CH_4 is formed exclusively from the methyl carbon of the acetate and not from the carboxyl carbon. Although this is more-or-less the case for most aquatic samples, it has been shown that strains of *Methanosarcina barkeri* are able to form some CH_4 from the carboxyl-C of acetate (Krzycki *et al.*, 1982) and that this proportion may decrease when syntrophic hydrogen-utilizing sulphate reducers are present (Phelps *et al.*, 1985). Furthermore, it must be noted that CO instead of CH_4 may be formed during methanogenic conversion of CH_4 precursors (Conrad and Thauer, 1983; Bott *et al.*, 1985; Eikmanns *et al.*, 1985).

12.4.2 Comparison of turnover rates

The principle of this method is the comparison of the utilization rates of the different immediate CH_4 precursors with the production rate of CH_4, both under *in situ* conditions. Hence, this method compares kinetics rather than specific radioactivities. The CH_4-production rate should finally be balanced by the stoichiometric CH_4 production rates calculated from the utilization rates of the individual CH_4 precursors. However, usually the main goal of studies applying this technique is the determination of organic carbon transformation rather than the determination of contribution of immediate CH_4 precursors to total methanogenesis (Lovley and Klug, 1982, 1983a; King *et al.*, 1983; Phelps and Zeikus, 1984, 1985). The reason for this is because of the relatively high experimental investment for the determination of substrate utilization rates.

The *in situ* rates of substrate utilization (v) are usually determined by measuring the turnover time (t_d) and the concentration (C) of the substrate pool. If steady state conditions exist, i.e. turnover rate = production rate = utilization rate, the substrate utilization rate is given by

$$v = C/t_d$$

The turnover time of a substrate is usually measured by addition of low amounts of labelled substrate to the aquatic sample, avoiding significant change of the substrate pool concentration. Then, the time course of decrease of the label in the available substrate pool is determined and the turnover time is calculated from its logarithmic decrease (e.g., Lovley and

Klug, 1982; Phelps and Zeikus, 1984). Basically, the turnover time of a substrate may be determined by addition of unlabelled substrate, and measuring the first order rate constant (k) of its utilization. However, it must be verified that the simultaneous production of the substrate is negligible during the incubation, and, thus, the measured rate is due to gross and not to net utilization. This precondition is met when the experiment is started with a substrate concentration that is significantly higher than the steady state pool concentration, but is sufficiently low that the substrate is utilized by a first order reaction. Otherwise, the steady state pool concentration must be considered in calculating the turnover time from the logarithmic decrease of substrate concentration (e.g. Conrad and Seiler, 1980):

$$k = 1/t_d = \frac{1}{t} \ln \frac{C_o - C_s}{C_t - C_s}$$

where $C_0 = C$ at time zero, $C_t = C$ at time t, and C_s = steady state concentration.

The resulting value gives the turnover time of the substrate, which is biologically available. If there are pools of different availability as in the case of acetate turnover in marine sediments (Ansbaek and Blackburn, 1980) the utilization rate will be overestimated when calculated by dividing the concentration of the total substrate pool rather than the available substrate pool by the measured turnover time. Unfortunately, there is no easy way to determine the available substrate pool. In fact, the existence of different substrate pools is only perceived, when substrate turnover rates do not match rates of CH_4 production or sulphate reduction. Determining hydrogen turnover rates in Lake Mendota sediments, Conrad *et al.* (1985; 1987) observed that the turnover of the hydrogen pool was much too low to account for methanogenesis. Apparently, an additional pool was involved in hydrogen turnover but was not detected during the measurements. This hydrogen turnover was most probably due to interspecies hydrogen transfer between juxta-positioned syntrophic hydrogen-producing and methanogenic bacteria.

These examples show that the actual utilization rate of a substrate may be over- or underestimated when pools with different availabilities exist. Another drawback of the turnover method is the necessity of determining to what extent other transformation reactions than methanogenesis contribute to the consumption of the CH_4 precursor. In addition to sulphate reduction, which is predominant in marine sediments (Nedwell, 1984) and may also be highly significant in freshwater sediments (Smith and Klug, 1981; Ingvorsen and Brock, 1982; Lovley and Klug, 1983b), CH_4 precursors may be utilized by bacterial reduction of ferric iron (Soerensen,

1982; Jones et al., 1983a; Lovley and Phillips, 1986). Therefore, it is necessary to test the contribution of other mineralization reactions to the turnover of CH_4 precursors by determining the RI coefficient.

12.4.3 Stimulation experiments

In these experiments, the stimulation of CH_4 production after addition of potential CH_4 precursors is measured. Stimulation experiments may be carried out easily. However, they can give only a qualitative impression of which CH_4 precursors might be important under in situ conditions. Stimulation experiments do not allow any quantitative evaluation of contribution of substrates to methanogenesis, and even qualitative interpretations have to be made with great care.

Addition of a potential CH_4 precursor increases the pool of this substrate in the aquatic sample. This increase may have different effects, as follows:

1. It stimulates the activity of resident methanogens, which also utilize this particular CH_4 precursor under in situ conditions.
2. It stimulates metabolically active methanogens to induce the catabolic enzymes required for utilization of this particular CH_4 precursor.
3. It stimulates growth of dormant methanogens having the potential to utilize the particular CH_4 precursor.

Hence, stimulation experiments will either show if a substrate is a CH_4 precursor under in situ conditions (case 1) or will demonstrate the potential capacity of metabolically active and dormant methanogens to utilize the substrate (case 2 and 3). Despite the chance of ambiguous interpretation, stimulation experiments may reveal much about carbon flow and methanogenesis, especially when combined with tracer, inhibition, and kinetic experiments (e.g., Winfrey et al., 1977; Zinder and Brock, 1978; Oremland and Polcin, 1982; Winfrey and Ward, 1983; King, 1984a).

12.5 CARBON FLOW ANALYSIS

Methanogenesis is just one of several terminal reactions in anaerobic mineralization of organic matter. Depending on the environmental conditions, methanogens have, for example, to compete with sulphate-reducing bacteria (Winfrey and Zeikus, 1977) or with (ferric) iron-reducing bacteria (Lovley and Phillips, 1986) for common substrates. This competition is a very important factor in direction of carbon flow and the biogeochemical role of methanogenic bacteria. Another important factor is syntrophy of methanogens to the bacteria which provide the immediate CH_4 precursors by metabolizing the more complex organic compounds

(Zehnder, 1978). Recent evidence indicates that the biogeochemistry of CH_4 in anoxic aquatic environments is even more complicated for CH_4 is not necessarily the end product of the mineralization chain, but may be re-oxidized by processes which are as yet incompletely understood. Re-oxidation of CH_4 apparently is taking place in the rhizosphere of submerged anoxic paddy soils (Holzapfel *et al.*, 1985, 1986), and, most probably, in vegetated littoral sediments. Even more amazingly, CH_4 oxidation seems to be a significant process in the anoxic sulphate reduction zone of marine sediments indicating the operation of anaerobic CH_4 oxidation processes (Devol, 1983; Alperin and Reeburgh, 1984; 1985; Iversen and Joergensen, 1985). For the study of competition, syntrophy and carbon flow in methanogenic aquatic environments a number of techniques may be useful, namely tracer experiments, kinetic experiments and $\delta^{13}C$ analysis.

12.5.1 Tracer experiments

An easy and meaningful approach to study carbon flow in aquatic environments is the measurement of the respiratory index (RI) of a ^{14}C-labelled substrate:

$$RI = {}^{14}CO_2/({}^{14}CO_2 + {}^{14}CH_4)$$

In analogy, one can define a methanogenic index (MI) according to:

$$MI = 1 - RI = {}^{14}CH_4/({}^{14}CO_2 + {}^{14}CH_4)$$

The method of determination of the RI value has been introduced by Cappenberg *et al.* (1978), and has subsequently been used by many researchers in analysis of environmental samples (see Winfrey and Zeikus, 1979a; 1979b; Winfrey *et al.*, 1981; Winfrey and Ward, 1983, Mountford *et al.*, 1980; Mountford and Asher, 1981; Smith and Klug, 1981; Lovley and Klug 1982; Sansone and Martens, 1981; Sleat and Robinson, 1983; Phelps and Zeikus, 1985). The RI value gives the percentage flow of labeled carbon atoms to CO_2 versus CH_4 and, thus, should be preferred to the use $^{14}CO_2/^{14}CH_4$ ratios which only indicate the relative amount of the gaseous products (King and Wiebe, 1980; King *et al.*, 1983).

The RI or MI value of a labelled substrate is determined by measuring the radioactivity in the CH_4 and CO_2 fractions. It is not necessary to measure concentrations of substrate, i.e. CH_4 or CO_2, and, thus, the RI value is much more easily determined than the specific radioactivities required for analysis of contribution of precursors to CH_4 production. Since CH_4 is only produced by methanogens but not by other anaerobic micro-organisms, which only form CO_2 during metabolism, the RI and MI values give a quantitative figure for the involvement of methanogens in the mineralization process. However, the contribution of methanogens to total mineralization must be interpreted with respect to the kind of substrate and to the position which is labelled with ^{14}C. The following overview shows the RI and MI values which would be obtained if the substrates had been exclusively metabolized by methanogens:

$$^{14}CH_3 - COOH \quad = \quad ^{14}CH_4 + CO_2; \qquad RI = 0 \qquad ; \ MI = 1.0$$
$$CH_3 - {}^{14}COOH \quad = \quad ^{14}CH_4 + CO_2; \qquad RI = 1.0 \quad ; \ MI = 0$$
$$^{14}CH_3 - {}^{14}COOH \quad = \quad ^{14}CH_4 + {}^{14}CO_2; \qquad RI = 0.5 \quad ; \ MI = 0.5$$
$$4 \ {}^{14}CO + H_2O \quad = \quad ^{14}CH_4 + 3 \ {}^{14}CO_2; \qquad RI = 0.75 \ ; \ MI = 0.25$$
$$4 \ {}^{14}CH_3 OH \quad = 3 \ {}^{14}CH_4 + {}^{14}CO_2 + 2 H_2O; \quad RI = 0.25 \ ; \ MI = 0.75$$

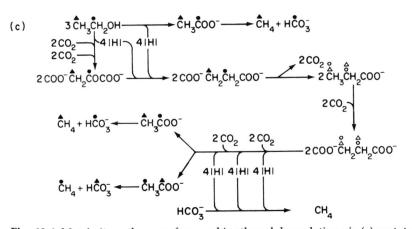

Fig. 12.4 Metabolic pathways of anaerobic ethanol degradation via (a) acetate, (b) butyrate or (c) propionate. Labelled carbon atoms are indicated by symbols: •. C-1; ▲, C-2 (original positions of the labels in ethanol); ○ and △ are used when there are two possible positions for the label (from Schink *et al.* (1985) and reproduced by permission of The Society for General Microbiology, Reading).

Any contribution of sulphate reducers or other end-mineralizing bacteria in utilizing the potential CH_4 precursors would tend to increase the RI and decrease the MI value according to the contribution.

Interpretation of RI and MI values is being complicated by application of position-labelled substrates which are more complex than the immediate CH_4 precursors. In this case, the MI or RI values do not only indicate the contribution of the methanogens to the mineralization process of the labelled carbon atom, but also give an idea by which pathway the substrate is degraded to CH_4 precursors. Therefore, degradation pathways of complex substrates are best studied in samples where methanogenesis is the only significant end-mineralizing process. The rational for elucidation of anaerobic degradation pathways is illustrated in Fig. 12.4 for the degradation of ethanol (Schink *et al.*, 1985). The distribution of label allows to distinguish between metabolic routes of ethanol-carbon via acetate, butyrate or propionate. The same approach has been used for analysis of the degradation pathways of propionate and succinate in aquatic environments (Schink, 1985a). The information provided by the RI or MI values is usually supplemented by analysing radioactivity in the pools of potential intermediates and by doing kinetic and inhibition experiments (e.g. Culbertson *et al.*, 1981; Lovley and Klug, 1982; Soerensen *et al.*, 1981).

12.5.2 Kinetic experiments

Kinetic experiments have been used by many investigators to study competition between methanogenic bacteria and sulphate reducers for common substrates. Most of these studies have been done by applying inhibitors specific for sulphate reducers or methanogens or by using inhibitory substrate analogues, in order to see how substrate utilization and product formation are affected. Molybdate (2–20 mM) is regarded as an inhibitor for sulphate reducers; choloroform (0.003–0.5%, v/v), bromoethane sulfonate (2–100 mM), acetylene or ethylene (1–20%) as inhibitors for methanogens; and fluoroacetate (1–50 mM) as an inhibitor for acetate utilization (Cappenberg, 1974; Banat *et al.*, 1981, 1983; Nedwell and Banat, 1981; Jones *et al.*, 1982; Winfrey and Ward, 1983; King, 1984a; Dicker and Smith, 1985; Raimbault, 1975; Oremland and Taylor, 1975; Knowles, 1979; Schink, 1985b). Care must be taken, however, since some natural samples need higher inhibitor concentrations than others for full efficiency and, sometimes, unspecific effects may be caused by an inhibitor. Acetylene is degradable under anoxic conditions (Culbertson *et al.*, 1981; Schink, 1985c) and, thus, may loose its inhibitory effect during incubation.

Kinetic experiments are usually carried out to determine the kinetic parameters of substrate utilization, i.e. maximum velocity (V_{max}), first order rate constant (k) of utilization, and the K_m value. Usually it is assumed that the substrate is utilized according to Michaelis–Menten kinetics and that no growth is taking place. Then, the following equations apply:

generally: $v = V_{max}C/(K_m + C)$
for $C << K_m$: $v = kC = C/t_d$

V_{max} is determined at saturating substrate concentrations that decrease linearly with time; and k is determined at rate-limiting substrate concentrations that decrease logarithmically with time. For determination of k, it must be insured that the simultaneous production of substrate is negligible, or is accounted for. The K_m can be approximated by

$$K_m = V_{max}/k$$

Usually, however, V_{max} and K_m are determined from progress kinetic experiments. The data are analysed by determining v at different C reached during progress of substrate utilization, by non-linear regression of C versus time to the integrated Michaelis–Menten equation (Robinson, 1985) or by regression to the linearized integrated Michaelis–Menten equation (Strayer and Tiedje, 1978b):

$$\frac{\ln C_o/C_t}{t} = \frac{1}{K_m}\frac{C_o - C_t}{t} + \frac{V_{max}}{K_m}$$

By knowing V_{max} and K_m of substrate utilization in a particular aquatic environment, the actual substrate utilization rate can be calculated for any given substrate concentration. This allows modelling of competition between different micro-organisms for the substrate and, finally, helps to understand electron and carbon flow in the methanogenic community as well as adaptation to environmental conditions.

Kinetic analysis of aquatic sediments has predominantly been used for study of hydrogen utilization by the methanogenic and sulphate-reducing microbial flora (Strayer and Tiedje, 1978a,b; Robinson and Tiedje, 1982; Lovley et al., 1982; Lovley and Klug, 1983b; Conrad et al., 1985; 1986a; 1987). However, the kinetic analysis of hydrogen consumption data is biased, if hydrogen is measured in a gaseous headspace. In this case, the rate-limiting step in hydrogen utilization may be the transfer of hydrogen from the headspace into the solution of the aquatic sample, rather than uptake by the utilizing bacteria (Robinson and Tiedje, 1982; Dolfing, 1985). Gas transfer limitation exists, if the utilization rate is not proportional to the amount of active sample and, then, the K_m may be greatly overestimated. Using headspace analysis, gas transfer limitation can only be avoided by dilution of the sample which, however, alters the concentrations of all compounds in the sample and may give unrealistic results. Another disadvantage of headspace analysis, in case of the poorly soluble hydrogen is the slow response time due to the relatively large hydrogen reservoir in the headspace. This requires relatively long incubation periods with increasing danger that the kinetics are influenced by growth of the hydrogen-utilizing microbial population., A further complication is the necessity to account for the simultaneous production of hydrogen which takes place in the aqueous phase but is only detected in the gaseous phase (Robinson and Tiedje, 1982).

Brief incubation experiments without headspace avoid all these problems, but require sensitive hydrogen probes. Polarographic hydrogen electrodes (e.g. Hanus *et al.*, 1980; Sweet *et al.*, 1980) have been used in pure culture studies of aerobic hydrogen-producing or consuming microorganisms. These electrodes, however, are not very suitable for measuring kinetic parameters in methanogenic systems. They are desensitized by sulphide, are sensitized at high hydrogen concentrations and generally exhibit a relatively unstable sensitivity which makes absolute calibration rather difficult. Furthermore, their response time increases at decreasing hydrogen concentrations, which precludes the determination of progress kinetics. In contrast to polarographic hydrogen electrodes, membrane inlet mass spectrometry provides a new exact technique to measure gases in dissolved state. Hydrogen or deuterated hydrogen can be measured directly in the aqueous phase simultaneously with other gas species (e.g. CH_4) (Hillman *et al.*, 1985; Scott *et al.*, 1983). The detection limit of electrodes or the mass spectrometer is in the range of $0.1\,\mu M$ hydrogen and, thus, is sufficient for analysis of the kinetic parameters V_{max} and K_m in methanogenic samples.

However, the detection limit is not sufficient for analysing *in situ* concentrations of dissolved hydrogen in lake sediments or anoxic hypolimnetic water which range between 5 and $50\,nM$ (Goodwin *et al.*, 1987; Conrad *et al.*, 1987) and $<3.5\,nM$ (Conrad *et al.*, 1983; Scranton *et al.*, 1984), respectively. Hydrogen at this low concentration can only be measured after extraction by using a hydrogen analyser based on the HgO-to-Hg vapour conversion technique (Schmidt and Seiler, 1970) which allows the detection of $>0.04\,nM$ H_2 (Conrad *et al.*, 1983; Scranton *et al.*, 1984). Headspace-free kinetic measurements at this low hydrogen concentration range may be done by incubation of samples in a syringe, from which subsamples are taken subsequently and extracted for analysis of hydrogen (Conrad *et al.*, 1985). Beyond that, the extraction technique can generally be applied for measuring kinetics of gas reactions in aqueous solution. The scheme of the procedure for extraction and analysis is shown in Fig. 2.5.

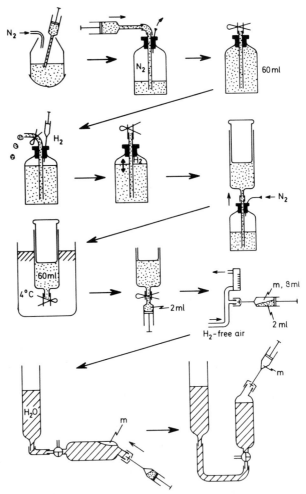

Fig. 12.5. Procedure for extraction and analysis of dissolved gases for kinetic studies without gas headspace. The technique is based on the measurement of the mixing ratio (m) of the gas after extraction from the liquid phase (from Conrad *et al.* (1985) and reproduced by permission of the American Society for Microbiology, Washington DC).

12.5.3 δ^{13}C Analysis

Analysis of δ^{13}C in aquatic pools of CH_4, CO_2 or organic carbon has the great advantage of providing information about biogeochemical processes under *in situ* conditions, without disturbance of the environment by adding substrates or retrieving samples for incubation. To date, $\delta^{13}C$ analysis is the only existing non-destructive technique to study biogeochemistry of carbon under *in situ* conditions. The natural samples are simply analysed for their ^{13}C and ^{12}C ratios using mass spectrometry. Carbon isotope ratios ($\delta^{13}C$) of samples are expressed as:

$$\delta^{13}C \ (\%) = \left(\frac{(^{13}C/^{12}C)\,Sample}{(^{13}C/^{12}C)\,PDB} - 1 \right) 1000$$

where PDB is the Peedee Belemnite carbonate standard whose $\delta^{13}C$ is unity, by definition.

Biological enzyme-catalysed reactions tend to transform ^{12}C-containing compounds more rapidly than ^{13}C-containing compounds, so that the product of a biological reaction will have more negative $\delta^{13}C$ values than the substrate. Hence, the pool of CO_2 becomes heavier of ^{13}C, when organic matter is synthesized, or CH_4 is formed from CO_2. Conversely, biogenic CH_4, will be much lighter than CO_2, because methanogenic bacteria fractionate against $^{13}CO_2$ by a fractionation factor of 1.02–1.06 (Heyer *et al.*, 1976; Games *et al.*, 1978; Belyaev *et al.*, 1983). The $\delta^{13}C$ values of CH_4 in aquatic sediments depend on the $\delta^{13}C$ values of the CH_4 precursors (e.g. CO_2), but usually are in the range of -40% to -94% (Claypool and Kvenvolden, 1983). Such negative values are indicative for the biogenic origin of CH_4 in contrast to thermogenic origin as in some deep sea sediments.

Beyond the indication of biogenic origin, however, the $\delta^{13}C$ analysis may be used to study the process of CH_4 biogeochemistry in more detail. In the past, $\delta^{13}C$ analysis of CH_4 in vertical profiles in aquatic environments have been used as an indication of anaerobic CH_4 oxidation (Alperin and Reeburgh, 1984; Oremland and DesMarais, 1983). In future, more information may be obtained when $\delta^{13}C$ values are determined simultaneously in the different carbon pools, e.g. CO_2, CH_4, organic carbon and are supplemented by study of δ^2H and $\delta^{18}O$ (Claypool and Kvenholden, 1983; Schoell, 1980; Woltemate *et al.*, 1984; Schidlowski *et al.*, 1985) and, most importantly, when we have a better knowledge of isotope fractionation by the different bacteria involved in biogeochemistry of carbon. Analysis of stable isotopes certainly is a powerful technique, which presently is not yet used to the extent deserved for environmental research.

Acknowledgement

We thank Professor D. N. Pfennig for critically reading the manuscript.

12.6 REFERENCES

Ahring, B. K. and Westermann, P. (1984). Isolation and characterization of a thermophilic, acetate-utilizing methanogenic bacterium. *FEMS Microbiology Letters*, **25**, 47–52.

Ahring, B. K. and Westermann, P. (1985). Methanogenesis from acetate: physiology of a thermophilic, acetate-utilizing methanogenic bacterium. *FEMS Microbiology Letters*, **28**, 15–19.

Alperin, M. J. and Reeburgh, W. S. (1984). Geochemical observations supporting anaerobic methane oxidation. In *Microbial Growth on C1 Compounds*, R. L. Crawford and R. S. Hanson (Eds), American Society for Microbiology, Washington, DC, pp. 282–289.

Alperin, M. J. and Reeburgh W. S. (1985). Inhibition experiments on anaerobic methane oxidation. *Applied and Environmental Microbiology*, **50**, 940–945.

Ansbaek, J. and Blackburn, T. H. (1980). A method for the analysis of acetate turnover in a coastal marine sediment. *Microbial Ecology*, **5**, 253–264.

Balch, W. E. and Wolfe, R. S. (1976). New approach to the cultivation of methanogenic bacteria: 2-mercaptoethanesulfonic acid(HS-CoM)-dependent growth of *Methanobacterium ruminantium* in a pressurized atmosphere. *Applied and Environmental Microbiology*, **32**, 781–791.

Balch, W. E., Fox, G. E., Magrum, L. J., Woese, C. R. and Wolfe, R. S. (1979). Methanogens: reevaluation of a unique biological group. *Microbiological Reviews*, **43**, 260–296.

Banat, I. M., Lindström, E. B., Nedwell, D. B. and Balba, M. T. (1981). Evidence for coexistence of two distinct functional groups of sulfate-reducing bacteria in salt marsh sediment. *Applied and Environmental Microbiology*, **42**, 985–992.

Banat, I. M., Nedwell, D. B. and Balba, M. T. (1983). Stimulation of methanogenesis by slurries of saltmarsh sediment after the addition of molybdate to inhibit sulphate-reducing bacteria. *Journal of General Microbiology*, **129**, 123–129.

Belay N., Sparling, R. and Daniels, L. (1984). Dinitrogen fixation by a thermophilic methanogenic bacterium. *Nature*, **312**, 286–288.

Belyaev, S. S. Finkelstein, Z. I. and Ivanov, M. V. (1975). Intensity of bacterial methane formation in ooze deposits of certain lakes. *Mikrobiologyia*, **44**, 272–275.

Belyaev, S. S., Wolkin, R. Kenealy, W. R., DeNiro, M. J., Epstein, S. and Zeikus, J. G. (1983). Methanogenic bacteria from Bondyuzhskoe oil field: general characterization and analysis of stable-carbon isotopic fractionation. *Applied and Environmental Microbiology*, **45**, 691–697.

Bhatnagar, L., Jain, M. K., Aubert, J.-P. and Zeikus, J. G. (1984). Comparison of assimilatory organic nitrogen, sulfur, and carbon sources for growth of *Methanobacterium* species. *Applied and Environmental Microbiology*, **48**, 785–790.

Blake, D. R. and Rowland, F. S. (1986). World-wide increase in tropospheric methane, 1978–1983. *Journal of Atmospheric Chemistry*, **4**, 43–62.

Blotevogel, K. H., Fischer, U., Mocha, M. and Jannsen, S. (1985). *Methanobacterium thermoalkaliphilium* spec. nov., a new moderately alkaliphilic and thermophilic autotrophic methanogen. *Archives of Microbiology*, **142**, 211–217.

Blotevogel, K. H., Fischer, U. and Lüpkes, K. H. (1986). *Methanococcus frisius* sp. nov., a new methlotrophic marine methanogen. *Canadian Journal of Microbiology*, **32**, 127–131.

Böck, A and Kandler, O. (1985). Antibiotic sensitivity of Archaebacteria. In *The Bacteria*, vol. 8, *Archaebacteria*, C. R. Woese and R. S. Wolfe (Eds), Academic Press, Orlando, FL, pp. 525–544.

Bomar, M., Knoll, K., and Widdel, F. (1985). Fixation of molecular nitrogen by *Methanosarcina barkeri*. *FEMS Microbiological Ecology*, **31**, 47–55.

Boone, D. R. and Bryant, M. P. (1980). Propionate-degrading bacterium, *Synthrophobacter wolinii* sp. nov. gen. nov. from methanogenic ecosystems. *Applied and Environmental Microbiology*, **40**, 626–632.

Boone, D. R., Worakit, S., Mathrani, I. M. and Mah, R. A. (1986). Alkaliphilic methanogens from high-pH lake sediments. *Systematic and Applied Microbiology*, **7**, 230–234.

Bott M. H., Eikmanns, B. and Thauer, R. K. (1985). Defective formation and/or utilization of carbon monoxide in H_2/CO_2 fermenting methanogens dependent on acetate as carbon source. *Archives of Microbiology*, **135**, 266–269.

Braun, M., Schoberth, S. and Gottschalk, G. (1979). Enumeration of bacteria forming acetate from hydrogen and carbon dioxide in anaerobic habitats. *Archives of Microbiology*, **120**, 201–204.

Bryant, M. P., Wolin, E. A., Wolin, M. J. and Wolfe, R. S. (1967). *Methanobacillus omelianskii*, a symbiontic association of two species of bacteria. *Archives of Microbiology*, **59**, 20–31.

Burke Jr., R. A., Reid, D. F., Brooks, J. M. and Lavoie, D. M. (1983). Upper water column methane geochemistry in the eastern tropical North Pacific. *Limnology and Oceanography*, **28**, 19–32.

Cappenberg, T. E. (1974). Interrelations between sulfate-reducing and methane-producing bacteria in bottom deposits of a fresh-water lake. 2. Inhibition experiments. *Antonie van Leeuwenhoek*, **40**, 297–306.

Cappenberg, T. E. and Prins, R. A. (1974). Interactions between sulfate-reducing and methane-producing bacteria in bottom deposits of a freshwater lake. 3. Experiments with ^{14}C-labeled substrates. *Antonie van Leeuwenhoek*, **40**, 457–469.

Cappenberg, T. E., Jongejan, E. and Kaper, J. (1978). Anaerobic breakdown processes of organic matter in freshwater sediments. In *Microbial Ecology*, M.'W. Loutit and J. A. R. Miles (Eds), Springer, Berlin, pp. 91–99.

Cheeseman, P., Toms-Wood, A. and Wolfe, R. S. (1972). Isolation and properties of a fluorescent compound factor[420] from *Methanobacterium* strain M.o.H. *Journal of Bacteriology*, **112**, 527–531.

Christensen, D. and Blackburn, T. H. (1982). Turnover of ^{14}C-labelled acetate in marine sediments. *Marine Biology*, **71**, 113–119.

Cicerone, R. J. and Shetter, J. D. (1981). Sources of atmospheric methane: measurments in rice paddies and a discussion. *Journal of Geophysical Research*, **86**, 7203–7209.

Claypool, G. E. and Kvenvolden, K. A. (1983). Methane and other hydrocarbon gases in marine sediments. *Annual Review of Earth and Planetary Science*, **11**, 299–327.

Conrad, R. and Seiler, W. (1980). Role of microorganisms in the consumption and production of atmospheric carbon monoxide by soil. *Applied and Environmental Microbiology*, **40**, 437–445.

Conrad, R. and Thauer, R. K. (1983). Carbon monoxide production by *Methanobacterium thermoautotrophicum*. *FEMS Microbiology Letters*, **20**, 229–232.

Conrad, R., Aragno, M. and Seiler, W. (1983). Production and consumption of hydrogen in a eutrophic lake. *Applied and Environmental Microbiology*, **45**, 502–510.

Conrad, R., Lupton, F. S. and Zeikus, J. G. (1987). Hydrogen metabolism and sulfate-dependent inhibition of methanogenesis in a eutrophic lake sediment (Lake Mendota). *FEMS Microbiological Ecology*, **45**, 107–115.

Conrad, R., Phelps, T. J. and Zeikus, J. G. (1985). Gas metabolism evidence in support of juxtapositioning between hydrogen producing and methanogenic bacteria in sewage sludge and lake sediments. *Applied and Environmental Microbiology*, **50**, 595–601.

Corder, R. E., Hook, L. A., Larkin, J. M. and Frea, J. I. (1983). Isolation and characterization of two new methane producing cocci: *Methanogenium olentangyi*, sp. nov., and *Methanococcus deltae*, sp. nov. *Archives of Microbiology*, **134**, 28–32.

Culbertson, C. W., Zehnder, A. J. B. and Oremland, R. S. (1981). Anaerobic oxidation of acetylene by estuarine sediments and enrichment cultures. *Applied and Environmental Microbiology*, **41**, 396–403.

Dacey, J. W. H. and Klug, M. J. (1979). Methane efflux from lake sediments through water lilies. *Science*, **203**, 1253–1255.

Daniels, L., Sparling, R. and Sportt, G. D. (1984). The bioenergetics of methanogenesis. *Biochimica et Biophysica Acta*, **768**, 113–163.

Daniels, L., Belay, N. and Rajagopal, B. S. (1986). Assimilatory reduction of sulfate and sulfite by methanogenic bacteria. *Applied and Environmental Microbiology*, **51**, 703–709.

Daughton, C. G., Cook, A. M. and Alexander, M. (1979). Biodegradation of phosphonate toxicants yields methane or ethane on cleavage of the C-P bond. *FEMS Microbiology Letters*, **5**, 91–93.

DeLaune, R. D., Smith, C. J. and Patrick, W. H. (1983). Methane release from gulf coast wetlands. *Tellus*, **35B**, 8–15.

Devol, A. H. (1983). Methane oxidation rates in the anaerobic sediments of Saanich Inlet. *Limnology and Oceanography*, **28**, 738–742.

Dicker, H. J. and Smith, D. W. (1985). Metabolism of low molecular weight organic compounds by sulfate-reducing bacteria in a Delaware salt marsh. *Microbial Ecology*, **11**, 317–336.

Doddema, H. J. and Vogels, G. D. (1978). Improved identification of methanogenic bacteria by fluorescense microscopy. *Applied and Environmental Microbiology*, **36**, 752–754.

Dolfing, J. (1985). Kinetics of methane formation by granular sludge at low substrate concentrations: the influence of mass transfer limitation. *Applied Microbiology and Biotechnology*, **22**, 77–81.

Dolfing, J. and Mulder, J.-W. (1985). Comparison of methane production rate and coenzyme F_{420} content of methanogenic consortia in anaerobic granular sludge. *Applied and Environmental Microbiology*, **49**, 1142–1145.

Dubach, A. C. and Bachofen, R. (1985). Methanogens: a short taxonomic review. *Experientia*, **41**, 441–445.

Edwards, T. and McBride, B. C. (1975). New method for the isolation and identification of methanogenic bacteria. *Applied Microbiology*, **29**, 540–545.

Ehhalt, D. H. (1979). Der atmosphärische Kreislauf von Methan. *Naturwissenschaften*, **66**, 307–311.

Eichler, B. and Schink, B. (1984). Oxidation of primary aliphatic alcohols by *Acetobacterium carbinolicum* sp. nov., a homoacetogenic anaerobe. *Archives of Microbiology*, **140**, 147–152.

Eikmanns, B., Fuchs, G. and Thauer, R. K. (1985). Formation of carbon monoxide from CO_2 and H_2 by *Methanobacterium thermoautotrophicum*. *European Journal of Biochemistry*, **146**, 149–154.

Eirich, L. D., Vogels, G. D. and Wolfe, R. S. (1978). Proposed structure for coenzyme F_{420} from *Methanobacterium*. *Biochemistry*, **17**, 4583–4593.

Eker, A. P. M. and Pol, A., van der Meijden, P. and Vogels, G. D. (1980). Purification and properties of 8-hydroxy-5-deazaflavin derivatives from *Streptomyces griseus*. *FEMS Microbiology Letters*, **8**, 161–165.

Farquhar, G. J. and Rovers, F. A. (1973). Gas production during refuse decomposition. *Water, Air and Soil Pollution*, **2**, 483–495.

Fathepure, B. Z. (1983). Isolation and characterization of an aceticlastic methanogen from biogas digestor. *FEMS Microbiology Letters*, **19**, 151–165.

Fallon, R. D., Harrits, S., Hanson, R. S. and Brock, T. D. (1980). The role of methane in internal carbon cycling in Lake Mendota during summer stratification. *Limnology and Oceanography*, **25**, 357–360.

Ferguson, T. J. and Mah, R. E. (1983). Effect of $H_2–CO_2$ on methanogenesis from acetata or methanol in *Methanosarcina* spp. *Applied and Environmental Microbiology*, **46**, 348–355.

Fuchs, G. and Stupperich, E. (1985). Evolution of autotrophic CO_2 fixation. In *Evolution of Prokaryotes*, K. H. Schleifer and E. Stackebrandt (Eds) Academic Press, London, pp. 235–251.

Games, L. M., Hayes, J. M. and Gunsalus, R. P. (1978). Methane-producing bacteria: natural fractionations of stable carbon isotopes. *Geochimica et Cosmochimica Acta*, **42**, 1295–1297.

Giani, D., Giani, L., Cohen, Y. and Krumbein, W. E. (1984). Methanogenesis in the hypersaline Solar Lake (Sinai). *FEMS Microbiology Letters*, **25**, 219–224.

Goodwin, S., Conrad, R. and Zeikus, J. G. (1987). Ecophysiological adaptation of anaerobic bacteria to low pH: relation of pH and hydrogen metabolism in diverse sedimentary ecosystems. *Applied and Environmental Microbiology* (submitted).

Hanus, F. J., Carter, K. R. and Evans, H. J. (1980). Technique for measurement of hydrogen evolution by nodules. In *Methods in Enzymology*, Vol. 69, A. SanPietro (Ed.), Academic Press, New York, pp. 731–739.

Harris, J. E., Pinn, P. A. and Davis, R. P. (1984). Isolation and characterization of a novel thermophilic, freshwater methanogen. *Applied and Environmental Microbiology*, **48**, 1123–1128.

Harriss, R. C. and Sebacher, D. J. (1981). Methane flux in forested freshwater swamps of the southeastern United States. *Geophysical Research Letters*, **8**, 1002–1004.

Harriss, R. C., Gorham, E., Sebacher, D. I., Bartlett, K. B. and Flebbe, P. A. (1985). Methane flux from northern peatlands. *Nature*, **315**, 652–654.

Harriss, R. C., Sebacher, D. J. and Day, F. P. (1982). Methane flux in the Great Dismal Swamp. *Nature*, **297**, 673–674.

Hermann, M., Noll, K. M. and Wolfe, R. S. (1986). Improved agar bottle plate for isolation of methanogens or other anaerobes in a defined gas atmosphere. *Applied and Environmental Microbiology*, **51**, 1124–1126.

Hesslein, R. (1976). An *in-situ* sampler for close interval pore water studies. *Limnology and Oceanography*, **21**, 912–914.

Heukelikian, H. and Heinemann, B. (1939). Studies on the methane producing bacteria. I. Development of a method for enumeration. *Sewage Works Journal*, **11**, 426–435.

Heyer, J., Hübner, H. and Maass, I. (1976). Isotopenfraktionierung des Kohlenstoffs bei der mikrobiellen Methanbildung. *Isotopenpraxis*, **12**, 202–205.

Hillman, K., Lloyd, D. and Williams, A. G. (1985). Use of a portable quadropole mass spectrometer for the measurement of dissolved gas concentrations in ovine rumen liquor *in situ*. *Current Microbiology*, **12**, 335–340.

Hippe, H., Caspari, D., Fiebig, K. and Gottschalk, G. (1979). Utilization of trimethylamine and other N-methyl compounds for growth and methane formation by *Methanosarcina barkeri*. *Proceedings of the National Academy of Sciences of the USA*, **76**, 494–498.

Holland, H. D., Lazar, B. and McCaffrey, M. (1986). Evolution of the atmosphere and oceans. *Nature*, **320**, 27–33.

Holzapfel-Pschorn, A. and Seiler, W. (1986). Methane emission during a vegetation period from an Italian rice paddy. *Journal of Geophysical Research*, **91**, 11803–11814.

Holzapfel-Pschorn, A., Conrad, R. and Seiler, W. (1985). Production, oxidation and emission of methane in rice paddies. *FEMS Microbiological Ecology*, **31**, 343–351.

Holzapfel-Pschorn, A., Conrad, R. and Seiler, W. (1986). Effects of vegetation on the emission of methane by submerged paddy soil. *Plant and Soil*, **92**, 223–233.

Howes, B. L., Dacey, J. W. H. and Wakeham, S. G. (1985). Effects of sampling technique on measurements of porewater constituents in salt marsh sediments. *Limnology and Oceanography*, **30**, 221–227.

Hungate, R. E. (1966). *The Rumen and its Microbes*, Academic Press, New York.

Hungate, R. E. (1969). A roll tube method for the cultivation of strict anaerobes. In *Methods in Microbiology*, Eds. J. R. Norris and D. W. Ribbons (Eds), Academic Press, London and New York, pp. 117–132.

Huser, B. A., Wuhrmann, K. and Zehnder, A. J. B. (1982). *Methanothrix soehngenii* gen. nov. sp. nov., a new acetotrophic non-hydrogen-oxidizing methane bacterium. *Archives of Microbiology*, **132**, 1–9.

Ingvorsen, K. and Brock, T. D. (1982). Electron flow via sulfate reduction and methanogenesis in the anaerobic hypolimnion of Lake Mendota. *Applied and Environmental Microbiology*, **27**, 559–564.

Iversen, N. and Blackburn, T. H. (1981). Seasonal rates of methane oxidation in anoxic sediments. *Applied and Environmental Microbiology*, **41**, 1295–1300.

Iversen, N. and Joergensen, B. B. (1985). Anaerobic methane oxidation rates at the sulfate methane transition in marine sediments from Kattegat and Skagerak (Denmark). *Limnology and Oceanography*, **30**, 944–955.

Joergensen, B. B. (1977). The sulfur cycle of a coastal marine sediment (Limfjorden, Denmark), *Limnology and Oceanography*, **22**, 814–832.

Joergensen, B. B. and Revsbech, N. P. (1985). Diffusive boundary layers and the oxygen uptake of sediment and detritus. *Limnology and Oceanography*, **30**, 111–122.

Jones, J. G. and Simon, B. M. (1985). Interaction of acetogens and methanogens in anaerobic freshwater sediments. *Applied and Environmental Microbiology*, **49**, 944–948.

Jones, J. G., Simon, B. M. and Gardener, S. (1982). Factors affecting methanogenesis and associated anaerobic processes in the sediments of a stratified eutrophic lake. *Journal of General Microbiology*, **128**, 1–11.

Jones, J. G., Gardener, S. and Simon, B. M. (1983a). Bacterial reduction of ferric iron in a stratified eutrophic lake. *Journal of General Microbiology*, **129**, 131–139.

Jones, W. J., Leigh, J. A., Mayer, F., Woese, C. R. and Wolfe, R. S. (1983b). *Methanococcus jannaschii* sp. nov., an extremely thermophilic methanogen from a submarine hydrothermal vent. *Archives of Microbiology*, **136**, 254–261.

Jones, W. J., Whitman, W. B., Fields, R. D. and Wolfe, R. S. (1983c). Growth and plating efficiency of Methanococci on agar media. *Applied and Environmental Microbiology*, **46**, 220–226.

Kandler, O. (1979). Zellwandstrukturen bei Methanbakterien. Zur Evolution der Prokaryonten. *Naturwissenschaften*, **66**, 95–105.

Kandler, K. O. and Zillig, W. (Eds) (1985). *Archaebacteria'85*, Fischer, Stuttgart.

Kiene, R. P. and Capone, D. G. (1985). Degassing of pore water methane during sediment incubation. *Applied and Environmental Microbiology*, **49**, 143–147.

Kiener, A. and Leisinger, T. (1983). Oxygen sensitivity of methanogenic bacteria. *Systematic and Applied Microbiology*, **4**, 305–312.

King, G. M. (1984a). Metabolism of trimethylamine, choline, and glycine betaine by sulfate-reducing and methanogenic bacteria in marine sediments. *Applied and Environmental Microbiology*, **48**, 719–725.

King, G. M. (1984b). Utilization of hydrogen, acetate and 'noncompetitive' substrates by methanogenic bacteria in marine sediments. *Geomicrobiological Journal*, **3**, 275–306.

King, G. M. and Wiebe, W. J. (1978). Methane release from soils of a Georgia salt marsh. *Geochimica et Cosmochimica Acta*, **42**, 343–348.

King, G. M. and Wiebe, W. J. (1980). Tracer analysis of methanogenesis in salt marsh soils. *Applied and Environmental Microbiology*, **39**, 877–881.

King, G. M., Klug, M. J. and Lovley, D. R. (1983). Metabolism of acetate, methanol, and methylated amines in intertidal sediments of Lowes Cove, Maine. *Applied and Environmental Microbiology*, **45**, 1848–1853.

Kipphut, G. W. and Martens, C. S. (1982). Biogeochemical cycling in an organic-rich coastal marine basin. 3. Dissolved gas transport in methane-saturated sediments. *Geochimica et Cosmochimica Acta*, **46**, 2049–2060.

Knowles, R. (1979). Denitrification, acetylene reduction and methane metabolism in lake sediments exposed to acetylene. *Applied and Environmental Microbiology*, **38**, 486–493.

König, H. and Stetter, K. O. (1982). Isolation and characterization of *Methanolobus tindarius* sp nov., a coccoid methanogen growing only on methanol and methylamines. *Zentralblatt für Bakteriologie, Mikrobiologie und Hygiene, Abteilung 1, Originale C*, **3**, 478–490.

Krzycki, J. A. and Zeikus, J. G. (1980). Quantification of corrinoids in methanogenic bacteria. *Current Microbiology*, **3**, 243–245.

Krzycki, J. A., Wolkin, R. H. and Zeikus, J. G. (1982). Comparison of unitrophic and mixotrophic substrate metabolism by an acetate-adapted strain of *Methanosarcina barkeri*. *Journal of Bacteriology*, **149**, 247–254.

Kuenen, J. G. and Harder, W. (1982). Microbial competition in continuous culture. In *Experimental Microbial Ecology*, R. G. Burns and J. H. Slater (Eds), Blackwell, Oxford, pp. 342–367.

Lauerer, G., Kristjansson, J. K., Langworthy, T. A., König, H. and Stetter, K. O. (1986). *Methanothermus sociabilis* sp. nov., a second species within the Methanothermaceae growing at 97 °C. *Systematic and Applied Microbiology*, **8**, 100–105.

Leigh, J. A., Rinehart Jr., K. L. and Wolfe, R. S. (1985). Methanofuran (carbon dioxide reduction factor), a formyl carrier in methane production from carbon dioxide in *Methanobacterium*. *Biochemistry*, **24**, 995–999.

Lein, A. Y., Namsaraev, B. B., Trotsyuk, V. Y. and Ivanov, M. V. (1981). Bacterial methanogenesis in Holocene sediments of the Baltic Sea. *Geomicrobiological Journal*, **2**, 299–315.

Lilley, M. D., Baross, J. A. and Gordon, L. I. (1982). Dissolved hydrogen and methane in Saanich Inlet, British Columbia. *Deep-Sea Research*, **29**, 1471–1484.

Lloyd, D., Davies, R. J. P. and Boddy, L. (1986). Mass spectrometry as an ecological tool for *in-situ* measurement of dissolved gases in sediment systems. *FEMS Microbiological Ecology*, **38**, 11–17.

Lovley, D. R. and Klug, M. J. (1982). Intermediary metabolism of organic matter in the sediments of a eutrophic lake. *Applied and Environmental Microbiology*, **43**, 552–560.

Lovley, D. R. and Klug, M. J. (1983a). Methanogenesis from methanol and methylamines and acetogenesis from hydrogen and carbon dioxide in the sediments of a eutrophic lake. *Applied and Environmental Microbiology*, **45**, 1310–1315.

Lovley, D. R., and Klug, M. J. (1983b). Sulfate reducers can outcompete methanogens at freshwater sulfate concentrations. *Applied and Environmental Microbiology*, **45**, 187–192.

Lovley, D. R. and Phillips, E. J. P. (1986). Organic matter mineralization with reduction of ferric iron in anaerobic sediments. *Applied and Environmental Microbiology*, **51**, 683–689.

Lovley, D. R., Dwyer, D. F. and Klug, M. J. (1982). Kinetic analysis of competition between sulfate reducers and methanogens for hydrogen in sediments. *Applied and Environmental Microbiology*, **43**, 1373–1379.

Lynd, L., Kerby, R. and Zeikus, J. G. (1982). Carbon monoxide metabolism of the methylotrophic acidogen *Butyribacterium methylotrophicum*. *Journal of Bacteriology*, **149**, 255–263.

Mah, R. A. (1980). Isolation and characterization of *Methanosarcina mazei*. *Current Microbiology*, **3**, 321–326.

Mah, R. A. and Smith, M. R. (1981). The methanogenic bacteria. In *The Prokaryotes*, vol. 1, M. P. Starr, H. Stolp, H. G. Trüper, A. Balows and H. G. Schlegel (Eds) Springer, Berlin, pp. 948–977.

Mah, R. A., Ward, D. M., Baresi, L. and Glass. T. L. (1977). Biogenesis of methane. *Annual Review of Microbiology*, **31**, 309–341.

Martens, C. S. and Val Klump, J. V. (1980). Biogeochemical cycling in an organic-rich coastal marine basin. I. Methane sediment–water exchange process. *Geochimica et Cosmochimica Acta*, **44**, 471–490.

Martz, R. F., Sebacher, D. I. and White, D. C. (1983). Biomass measurement of methane forming bacteria in environmental samples. *Journal of Microbiological Methods*, **1**, 53–61.

Matthess, G. (1985). Geochemical conditions in the ground water environment. In *Planetary Ecology*, D. E. Caldwell, J. A. Brierley and C. L. Brierley (Eds), Van Nostrand Reinhold, New York, pp. 347–355.

McInerney, M. J., Bryant, M. P. and Pfennig, N. (1979). Anaerobic bacterium that degrades fatty acids in syntrophic association with methanogens. *Archives of Microbiology*, **122**, 129–135.

McInerney, M. J., Bryant, M. P., Hespeel, R. B. and Costerton, J. W. (1981). *Syntrophomonas wolfei*, gen. nov. spec. nov. and anaerobic, syntrophic, fatty-acid oxidizing bacterium. *Applied and Environmental Microbiology*, **41**, 1029–1039.

Miller, T. L. and Wolin, M. J. (1974). A serum bottle modification of the Hungate technique for cultivating obligate anaerobes. *Applied Microbiology*, **27**, 985–987.

Miller, T. L. and Wolin, M. J. (1982). Enumeration of *Methanobrevibacter smithii* in human feces. *Archives of Microbiology*, **131**, 14–18.

Miller, T. L. and Wolin, M. J. (1983). Oxidation of hydrogen and reduction of methanol is the sole energy source for a methanogen isolated from human feces. *Journal of Bacteriology*, **153**, 1051–1055.

Miller, T. L. and Wolin, M. J. (1985). *Methanosphaera stadtmaniae*, gen. nov., sp. nov.: a species that forms methane by reducing methanol with hydrogen. *Archives of Microbiology*, **141**, 116–122.

Mink, R. W. and Dugan, P. R. (1977). Tentative identification of methanogenic bacteria by fluorescense microscopy. *Applied and Environmental Microbiology*, **33**, 713–717.

Mountford, D. O. and Asher, R. A. (1981). Role of sulfate reduction versus methanogenesis in terminal carbon flow in polluted intertidal sediment of Waimea Inlet, Nelson, New Zealand. *Applied and Environmental Microbiology*, **42**, 252–258.

Mountford, D. O. and Bryant, M, O, (1982). Isolation and characterization of an anaerobic syntrophic benzoate-degrading bacterium from sewage sludge. *Archives of Microbiology*, **133**, 249–256.

Mountford, D. O., Asher, R. A., Mays, E. L. and Tiedje, J. M. (1980). Carbon and electron flow in mud and sandflat intertidal sediments at Delaware inlet, Nelson, New Zealand. *Applied and Environmental Microbiology*, **39**, 686–694.

Murray, P. A. and Zinder, S. H. (1984). Nitrogen fixation by a methanogenic archaebacterium. *Nature*, **314**, 284–286.

Nedwell, D. B. (1984). The input and mineralization of organic carbon in anaerobic aquatic sediments. *Advances in Microbial Ecology*, **7**, 93–131.

Nedwell, D. B. and Banat, I. M. (1981). Hydrogen as an electron donor for sulfate-reducing bacteria in slurries of salt marsh sediment. *Microbial Ecology*, **7**, 305–313.

Nelson, D. R. and Zeikus, J. G. (1974). Rapid method for the radioisotopic analysis of gaseous end products of anaerobic metabolism. *Applied Microbiology*, **28**, 258–261.

Oremland, R. S. (1975). Methane production in shallow-water, tropical marine sediments. *Applied Microbiology*, **30**, 602–608.

Oremland, R. S. (1979). Methanogenic activity in plankton samples and fish intestines: a mechanism for in situ methanogenesis in oceanic surface waters. *Limnology and Oceanography*, **24**, 1136–1141.

Oremland, R. S. and DesMarais, D. J. (1983). Distribution, abundance and carbon isotopic composition of gaseous hydrocarbons in Big Soda Lake, Nevada; an alkaline, meromictic lake. *Geochimica et Cosmochimica Acta*, **47**, 2107–2114.

Oremland, R. S. and Polcin, S. (1982). Methanogenesis and sulfate reduction: competitive and noncompetitive substrates in estuarine sediments. *Applied and Environmental Microbiology*, **44**, 1270–1276.

Oremland, R. S. and Taylor, B. F. (1975). Inhibition of methanogenesis in marine sediments by acetylene and ethylene: validity of the acetylene reduction assay for anaerobic microcosms. *Applied Microbiology*, **30**, 707–709.

Oremland, R. S., Marsh, L. and DesMarais, D. J. (1982). Methanogenesis in Big Soda Lake Nevada: an alkaline, moderately hypersaline desert lake. *Applied and Environmental Microbiology*, **43**, 462–468.

Pace, N. R., Stahl, D. A., Lane, D. J. and Olsen, G. J. (1986). The analysis of natural microbial populations by ribosomal RNA sequences. *Advances in Microbial Ecology*, **9**, 1–56.

Parkes, R. J., Taylor, J. and Joerck-Ramberg, D. (1984). Demonstration, using *Desulfobacter* sp., of two pools of acetate, with different biological availabilities in marine pore water. *Marine Biology*, **83**, 271–276.

Patel, G. B. (1984). Characterization and nutritional properties of *Methanothrix concilii* sp. nov., a mesophilic, aceticlastic methanogen. *Canadian Journal of Microbiology*, **30**, 1383–1396.

Paterek, J. R. and Smith, P. H. (1985). Isolation and characterization of a halophilic methanogen from Great Salt Lake. Applied and Environmental Microbiology, **50**, 877–881.

Pfennig, N. (1978). *Rhodocyclus purpureus*, gen. nov. and sp. nov., a ring-shaped, vitamin B12-requiring member of the family Rhodospirillaceae. *International Journal of Systematic Bacteriology*, **28**, 283–288.

Pfennig, N. (1984). Microbial behaviour in natural environments. In *The Microbe 1984*, Part 2, *Prokaryotes and Eukaryotes* D. P. Kelley and N. G. Carr (Eds), Cambridge University Press, Cambridge, pp. 23–50.

Phelps, T. J. and Zeikus, J. G. (1984). Influence of pH on terminal carbon metabolism in anoxic sediments from a mildly acidic lake. *Applied and Environmental Microbiology*, **48**, 1088–1095.

Phelps T. J. and Zeikus, J. G. (1985) Effect of fall turnover on terminal carbon metabolism in Lake Mendota sediments. *Applied and Environmental Microbiology*, **50**, 1285–1291.

Phelps, T. J., Conrad, R. and Zeikus, J. G. (1985). Sulfate-dependent interspecies H_2 transfer between *Methanosarcina barkeri* and *Desulfovibrio vulgaris* during coculture metabolism of acetate and methanol. *Applied and Environmental Microbiology*, **50**, 589–594.

Raimbault, M. (1975). Etude de l'influence inhibitrice de l'acétylene sur la formation biologique du méthane dans un sol de rizière. *Annales de Microbiologie de l' Institut Pasteur*, **126A**, 247–258.

Rasmussen, R. A. and Khalil, M. A. K. (1981). Atmospheric methane (CH_4): trends and seasonal cycles. *Journal of Geophysical Research*, **86**, 9826–9832.

Reeburgh, W. S. (1983). Rates of biogeochemical processes in anoxic sediments. *Annual Review of Earth and Planetary Science*, **11**, 269–298.

Robinson, J. A. (1985). Determining microbial kinetic parameters using nonlinear regression analysis: advantage and limitations in microbial ecology. *Advances in Microbial Ecology*, **8**, 61–114.

Robinson, J. A. and Tiedge, J. M. (1982). Kinetics of hydrogen consumption by rumen fluid, anaerobic digestor sludge, and sediment. *Applied and Environmental Microbiology*, **44**, 1374–1384.

Rönnow, P. H. and Gunnarson, L. A. H. (1981). Sulfide-dependent methane production and growth of a thermophilic methanogenic bacterium. *Applied and Environmental Microbiology*, **42**, 580–584.

Rönnow, P. H. and Gunnarson, L. A. H. (1982). Response of growth and methane production to limiting amounts of sulfide and ammonium in two thermophilic methanogenic bacteria. *FEMS Microbiology Letters*, **14**, 311–315.

Rudd, J. W. M. and Taylor, C. D. (1980). Methane cycling in aquatic environments. *Advances in Aquatic Microbiology*, **2**, 77–150.

Sansone, F. J. and Martens, C. S. (1981). Methane production from acetate and associated methane fluxes from anoxic coastal sediments. *Science*, **211**, 707–709.

Sayler, G. S., Shields, M. S., Tedford, E. T., Breen, A., Hooper, S. W., Sirotkin, K. M. and Davis, J. W. (1985). Application of DNA-DNA colony hybridization to the detection of catabolic genotypes in environmental samples. *Applied and Environmental Microbiology*, **49**, 1295–1303.

Scherer, P. and Sahm, H. (1981). Influence of sulphur-containing compounds on the growth of *Methanosarcina barkeri* in a defined medium. *European Journal of Applied Microbiology and Biotechnology*, **12**, 28–35.

Schidlowski, M. (1980). The atmosphere. In *The Handbook of Environmental Chemistry*, vol. 1A, O. Hutzinger (ed.), Springer, Berlin, pp. 1–16.

Schidlowski, M., Matzigkeit, U., Mook, W. G. and Krumbein, W. (1985). Carbon isotope geochemistry and ^{14}C ages of microbial mats from the Gavish Sabkha and the Solar Lake. In *Ecological Studies*, Vol. 53, *Hypersaline Ecosystems* G. M. Friedman and W. E. Krumbein (Eds), Springer, Berlin, pp. 381–401.

Schink, B. (1985a). Mechanisms and kinetics of succinate and propionate degradation in anoxic freshwater sediments and sewage sludge. *Journal of General Microbiology*, **131**, 643–650.

Schink, B. (1985b). Inhibition of methanogenesis by ethylene and other unsaturated hydrocarbons. *FEMS Microbiological Ecology*, **31**, 63–68.

Schink, B. (1985c). Fermentation of acetylene by an obligate anaerobe, *Pelobacter acetylenicus* sp. nov. *Archives of Microbiology*, **142**, 295–301.

Schink, B. (1988). Principles and limits of anaerobic degradation. Environmental and technological aspects. In *Environmental Microbiology of Anaerobes*, A. J. B. Zehnder (Ed.), Wiley, New York, in press.

Schink, B. and Pfennig, N. (1982a). *Propionigenium modestum* gen. nov. sp. nov. a new strictly anaerobic, nonsporing bacterium growing on succinate. *Archives of Microbiology*, **133**, 209–216.

Schink, B. and Pfennig, N. (1982b). Fermentation of trihydroxybenzenes by *Pelobacter acidigallici* gen. nov. sp. nov., a new strictly anaerobic, non-sporeforming bacterium. *Archives of Microbiology*, **133**, 195–201.

Schink, B., Phelps, T. J., Eichler, B. and Zeikus, J. G. (1985). Comparison of ethanol degradation pathways in anoxic freshwater environments. *Journal of Microbiology*, **131**, 651–660.

Schmidt, U. and Seiler, W. (1970). A new method for recording molecular hydrogen in atmospheric air. *Journal of Geophysical Research*, **75**, 1713–1716.

Schoell, M. (1980). The hydrogen and carbon isotopic composition of methane from natural gases of various origins. *Geochimica et Cosmochimica Acta*, **44**, 649–661.

Scott, R. I. Williams, T. N., Whitmore, T. N. and Lloyd, D. (1983). Direct measurement of methanogenesis in anaerobic digestors by membrane inlet mass spectrometry. *European Journal of Applied Microbiology and Biotechnology*, **18**, 236–41.

Scranton, M. I., Novelli, P. C. and Loud, P. A. (1984). The distribution and cycling of hydrogen gas in the waters of two anoxic marine environments. *Limnology and Oceanography*, **29**, 993–1003.

Sebacher, D. I. and Harriss, R. C. (1982). A system for measuring methane fluxes from inland and coastal wetland environments. *Journal of Environmental Quality*, **11**, 34–37.

Sebacher, D. J., Harriss, R. C. and Bartlett, K. B. (1985). Methane emissions to the atmosphere through aquatic plants. *Journal of Environmental Quality.*, **14**, 40–46.

Seiler, W. (1985). Increase of atmospheric methane: causes and impact on the environment. In *WMO-Special Environmental Report*, no. 16, World Meteorological Organization, Geneva, Switzerland, pp. 177–203,

Seiler, W. amd Schmidt, U. (1974). Dissolved nonconservative gases in seawater. In *The Sea*, vol. 5, E. D. Goldberg (Ed.), Wiley, New York, pp. 219–243.

Seiler, W., Holzapfel-Pschorn, A., Conrad, R. and Scharffe, D. (1984). Methane emission from rice paddies. *Journal of Atmospheric Chemistry*, **1**, 241–268.

Siebert, M. L. and Hattingh, W. H. J. (1967). Estimation of methane producing bacterial numbers by the most probable number (MPN) technique. *Water Research*, **1**, 13–19.

Sleat, R. and Robinson, J. P. (1983). Methanogenic degradation of sodium benzoate in profundal sediments from a small eutrophic lake. *Journal of General Microbiology*, 129, 141–152.

Smith, P. H. (1966). The microbial ecology of sludge methanogenesis. *Developments in Industrial Microbiology*, **7**, 156–161.

Smith, M. R. and Mah, R. A. (1978). Growth and methanogenesis by *Methanosarcina* strain 227 on acetate and methanol. *Applied and Environmental Microbiology*, **36**, 870–879.

Smith, R. L. and Klug, M. J. (1981). Electron donors utilized by sulfate-reducing bacteria in eutrophic lake sediments. *Applied and Environmental Microbiology*, **42**, 116–121.

Soerensen, J. (1982). Reduction of ferric iron in anaerobic, marine sediment and interaction with reduction of nitrate and sulfate. *Applied and Environmental Microbiology*, **43**, 319–324.

Soerensen, J., Christensen, D. and Joergensen, B. B. (1981). Volatile fatty acids and hydrogen as substrates for sulfate-reducing bacteria in anaerobic marine sediments. *Applied and Environmental Microbiology*, **42**, 5–11.

Sowers, K. R. and Ferry, J. G. (1983). Isolation and characterization of a methylotrophic marine methanogen, *Methanococcoides methylutens* gen. nov., sp. nov. *Applied and Environmental Microbiology*, **45**, 684–690.

Sowers, K. R., Baron, S. F. and Ferry, J. G. (1984). *Methanosarcina acetivorans* sp. nov., an acetotrophic methane-producing bacterium isolated from marine sediments. *Applied and Environmental Microbiology*, **47**, 971–978.

Stetter, K. O. and Gaag, G. (1983). Reduction of molecular sulfur by methanogenic bacteria. *Nature*, **305**, 309–311.

Stetter, K. O., Thomm, M., Winter, J., Wildgruber, G., Huber, H., Zillig, W., Jane-Covic, D., König, H., Palm, P. and Wunderl, S. (1981). *Methanothermus fervidus*, sp. nov., a novel extremely thermophilic methanogen isolated from an Icelandic hot spring. *Zentralblatt für Bakteriologie, Mikrobiologie und Hygiene, Abteiluag, 1, Originale C* **2**, 166–178.

Stieb, M., and Schink, B. (1985). Anaerobic oxidation of fatty acids by *Clostridium bryantii* sp. nov., a sporeforming, obligately syntrophic bacterium. *Archives of Microbiology*, **140**, 387–390.

Strayer, R. F. and Tiedje, J. M. (1978a). Application of the fluorescent-antibody technique to the study of a methanogenic bacterium in lake sediments. *Applied and Environmental Microbiology*, **35**, 192–198.

Strayer, R. F. and Tiedje, J. M. (1978b). Kinetic parameters of the conversion of methane precursors to methane in a hypereutrophic lake sediement. *Applied and Environmental Microbiology*, **36**, 330–340.

Svensson, B. H. (1984). Different temperature optima for methane formation when enrichments from acid peat are supplemented with acetate or hydrogen. *Applied and Environmental Microbiology*, **48**, 389–394.

Svensson, B. H. and Rosswall, T. (1984). *In situ* methane production from acid peat in plant communities with different moisture regimes in a subarctic mire. *Oikos*, **43**, 341–350.

Sweet, W. J., Houchins, J. P., Rosen, P. R. and Arp, D. J. (1980). Polarographic measurement of H_2 in aqeous solutions. *Analytical Biochemistry*, **107**, 337–340.

Takai, Y. (1070). The mechanism of methane fermentation in floaded paddy soil. *Soil Science and Plant Nutrition*, **16**, 238–244.

Taylor, J. and Parkes, R. J. (1985). Identifying different populations of sulfate-reducing bacteria within marine sediment systems, using fatty acid biomarkers. *Journal of General Microbiology*, **131**, 631–642.

Thauer, R. K., Brandis-Heep, A., Diekert, G., Gilles, H. H., Graf, E. G., Jaenchen, R. and Schönheit, P. (1983). Drei neue Nickelenzyme aus anaeroben Bakterien. *Naturwissenschaften*, **70**, 60–64.

Traganza, E. D., Swinnerton, J. W. and Cheek, C. H. (1979). Methane supersaturation and ATP-zooplankton blooms in near-surface waters of the Western Mediterranean and subtropical North Atlantic Ocean. *Deep-Sea Research*, **26A**, 1237–1245.

Van Bruggen, J. J. A., Stumm, C. K. and Vogels, G. D. (1983). Symbiosis of methanogenic bacteria and sapropelic protozoa. *Archives of Microbiology*, **136**, 89–95.

Van Bruggen, J. J. A., Stumm, C. K., Zwart, K. B. and Vogels, G. D. (1985).

Endosymbiontic methanogenic bacteria of the sapropelic amoeba *Mastigella*. *FEMS Microbiological Ecology*, **31**, 187–192.

Warford, A. L., Kosiur, D. R. and Doose, P. R. (1979). Methane production in Santa Barbara Basin (Southern California) sediments. *Geomicrobiological Journal*, **1**, 117–137.

Weimer, P. J. and Zeikus, J. G. (1978a). Acetate metabolism in *Methanosarcina barkeri*. *Archives of Microbiology*, **119**, 175–182.

Weimer, P. J. and Zeikus, J. G. (1978b). One carbon metabolism of methanogenic bacteria: cellular characterization and growth of *Methanosarcina barkeri*. *Archives of Microbiology*, **119**, 49–57.

Whitman, W. B. (1985). Methanogenic bacteria. In *The Bacteria*, vol. 8, Archaebacteria, C. R. Woese and R. S. Wolfe (Eds), Academic Press, Orlando, pp. 3–84.

Whitmore, T. N., Etheridge, S. P., Stafford, D. A., Leroff, U. E. A. and Hughes, D. (1986). The evaluation of anaerobic digestor performance by coenzyme F_{420} analyzis. *Biomass*, **9**, 29–35.

Widdel, F. (1983). Methods for enrichment and pure culture isolation of filamentous gliding sulfate-reducing bacteria. *Archives of Microbiology*, **134**, 282–285.

Widdel, F. (1986). Growth of methanogenic bacteria in pure culture with 2-propanol and other alcohols as hydrogen donor. *Applied and Environmental Microbiology*, **51**, 1056–1062.

Widdel, F. and Pfennig, N. (1981). Studies on dissimilatory sulfate-reducing bacteria that decompose fatty acids. I. Isolation of new sulfate-reducing bacteria enriched with acetate from saline environments. Description of *Desulfobacter postgatei* gen. nov. sp. nov. *Archives of Microbiology*, **129**, 395–400.

Widdel, F. and Pfennig, N. (1984). Dissimilatory sulfate- or sulfur-reducing bacteria. In *Bergey's Manual of Systematic Bacteriology*, vol. 1, N. R. Krieg and J. G. Holt (Eds), Williams and Wilkins Baltimore, pp. 663–679.

Williams, R. T. and Crawford, R. L. (1983). Microbial diversity of Minnesota peatlands. *Microbial Ecology*, **9**, 201–214.

Williams, R. T. and Crawford, R. L. (1984). Methane production in Minnesota peatlands. *Applied and Environmental Microbiology*, **47**, 1266–1271.

Williams, R. T. and Crawford, R. L. (1985). Methanogenic bacteria, including an acid-tolerant strain, from peatlands. *Applied and Environmental Microbiology*, **50**, 1542–1544.

Winfrey, M. R. and Ward, D. M. (1983). Substrates for sulfate reduction and methane production in intertidal sediments. *Applied and Environmental Microbiology*, **45**, 193–199.

Winfrey, M. R. and Zeikus, J. G. (1977). Effect of sulfate on carbon and electron flow during microbial methanogenesis in freshwater sediments. *Applied Environmental Microbiology*, **33**, 275–281.

Winfrey, M. R. and Zeikus, J. G. (1979a). Microbial methanogenesis, and acetate metabolism in a meromictic lake. *Applied and Environmental Microbiology* **37**, 213–221.

Winfrey, M. R. and Zeikus, J. G. (1979b). Anaerobic metabolism of immediate methane precursors in Lake Mendota. *Applied and Environmental Microbiology*, **37**, 244–253.

Winfrey, M. R., Marty, D. G., Bianchi, A. J. M. and Ward, D. M. (1981). Vertical distribution of sulfate reduction, methane production and bacteria in marine sediments. *Geomicrobiological Journal* **2**, 341–362.

Winfrey, M. R., Nelson, D. R., Klevickis, S. C. and Zeikus, J. G. (1977). Association of hydrogen metabolism with methanogenesis in Lake Mendota sediments. *Applied and Environmental Microbiology*, **33**, 312–318.

Winter, J., Lerp, C., Zabel, H. P., Wildenauer, F. X., König, H. and Schindler, F. (1984). *Methanobacterium wolfei*, a new tungsten-requiring, thermophilic, autotrophic methanogen. *Systematic and Applied Microbiology*, **5**, 457–466.

Woese, C. R. and Olsen, G. J. (1986). Archaebacterial phylogeny: perspectives on the urkingdoms. *Systematic and Applied Microbiology*, **7**, 161–177.

Woese, C. R. and Wolfe, R. S. (Eds) (1985). *The Bacteria*, vol. 8, *Archaebacteria*, Academic Press, Orlando.

Wolfe, R. S. (1979). Microbial biochemistry of methane—a study in contrasts. Part 1. Methanogenesis. *International Review of Biochemistry*, **21**, 267–300.

Wolfe, R. S. (1980). Respiration in methanogenic bacteria. In *Diversity of Bacterial Respiratory Systems*, vol. 1, C. J. Knowles (Ed.), CRC Press, Boca Raton, Florida, pp. 161–186.

Woltemate, I., Whiticar, M. J. and Schoell, M. (1984). Carbon and hydrogen isotopic composition of bacterial methane in a shallow freshwater lake. *Limnology and Oceanography*, **29**, 985–992.

Zaiss, U. (1981). Seasonal studies of methanogenesis and desulfurication in sediments of the River Saar. *Zentralblatt für Bakteriologie, Mikrobiologie und Hygiene, Abteilung 1, Originale C* **2**, 76–89.

Zehnder, A. J. B. (1978). Ecology of methane formation. In *Water Pollution Microbiology*, vol. 2, R. Mitchell (Ed.), Wiley, New York, pp. 349–376.

Zehnder, A. J. B., and Brock, T. D. (1979). Methane formation and methane oxidation by methanogenic bacteria. *Journal of Bacteriology*, **137**, 420–432.

Zehnder, A. J. B., Huser, B. and Brock, T. D. (1979). Measuring radioactive methane with the liquid scintillation counter. *Applied and Environmental Microbiology*, **37**, 897–899.

Zehnder, A. J. B., Ingvorsen, K. and Marti, T. (1982). Microbiology of methane bacteria. In *Anaerobic Digestion 1981* (Eds. D. E. Hughes, D. A. Stafford, B. I. Wheatley, W. Baader, G. Lettinga, E. Y. Nyns, W. Verstraete and R. L. Wentworth), pp. 45–68, Elsevier, Amsterdam.

Zeikus, J. G. (1977). The biology of methanogenic bacteria. *Bacteriological Reviews*, **41**, 514–541.

Zeikus, J. G. (1983). Metabolism of one-carbon compounds by chemotrophic anaerobes. *Advances in Microbial Physiology*, **24**, 215–299.

Zeikus, J. G. and Ward, J. C. (1974). Methane formation in living trees: a microbial origin. *Science*, **184**, 1181–1183.

Zeikus, J. G. and Winfrey, M. R. (1976). Temperature limitation of methanogenesis in aquatic sediments. *Applied and Environmental Microbiology*, **31**, 99–107.

Zeikus, J. G., Ben-Bassat, A. and Hegge, P. W. (1980). Microbiology of methanogenesis in thermal, volcanic environments. *Journal of Bacteriology*, **143**, 432–440.

Zhilina, T. N. (1983). New obligate halophilic methane-producing bacterium. *Mikrobiologyia*, **52**, 375–382.

Zhilina, T. N. (1986). Methanogenic bacteria from hypersaline environments. *Systematic and Applied Microbiology*, **7**, 216–222.

Zinder, S. H. and Brock, T. D. (1978). Production of methane and carbon dioxide from methane thiol and dimethyl sulphide by anaerobic lake sediments. *Nature*, **273**, 226–228.

Zinder, S. H. and Mah, R. A. (1979). Isolation and characterization of a thermophilic strain of *Methanosarcina* unable to use hydrogen-carbon dioxide for methanogenesis. *Applied and Environmental Microbiology*, **38**, 996–1008.

Zinder, S. H. and Koch, M. (1984). Non-aceticlastic methanogenesis from acetate: acetate oxidation by a thermophilic syntrophic coculture. *Archives of Microbiology*, **138**, 263–272.

Zinder, S. H., Sowers, K. R. and Ferry, J. G. (1985). *Methanosarcina thermophila* sp. nov., a thermophilic, acetotrophic, methane-producing bacterium. *International Journal of Systematic Bacteriology*, **35**, 522–523.

PART 4

Activity

Methods in Aquatic Bacteriology
Edited by B. Austin
© 1988 John Wiley & Sons Ltd.

13

Assessment of Bacterial Activity

K. O'Carroll

*Institute of Offshore Engineering, Heriot-Watt University, Riccarton,
Edinburgh EH14 4AS, Scotland*

13.1 INTRODUCTION

Heterotrophic bacteria, in general, perform several activities which are amenable to quantification. Some of these activities are listed below, together with the corresponding methods of measurement described in this chapter.

Cell division	Frequency of cell division
	Nalidixic acid cell enlargement
	Thymidine incorporation
	Increase in cell number
Substrate uptake	Microautoradiography
	Kinetic analysis
Electron transport	Tetrazolium dye reduction
ATP metabolism	Adenylate energy charge

The selection of an appropriate method for a particular study is important and the following suggestions are provided with this in mind.

Object of study	*Method(s)*
Relate uptake of known substrate to an individual bacterial strain	Microautoradiography
Assess proportion of active bacteria in a sample	Microautoradiography
	Nalidixic acid cell enlargement
	Tetrazolium dye reduction
Assess metabolic activity of	Tetrazolium dye reduction

347

identifiable species	Microautoradiography
Growth rate of single strains within a community	Microautoradiography Increase in cell numbers
Relate uptake of known substrates to heterotrophic community activity (heterotrophic potential)	Kinetic analysis Frequency of dividing cells Increase in cell number Adenylate energy charge

13.2 GENERAL PRINCIPLES

If we are seeking to measure the activity of a natural heterotrophic bacterial community, the method chosen must interfere with the normal functions of that community as little as possible. For example, in the methods which follow, reference is made to the use of *in situ* incubation temperatures. The simplest way of ensuring that environmental parameters, such as temperature, are identical to those in the natural habitat is to incubate the bacteria in the water body from which they have been sampled, i.e. immerse the container in the lake, stream or sea. If this is impossible, use a tank containing water from the sample site.

If transportation, or delay in processing the sample, is unavoidable experimenters should be aware of the 'bottle effect'. Storage of both freshwater and seawater samples in containers can increase cell activity and numbers by up to three orders of magnitude. A reduction of this effect can be obtained by minimizing the time of storage or the use of large volume containers ($>10^3$ litres; Menzel and Case, 1977). Since 1000 litre containers are normally impractical in use, preservation of samples should be for as short a period as possible. A figure of 10% increase in cell numbers over a 5-h preservation period has been quoted by Ferguson et al. (1984) and this can be taken as the absolute maximum period of preservation, even at low temperatures, before assessments of activity are performed.

A choice must be made before embarking on an activity assessment as to whether the presence or absence of predators, such as protozoans, is required. Pre-filtration of samples to exclude bacteriovores is sometimes carried out, using $3.0 \mu m$ filters. Although this process can give an indication of bacterial productivity in the absence of predation, it can alter bacterial activity in other ways. Injury to phytoplankton by the filtration process can release dissolved primary amines, and these have been shown to increase the activity and numbers of culturable cells (Ferguson et al., 1984).

If during the course of a measurement of bacterial activity, alteration of the normal growth conditions is unavoidable, be aware of the consequences when extrapolating results back to the natural environment. If incubations are carried out in the dark, will the oxygen content of the water be lower due to the lack of photosynthetic activity? Will this in turn lower the observed bacterial activity? If labelled glucose is added to an oligotrophic bacterial culture, will the concentration be too great for the organisms to respond? Conversely, will dormant organisms be triggered into growth and give a misleadingly high activity figure?

It is impossible to carry out an experiment without influencing its outcome in some way, but as long as an effort is made to take some account of the effects of the interference, useful results may be obtained.

The use of radioisotopes is common to several of the techniques presented here. A thorough grounding in radioactive work should be obtained before attempting these techniques, particularly as these will require modification to suit the particular laboratory equipment or isotopes being used.

The method of liquid scintillation counting (LSC) is probably the best available for counting weak beta emitters such as ^3H, ^{14}C and ^{35}S and consists, in essence, of combining a radioactive source with an organic scintillator, the emissions from which are picked up by a photomultiplier tube. The detector can usually be preset to count a fixed number of scintillations and note the time taken, or to count the number of scintillations within a fixed time interval. Detection efficiencies for the two most commonly used isotopes are typically 90% for ^{14}C and 40% for ^3H.

The organic scintillator used is dissolved in an organic solvent. The purpose of this solvent is to pass excitation energy by collision from molecule to molecule in the solution. It is important, therefore, that the solvent molecule has an excited molecular state with mean half-life long enough for this energy transmission to take place. The double-bonds of several aromatic compounds have been found suitable, with toluene, benzene and m-xylene being typical examples.

The scintillator molecule may emit light directly to the photomultiplier tube (primary scintillator) or via another scintillator molecule (secondary scintillator). These molecules are normally phenolic in nature and are known by acronyms, such as PPP (para-terphenyl) or PPO (2,5-diphenyloxazole).

Any process interferring with the creation or transmission of light in liquid scintillation counting is known as quenching.

Colour of solution: yellow substances absorb near the UV end of the spectrum and can cause severe quenching.

Chemical: a compound in the sample may have an excited state which absorbs stray energy without transmitting light, or may complex with the emitter or scintillator. The C=O group is particularly troublesome and even dissolved oxygen can be a significant quencher.

Physical: photons may be scattered, reflected or refracted. Clear solutions are therefore best for light transmission.

Three common methods of quench correction are:

1. *Internal standard.* A standard amount of radioactivity is added to the scintillation vial and the effect of the sample present is calculated. This ruins the sample for the purposes of a repeat reading and involves a correction for the volume increase of the scintillating solution.
2. *Channels ratio.* The detector counts emissions of light on several different channels and by comparing them corrects for quenching.
3. *Automatic quench correction.* Microprocessor control of the detector analyses the photon emission pattern and corrects for quench.

Since organic compounds are used as scintillators and solvents, aqueous samples may need preparation if they are to form a true solution. Blenders such as 2-ethoxy ethanol or 1,4-dioxane are commonly used, as is the detergent Triton-X100 (Rohm and Haas). Commercial scintillation preparations, such as Unisolve (Koch-Light) combine solvents, emulsifiers and scintillators, and are widely used.

Mention has already been made of the detection efficiencies of LSC for ^{14}C and ^{3}H. Some characteristics of the commonly used isotopes are given in Table 13.1.

The use of ^{3}H labelled substrates is often preferred in radioisotopic methods, since their higher specific activity enables smaller concentrations of labelled substrate to be used. This can allow an approach to *in situ* substrate concentrations to be made, thus minimizing perturbations to the system under study.

It can be seen from the short outline presented above that factors such as choice of isotope, scintillator, solvent, blender, quenching method and

TABLE 13.1 Characteristics of commonly used isotopes (after Faires and Boswell, 1981)

Radioisotope	E_{max} (keV)	$t_{1/2}$ (years)[1]	Max, specific activity (MBq/µg atom)
^{3}H	18	12.26	1079
^{14}C	156	5730	2.308
^{35}S	167	87.2 days	55 408

[1] Half-life.

detection equipment will influence the protocol of any experiment using LSC. The methods given in this chapter relate specifically to the relevant papers from which they were abstracted. It is important to realize that the LSC protocols given can be varied as necessary for a particular laboratory as long as the principles behind these protocols are understood. For a more detailed background to these techniques see Faires and Boswell (1981).

13.3 EPIFLUORESCENCE MICROSCOPY

This topic has been adequately dealt with in Chapter 2.

13.4 RADIOISOTOPIC METHODS

13.4.1 Microautoradiography

This method uses radioisotope labelled substrates which are incorporated only by metabolically active cells. Isotopes such as ^{14}C and ^{3}H are most commonly used, although others, such as ^{35}S, may be used in certain circumstances. The labelled and unlabelled cells are placed onto a photographic emulsion, where the radiation emitted by the labelled cells darkens the photosensitive grains. These can then be counted by light microscopy and related to metabolically active cell numbers. The following method, which is based on Tabor and Neihof (1982b), combines microautography with epifluorescence microscopy. It also eliminates the tedious counting of photosensitive grains which can lead to problems in interpretation (Meyer-Reil, 1978).

Ten millilitre volumes of the sample, in duplicate, are pipetted into 30 ml sterile amber glass bottles. ^{3}H-acetic acid (specific activity 800 mCi mmol^{-1}) is added to give a final activity of 0.2 µCi ml^{-1} and a final added concentration of 250 nM.

The sample is incubated at *in situ* temperatures for 2.5 h, with shaking. Formalin is added (37% (v/v) formaldehyde solution filtered through a 0.2 µm Nucleopore filter; final concentration in sample 2% (v/v) formaldehyde). This fixes the cells and stops substrate uptake. Control bottles are fixed within 15 min of collection, prior to the addition of the substrate as above.

Duplicate dilutions are prepared (1 ml sample in 9 ml sterile, particulate-free water of the same salinity as the sample) and filtered (0.2 µm Nucleopore membrane). The filter and sample are washed by passing two 10 ml volumes of diluent water through the membrane.

The filter is then cut in half, and one half is temporarily attached to an acid-cleaned slide by means of a drop of glycerin at the corner of the slide.

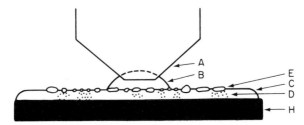

Fig. 13.1 Autoradiographic slide preparation. A: objective lens, B: immersion oil, C: NTB emulsion, D: microautoradiograms, E: microorganisms, H: microscope slide (from Tabor and Neihof, 1982b; reproduced by permission of the American Society for Microbiology).

The sample is facing away from the slide, and the filter should be attached by one corner only.

In a darkroom the filter is removed from the slide and, in total darkness, the slide is dipped into NTB-2 autoradiographic emulsion (Kodak), diluted 1:2 in distilled water. The slide is held at 43 °C to drain for 20 s. The filter is then applied, sample side down, to the emulsion film (momentary use of the safe light). The back of the slide is wiped clean of emulsion and stored in a cool (7 °C), light-tight container until the emulsion solidifies (Fig. 13.1).

After 3 days exposure to the sample at 18 °C, under vacuum and over silica gel, the slides are developed for 30 s at 20 °C in Kodak D-19 developer (diluted to a final concentration of 1:3 with tap water), fixed in 30% (w/v) sodium thiosulphate (or normal Kodak fixer) for 2 min, and washed in tap water for 15 min before drying in air. After drying, the slides are pre-soaked in citrate buffer (pH 6.6), stained with Acridine Orange (AO) (40 mg 100 ml^{-1} citrate buffer) for 7 min and destained for 6 min in each of two citrate buffers at pH 6.6 then 5.0, and finally for 10 min at pH 4.0. The slides are now rinsed in distilled water for 1 min, and examined to make sure that no visible AO remains in the developed emulsion or gelatin. Thence, the slides are dipped in 1% (w/v) glycerin for 1 min to ease the removal of the filter. Once the slide is completely dry, the outline of the filter is traced with a marker pen on the underside of the slide and the filter is carefully peeled off.

The slide is now ready for epifluorescence microscopy, using an oil immersion lens (no cover-slip is required). Organisms surrounded by blackened grains may be deemed to have taken up the labelled substrate. Moreover, it is possible to determine the proportion of cells which have taken up the radiolabel.

Other radioisotopes may be used, either singly or in combination. If heterotrophic community activity is being studied, it may be prudent to supply a mixture of labelled amino acids and perhaps ^{3}H-thymidine in addition to a carbon/energy source such as acetate.

The use of Lugol's Iodine as a fixative has been recommended (Staley and Konopka, 1985). However, it has been shown to be unsuccessful in combination with acridine orange, which is precipitated on contact with the sodium thiosulphate solution used for decolorization (Pomroy, 1984). An alternative to AO, such as DAPI (4′,6-diamidino-2-phenylindole) will solve this problem, but may not be as effective a staining agent as AO, although it is of greater use when staining sediment samples, because it does not stain non-microbial particulates (Tabor and Neihof, 1984).

The use of fluorescent antibodies (immunofluorescence) enables a measure of activity of individual strains, within a community, to be measured when combined with autoradiography (Fliermans and Schmidt, 1975).

Meyer-Reil (1978) investigated various parameters influencing the success of autoradiography and concluded that an exposure time of 2–4 h with 1–5 μCi of ^{3}H-glucose, followed by 3 days incubation at 7 °C, gave best results. He also obtained good correlations between uptake of glucose, as measured by tracer experiments, and number of actually metabolizing cells as determined by autoradiography.

13.4.2 Kinetic analysis

The kinetic analysis of uptake and mineralization of organic substrates has, for some time, been a frequently used method of assessing the hetero-trophic activity of natural populations of aquatic bacteria (e.g. Wright and Hobbie, 1966; Hobbie and Crawford, 1969). ^{3}H-acetate is a substrate which is easily available and in common use, and the method given here uses this substrate. Other labelled compounds could be substituted in this method with little alteration being necessary apart from selecting appropriate concentrations and incubation times. In any case, it may be sensible to try the effect of varying substrates, because different groups of organisms may utilize different substrates within the same population.

The following protocol is adapted from Stanley and Staley (1977): 0.01 ml of ^{3}H-acetate (615 mCi mmol^{-1}) and 0.01 of cold acetate of appropriate concentration are placed in a vial. Two millilitres of sample water is added, and the mixture is incubated for 4–80 min in the dark at the *in situ* temperature of the sample. Uptake is terminated by the addition of 0.01 ml of iodine solution (5% I$_2$, 10% KI in distilled water). For the zero time control, the 2 ml of sample water is added to a mixture of ^{3}H-acetate and iodine solution. In each experiment, uptake is measured as a function of time to ensure that the incubation period selected is within the linear portion of the uptake versus time curve (Wright, 1973). This also enables the minimum incubation time to be selected in order that low substrate concentrations, which approximate to *in situ* substrate concentrations, can be obtained without lengthening incubation time so that the 'bottle effect' is encountered. A range of times used by Fry and Ramsey (1977) for 22 °C incubation was 30, 60, 90 and 120 min. If the incubation temperature is lower, longer incubation periods may be necessary (6–12 h at 3 °C for eutrophic lake water; Overbeck, 1972).

The experiment needs to be repeated over a range of acetate concentrations. To measure uptake, 0.5 ml of the incubated sample is filtered onto a 0.22 µm Millipore membrane filter. The filter is washed three times with 3 ml of tapwater, and air dried. Counting methods will depend on the equipment available; this method is for a Beckman LS 100 counter using 2,5-diphenyloxazole (PPO)-toluene as a scintillation fluid.

Stanley and Staley (1977) corrected for quenching, by either of two

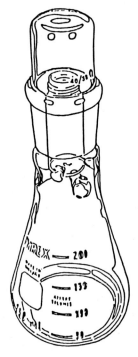

Fig. 13.2 Design of shaking flask for trapping $^{14}CO_2$ produced by heterotrophs in seawater (from Massie *et al.*, 1985; reproduced by permission of Elsevier Science Publisher Ltd).

methods. For filtered samples of less than 0.1 ml, disintegrations per minute per millilitre (dpm ml^{-1}) were constant. To determine quench correction 0.05 ml of labelled sample was filtered with and without the addition of 0.5 ml of unlabelled sample. The percentage reduction of dpm ml^{-1} with the addition of unlabelled sample was used to determine the quench factor. To convert from counts per minute (cpm) to dpm, it was assumed that the membrane filters did not cause a decrease in counting efficiency, which was found to be 53% for their machine using a ^{3}H-toluene standard.

To determine the percentage of substrate, which has been respired, 0.1 μCi of ^{14}C-acetate is placed in a 25 ml flask. The flask is closed by a serum stopper from which is suspended a 1 ml serum vial containing a fluted piece of Whatman No. 1 filter paper measuring approximately 78 × 30 mm. (Alternatively, a special flask, such as the one used by Massie *et al.* (1985), with 10% NaOH (0.2 ml) as CO_2 absorbant, could be used; Fig. 13.2.)

Five millilitres of sample is added, by syringe, through the stopper. The sample is incubated for 4–10 min, and the reaction stopped by the addition of 0.2 ml of 25% (v/v) glutaraldehyde. Phenethylamine (0.2 ml) is injected onto the filter paper. Quenching of filtered cells is determined by extrapolating to zero, from a range of acetate concentrations (Peng, 1964). These acetate concentrations are made by the addition of non-labelled acetate to the labelled acetate, already present. Quenching by the filter paper is determined by absorbing a constant amount of $^{14}CO_2$ onto filter papers which are 50, 67 and 100% of the size of the papers used in the uptake experiments (Peng, 1964). All incubations should be done on site if possible, placing the flasks in the water body itself, or into a container of freshly drawn water to ensure correct incubation conditions.

The *in situ* acetate concentration should be obtained. To do this, water samples should be kept on ice for as short a time as possible before measuring acetate levels in the laboratory. The sample is filtered (0.22 μm Millipore membrane), after centrifugation if necessary. The pH is adjusted to 9.0 with NaOH and duplicate 500 ml samples are evaporated to dryness in a rotary flash evaporator at 35 °C. The samples are re-suspended in distilled water, acidified with 50% H_2SO_4 and centrifuged. Acetic acid is analysed by gas chromatography. The equipment used by Stanley and Staley (1977) consisted of a 72 × ⅛ inch (183 × 0.32 cm) OD copper column packed with Porapak Q which was treated with 0.1% H_3PO_4. A Hewlett-Packard 5711A with flame ionization detector was fitted. Percentage recovery was calculated by adding a known amount of acetate to two aliquots of filtered sample water, immediately before the concentration step.

When the rates of substrate uptake are plotted against their concentrations a straight line graph should result, the slope of which is $1/V_{max}$, the X intercept $(K_t + S_n)$ and the Y intercept Tt, the turnover time. The equation used and terms defined are (Wright and Hobbie, 1966):

$$T/F = 1/V_{max}(A) + (K_t + S_n)/V_{max}$$

where T/F = velocity of substrate uptake (h)

V_{max} = maximal uptake velocity (h)
K_t = transport constant $(\mu g \, l^{-1})$
S_n = natural substrate concentration $(\mu g \, l^{-1})$
A = substrate concentration $(\mu g \, l^{-1})$

It is important to note that the term T/F (time over fraction of substrate taken up) includes the $^{14}CO_2$ respired, which is added to the assimilation result to give the true substrate uptake figure.

The term V_{max} when expressed as V_{max}/bacterium can be used as an indicator of metabolic activity, or stress (Goulder et al., 1979). The same authors compared this measure with %R, the percentage of total substrate uptake which is respired, and found V_{max}/bacterium to be a more useful indicator of copper-induced stress for an estuarine community. Fry and Ramsey (1977) also used V_{max}/bacterium as an indicator of substrate uptake by epiphytic bacteria from paraquat-treated water plants.

Stanley and Staley (1977) determined that not all strains of bacteria in the community took up labelled acetate, and, that within identifiable species, acetate uptake was a function of cell length (biomass). Ramsey (1974) reported that not all bacteria, in natural aquatic habitats will take up glucose. The use of autoradiography in conjuction with kinetic analysis will quantify the proportions of the community under study which take up the chosen substrate, and may also be used as a guide to a choice of substrate.

There is an increasing variety of labelled substrates available and this enables more specialized work to take place, such as that of Massie et al. (1985) who used labelled naphthalene, amino acids and benzo(a)pyrene to study bacterial activity near oil installations in the North Sea. Such substrates may be expensive, however, and it is recommended that non-labelled substrates be used initially for a trial run of the protocol in order to avoid expensive accidents! It is possible to purchase the more common labelled compounds in ampoules, which have predetermined concentrations, and may be directly added to flasks in the field. This facilitates handling, and helps avoid errors of measurment.

13.4.3 Tritiated thymidine incorporation

Utilization of exogenous thymidine is generally restricted to prokaryotic organisms (Grivell and Jackson, 1968). The detection of its incorporation into bacterial DNA is therefore a suitable method of assessment of bacterial activity. Unfortunately, not all ^{3}H-thymidine taken up by bacteria is used in building DNA. Thymidine can be subject to catabolism and the radioactive component distributed around the cell, or the labelled methyl group can be separated and incorporated into protein (Staley and Konopka, 1985). A significant amount of labelled thymidine may be incorporated into RNA. It is, therefore, important to purify the DNA fraction from the cells under investigation before assaying the level of radioactivity.

The protocol given here follows part of the modification of Fuhrman and Azam's method (1980) as used by Riemann and Søndergaard (1984).

Replicate samples (10 ml) are incubated with 2 to 20 nM of (methyl-^{3}H) thymidine (40 to 60 Ci mmol^{-1}). Incubation times can be from 30 to 120 min, but these must be checked against thymidine incorporation to ensure linear uptake is occurring. Shorter or longer incubation times may be required for the bacteria being studied.

Subsamples (3 to 10 ml) are chilled for 1 min in an ice water bath after incubation, and an equal volume of ice-cold 10% (w/v) trichloroacetic acid is added. After 5 min incubation on ice, the mixture is filtered through 25 mm HA membrane filter (0.45 μm nominal pore size; Millipore), and rinsed twice with 3 ml of ice-cold 5% (w/v) trichloroacetic acid. Subsequently, the filter is placed in a scintillation vial, and 1.0 ml of ethyl acetate added to dissolve the filter. This takes about 10 min. Ten millilitres of Aquasol-2 scintillation fluid is then added, and radioactivity is assayed by LSC.

Fuhrman and Azam (1980) used a Beckman LS-100C machine, using the external standard ratio method of quenching with Beckman quenched standards. Blanks are prepared using formalin (1% (v/v) formaldehyde, final concentration) which is added with the thymidine to the samples before incubation. This method may have the disadvantage that labelled thymidine is diluted by naturally occurring thymidine, especially in sediment samples (Staley and Konopka, 1985). The assumptions upon which Fuhrman and Azam (1980) based their method were that:

1. Only bacteria utilized the added thymidine at the low concentrations employed.
2. All the bacteria present in the sample were capable of utilizing exogenous thymidine.
3. The label found in DNA was 80% of that extracted.
4. There was little isotope dilution by natural thymidine.
5. Total bacterial DNA residues contained 25 mol% thymidilic acid residues.
6. The amount of DNA cell^{-1} ranged from 7.47×10^{-16} to 4.82×10^{-15} g.

Filtration (3 μm, gravity) can be performed before treating the sample to exclude predation (Fuhrman and Azam, 1980). Ducklow and Hill (1985) noted that a significant direct relationship between initial rates of ^{3}H-thymidine deoxyribose incorporation per cell and specific growth rates was obtained when incubation times were between 15 and 45 min. The tritiated thymidine method has been applied to coastal (Fuhrman and Azam, 1980; Riemann and Søndergaard, 1984), lacustrine (Riemann and Søndergaard, 1984) and oceanic (Ducklow and Hill, 1985) environments with some success. It has generally proved the most popular method of assessment of bacterial activity (Staley and Konopka, 1985) and one of the most precise (Riemann and Søndergaard, 1984).

13.5 METABOLIC INDICATORS

13.5.1 Nalidixic acid cell enlargement

If cell division of bacteria is inhibited, but growth continues, enlarged cells result. Some compounds, such as nalidixic acid, can cause such enlargement. This increase in size observed is generally longitudinal in rod- or vibrio-shaped organisms. Because nalidixic acid is only effective on Gram-negative organisms (and some Gram-negative bacteria are resistant to this compound) methods using a range of antibiotics have been developed. The protocol given here is based on Kogure et al. (1984) and uses a combination of nalidixic acid (NA), piromidic acid (PA) and pipemidic acid (PPA).

Piromidic acid (8-ethyl-5,8-dihydro-5-oxo-2-pyrrolidinopyrido-2,3-d pyrimidine-6-carboxylic acid) is effective against some Gram-positive taxa such as Staphylococcus spp., whereas pipemidic acid (8-ethyl-5,8-dihydro-5-oxo-2-(l-piperazinyl)-pyrido-2,3-d—pyrimidine-6-carboxylic acid trihydrate) is mainly effective on Gram-negative organisms. However, some taxa, such as Pseudomonas aeruginosa, are resistant to all three of these antimicrobial compounds, Nevertheless, the usefulness of the method lies in the fact that the great majority of aquatic bacteria will be affected by at least one of these agents. Another restriction of this method lies in its use of yeast extract as a substrate. Quite simply, if the organisms fail to use yeast extract, or if the concentration of yeast extract is too great for oligotrophs, reduced counts will be obtained. Conversely, dormant bacteria may be triggered into growth by the added nutrient supply and the figure obtained may represent living, rather than metabolically active, bacteria (Kogure et al., 1980). Despite these drawbacks, the general method has proved successful in comparisons with tetrazolium dye reduction (Maki and Remsen, 1981), and its simplicity recommends it. To carry out the procedure, duplicate 100 ml samples are placed in sterile cotton-wool plugged 250 ml capacity amber glass bottles and the following antibiotics added: NA (2.0 mg), PA (1.0 mg) and PPA (1.0 mg); 25 mg of yeast extract are also added.

The sample is then incubated at 20 °C for 8 h, before fixation with formalin (final concentration 2% (v/v) formaldehyde). Following filtration, the sample is examined by epifluorescence microscopy, using at least 10 random fields per filter. Only those cells, which have elongated or enlarged, are deemed as metabolically active. These cells also fluoresce a reddish-orange colour (Kogure *et al.*, 1980; Maki and Remson, 1981). The number of these fluorescing organisms is expressed as a proportion of all fluorescent organisms.

Nalidixic acid, as the sole growth inhibitor, has frequently been used, although other inhibitory compounds may be of use in specific circumstances. If NA is used exclusively, a shorter incubation time (6 h) should be used. The longer incubation period, i.e. 8 h, can give clearer results, due to greater cell enlargement without loss of DNA synthesis inhibition.

In comparison with viable counts, Tabor and Neihof (1984) noted that the NA method gave from 10 to 300 times the number of metabolically active bacteria from open ocean and coastal water samples. However, they also reported that the concentration of NA needed to be increased to 0.01% (w/v) when large numbers of phytoplankton were present in the sample. Prefiltration to remove phytoplankton is of little use because the reduction of efficacy of nalidixic acid is attributed to dissolved components associated with high phytoplankton numbers (Tabor and Neihof, 1984).

13.5.2 Tetrazolium dye reduction

Virtually all bacteria possess electron transport systems (ETS). The use of artificial electron acceptors as indicators of ETS activity has much to commend it. No substrate is required and the presence of artificial electron acceptors does not appear to induce ETS activity (Zimmermann *et al.*, 1978). This method does not, therefore, involve any significant alteration of existing environmental parameters. The particular protocol recommended here is that of Tabor and Neihof (1982a), and has similarities in technique to their autoradiography method. The electron acceptor used is 2-(p-iodophenyl)-3-(*p*-nitrophenyl)-5-phenyl tetrazolium chloride (INT). It is relevant to note that this is the electron acceptor used by most other workers (e.g. Maki and Remsen, 1981; Baker and Mills, 1982). INT is reduced to INT-formazan which forms optically dense crystals within the cells. The method of Tabor and Neihof (1982a) combines epifluorescence microscopy with INT reduction. Two advantages of the method are that it avoids direct contact of immersion oil with stained organisms, and that fluorescing non-microbial particles are destained as part of the protocol when the gelatin is destained. (Direct contact of immersion oil can dissolve INT formazan crystals, resulting in a lowering of INT reducing organisms counted, by more than 70%; Tabor and Neihof, 1984.)

To carry out the procedure, duplicate 10 ml samples are pipetted into 30 ml capacity sterile, particle-free amber glass bottles. One millilitre of 0.2% (w/v) INT dye is added to each of the samples, which are incubated at *in situ* temperature with shaking for 45 min. The samples are fixed with formalin (final concentration 2% (v/v) formaldehyde). The controls are fixed with formalin on collection, and are shaken for 1 h at the same *in situ* temperature before the addition of INT. Duplicate 1 ml volumes of dyed samples are diluted to 10 ml in sterile, particulate-free water (pH and salinity as for the sample), and are filtered through Nucleopore filters (pore size 0.2 μm, diameter 25 mm). An acid-cleaned coverslip (no. 1 thickness, 24 × 50 mm) is dipped into a filtered (0.22 μm membrane, Millipore) 5% (w/v) gelatin-0.05% (w/v) KCr $(SO_4)_2 \cdot 12H_2O$ solution maintained at 43 °C. One surface of the cover slip is wiped clean, before it is drained vertically for 20 s to obtain a thin gelatin film. The filter retaining the sample is cut in half, and one half is applied to the gelatin film with the organisms in direct contact with the protein film. The cover slip is immediately placed on a cold tray to solidify. Subsequently, it is dried in a desiccator over silica gel. The other half of the filter is treated in the same manner. The coverslip is stained with acridine orange, and destained in citrate buffers exactly as described for the autoradiography method given above.

Once the filter has been removed, the gelatin surface of the cover slip is sprayed with a fine mist of 2% (w/v) gelatin-0.05% (w/v) KCr $(SO_4)_2 \cdot H_2O$ solution from an aerosol sprayer. It is immediately air dried, a procedure which completely embeds the organisms and obliterates the impression of the pore structure of the Nucleopore filter which would otherwise interfere with microscopic examination. One millimetre spacers of adhesive tape are fixed to each end of a microscope slide, and the cover slip is placed onto these spacers, gelatin side facing the slide (Fig. 13.3). The sample is now viewed through the cover glass with an oil immersion objective by epifluorescene microscopy. Organisms fluorescing and containing dense formazan deposits are counted as being metabolically active, and are compared to a total count of all fluorescing organisms. The controls provide an indication of particulate fluorescence or non-biological INT reduction. Formazan deposits frequently exhibit a round, dark-red appearance, and may not be in the focal plane of the cell. Fine focus adjustment in combination with bright-field transmitted and incident fluorescent light provides the best results. It is possible to use a dark-stained Nucleopore filter (Irgalan black or Sudan black B), and examine the organisms against this background by placing a coverslip onto the filter itself, which is in turn mounted on a slide (Maki and Remsen, 1981). The formazan deposits may not be easy to observe with this simpler method, however, and small

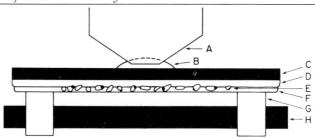

Fig. 13.3 INT slide preparation. A: objective lens, B: immersion oil, C: cover-slip, D: gelatin film, E: filtered micro-organism (stained with AO containing intracellular INT-formazan deposits) embedded in a gelatin matrix, (F) gelatin film, (G) adhesive spacers applied to microscope slide, (H) microscope slide (from Tabor and Neihof, 1982a; reproduced by permission of the American Society for Microbiology).

organisms in particular may be 'lost' against the background of the filter. It is possible to combine nalidixic acid cell enlargement with INT reduction although Maki and Remsen (1981) found little to commend this technique.

13.5.3. Adenylate energy charge

The adenylate energy charge (EC) is calculated by assaying the ATP, ADP and AMP concentrations of a bacterial population. The EC is derived as shown:

$$EC = [ATP] + \tfrac{1}{2}[ADP]/[ATP] + [ADP] + [AMP]$$

The numerator contains the high energy phosphate compounds, the denominator the total adenylate pool. Atlas and Bartha (1981) give the following EC rations:

	EC
Actively growing cells	0.8–0.95
Stationary phase cells	0.6
Senescent or resting cells	<0.5

The protocol is explained in chapter 2.

13.6 INCREASE IN BACTERIAL ABUNDANCE

13.6.1 Frequency of dividing cells

In a growing bacterial culture, a given proportion of the cells will be dividing at any one time. Microscopic examination of a series of subsamples of a bacterial population can provide an estimate of the frequency of dividing cells (FDC), Hagström et al. (1979) suggested that measurements of FDC in natural samples can be used to estimate specific growth rates of populations. This method does, however, present some problems. Firstly, it assumes that all cells in the population spend a similar time interval undergoing the process of division. Division time may vary, in fact, between species and with growth rate. Such variations may also be most pronounced at the low growth rates which are often measured in natural heterotrophic populations (Staley and Konopka, 1985). Secondly, recognition of dividing cells is difficult by light microscopy (particularly in small cells) and the use of scanning electron microscopy does not lend itself to routine analysis. If experimental evidence can be found that the time taken for cell division is constant for the population under study, FDC may be a useful measurement to perform. Riemann and Søndergaard (1984) considered FDC to be intermediate in comparisons of expected error with the methods of thymidine uptake and $^{14}CO_2$ dark uptake. The following protocol is adapted from Hagström et al. (1979) and Riemann and Søndergaard (1984):

Five millilitres of water samples are taken and directly preserved and stained by the addition of 0.4 ml of filtered (0.22 μm), buffered (hexamethylene-diamine 20g/100 ml, pH 7.2) formaldehyde (20% w/w) containing acridine orange (0.125g 100 ml^{-1}). Final concentrations of formaldehyde and acridine orange should be 1.5% (v/v) and 0.01% (w/v), respectively. To ensure even filtration, subsamples of 0.1 to 0.6 ml are mixed with 5 ml of particle-free water in a 13 mm stainless steel funnel (× × 30 01240, Millipore) fitted with a 0.2 μm Nucleopore filter (dyed as previously described), placed on a pre-filter (AP20 013 00, Millipore). The filters are rinsed with 3 × 5 ml of particle-free water. To avoid the less effectively rinsed filter edge, small square should be cut from the centre of the filter (after drying), and mounted on a glass slide using cinnamaldehyde and eugenol (2:1). Place a cover slip over the specimen and seal the edges with nail varnish. The organisms are then counted by epifluorescence microscopy.

A total of 300 bacteria are counted for each slide unless the number of dividing bacteria counted in the sample is less than 30. In this case, a maximum of 20 additional fields are examined. Bacteria showing invagination, but no clear zone between cells, are classed as dividing. This protocol could be altered to use the tetrazolium reduction method of Tabor and Neihof (1982a). FDC values reported by Hagström et al. (1979) range from 0.6 to 6% for bacterial populations in a coastal area of the Baltic Sea.

13.6.2 Dilution method

This method relies upon the measurement of the increase of cell numbers in an isolated water sample over a relatively short period of time. To prevent substrate exhaustion during the incubation period, the original sample is diluted with a known quantity of filter sterilized water from the sample site. To prevent predation, the original sample may be prefiltered (3 μm). The diluted sample is incubated *in situ*, with epifluorescence microscopy counts being made at the beginning and end of the incubation period. The chief disadvantage of this method is the 'bottle effect'. If incubation times are extended to more than a few hours, unnatural growth may take place. Oligotrophic samples can be particularly difficult to assess by this method, because the low bacterial numbers present in the original sample are further diluted, and long incubation times may be necessary before a significant increase in cell numbers can be observed.

One solution to this problem is to assess the bottle effect by experiment. Containers of different surface to volume ratios may be employed, and cell numbers measured at fixed intervals to assess the importance of the effect to the population under study. Fuhrman and Azam (1980) used the following protocol:

Water samples are collected in acid-washed, autoclaved flasks and kept at *in situ* temperatures. One hundred ml subsamples are gently filtered (gravity only) through 47 mm, 3 μm Nucleopore filters. (Harsh filtration will lyse cells and artificially raise the substrate levels in the incubation water). Half the samples are diluted 10 times using sample water, which has been filter sterilized (0.22 μm Millipore membrane—care, as above). Both sets of samples are incubated in containers of the same size (100 ml in this protocol, but see above) at *in situ* conditions, normally by suspending the bottles in the water body. Subsample the incubation bottles and count by epifluorescence microscopy periodically. If samples cannot be counted immediately they may be preserved in borate-buffered 1% (v/v) formaldehyde.

13.6.3 Micro-colony method

In the absence of predation and substrate limitation, bacterial numbers may be expected to increase, as in the previous method. The microcolony method involves filtration of the sample and incubation of the bacteria on the filters themselves. A simple method is described by Straskrabova (1972):

A water sample is obtained and immediately filtered through six parallel filters (0.2 μm Irgalan-stained Nucleopore). The volume filtered will depend on the size of the bacterial population sampled and can be arrived at by experiment.

Three of the filters are immediately placed on pads of Whatman chromatography paper soaked in filtered sample water (organisms uppermost). The pads are prepared by boiling in distilled water, drying at 40–50 °C, and autoclaving. Both pads and filters are placed in petri dishes to minimise evaporation. The three remaining filters are placed on pads soaked with 3% (w/v) formaldehyde. The six filters are incubated for 3 h at *in situ* temperatures, then dried and stained for counting by epifluorescence microscopy, as previously described.

Bacterial generation time (g) may be calculated as follows:

$$g \text{ (hours)} = t \times 0.693/\ln \ Nf - \ln \ Ni$$

where t = incubation time in hours

ln Nf = natural logarithm of final population
ln Ni = natural logarithm of initial population
(derived from formaldehyde fixed filters).

An improved method is given by Meyer-Reil (1977). This method substitutes a continuous flow of water from the sampling site for the filter pads used by Straskrabova. The filters float in buoyant rings on the surface of the water, which is magnetically stirred. This method prevents substrate/oxygen depletion of the incubation water. Both methods can be used to estimate bacterial production if cell volumes are measured and biomass calculated, and the growth rate of the cultured cells (μ) measured by:

$$\mu = 0.693/g.$$

13.7 REFERENCES

Atlas, R. M. and Bartha, R. (1981). Measurement of microbial metabolism. In *Microbial Ecology*, R. M. Atlas and R. Bartha (Eds), Addison-Wesley, Philippines, pp. 113–124.

Baker, K. H. and Mills, A. L. (1982). Determination of the number of respiring *Thiobacillus ferrooxidans* cells in water samples by using combined fluorescent antibody-2-(p-iodophenyl)-3-(p-nitrophenyl)-5-phenyltetrazolium chloride staining. *Applied and Environmental Microbiology*, **43**, 338–344.

Ducklow, H. W. and Hill, S. M. (1985). Tritiated thymidine incorporation and the growth of heterotrophic bacteria in warm core rings. *Limnology and Oceanography*, **30**, 260–72.

Faires, R. A. and Boswell, G. G. J. (1981). *Radioisotope Laboratory Techniques*, 4th edn, Butterworths, London.

Ferguson, R. L., Buckley, E. N. and Palumbo, A. V. (1984). Response of marine bacterioplankton to differential filtration and confinement, *Applied and Environmental Microbiology*, **47**, 49–55.

Fliermans, C. B. and Schmidt, E. L. (1975). Autoradiography and immunofluorescence combined for autecological study of single cell activity with *Nitrobacter* as a model system. *Applied Microbiology*, **30**, 676–684.

Fry, J. C. and Ramsey, A. J. (1977). Changes in the activity of epiphytic bacteria of *Elodea canadensis* and *Chara vulgaris* following treatment with the herbicide Paraquat. *Limnology and Oceanography*, **22**, 556–562.

Fuhrman, J. A. and Azam, F. (1980). Bacterioplankton secondary production estimates for coastal water of British Columbia, Antarctica and California. *Applied and Environmental Microbiology*, **39**, 1085–1095.

Goulder, R., Blanchard, A. S., Sanderson, P. L. and Wright, B. (1979). A note on the recognition of pollution stress in populations of estuarine bacteria. *Journal of Applied Bacteriology*, **46**, 285–289.

Grivell, A. R. and Jackson, J. F. (1968). Thymidine kinase: evidence for its absence from *Neurospora crassa* and some other microorganisms, and the relevence of this to the specific labelling of deoxyribonucleic acid. *Journal of General Microbiology*, **54**, 307–317.

Hagström, A., Larsson, U., Hörstedt, P. and Normark, S. (1979). Frequency of dividing cells, a new approach to the determination of bacterial growth rates in aquatic environments. *Applied and Environmental Microbiology*, **37**, 805–812.

Hobbie, J. E. and Crawford, C. C. (1969). Respiration corrections for bacterial uptake of dissolved organic compounds in natural waters. *Limnology and Oceanography*, **14**, 528–532.

Kogure, K., Simidu, U. and Taga, N. (1980). Distribution of viable marine bacteria in neritic seawater around Japan, *Canadian Journal of Microbiology*, **26**, 318–323.

Kogure, K., Simidu, U. and Taga, N. (1984). An improved direct viable count method for aquatic bacteria. *Archives for Hydrobiology*, **102**, 117–122.

Maki, J. S. and Remsen, C. C. (1981). Comparison of two direct-count methods for determining metabolizing bacteria in freshwater. *Applied and Environmental Microbiology*, **41**, 1132–1138.

Massie, C. C., Ward, A. P. and Davies, J. M. (1985). The effects of oil exploration and production in the northern North Sea: Part 2—Microbial biodegradation of hydrocarbons in water and sediments, 1978–1981. *Marine Environmental Research*, **15**, 235–262.

Menzel, D. W. and Case, J. (1977). Concept and design: controlled ecosystem pollution experiment. *Bulletin of Marine Science*, **271**, 1–7.

Meyer-Reil, L.-A. (1977). Bacterial growth rates and biomass production. In *Microbial Ecology of a Brackish Water Environment*, G. Rheinheimer (Ed.), Springer-Verlag, Berlin, pp. 223–236.

Meyer-Reil, L.-A. (1978). Autoradicgraphy and epifluorescence microscopy combined for the determination of number and spectrum of actively metabolizing bacteria in natural waters. *Applied and Environmental Microbiology*, **36**, 506–512.

Overbeck, J. (1972). Measurement of uptake of organic matter by micro-organisms. In *Microbial Production and Decomposition in Fresh Waters*, IBP Handbook No. 23, Y. I. Sorokin and H. Kadota (Eds), IBP/Blackwell Scientific, Oxford, pp. 20–22.

Peng, C. T. (1964). Correction of quenching in liquid scintillation counting of homogenous samples containing both carbon-14 and tritium by extrapolation method. *Analytical Chemistry*, **36**, 2456–2461.

Pomroy, A. J. (1984). Direct counting of bacteria preserved with Lugol iodine solution. *Applied and Environmental Microbiology*, **47**, 1191–1192.

Ramsey, A. J. (1974). The use of autoradiography to determine the proportion of bacteria metabolising in an aquatic habitat. *Journal of General Microbiology*, **80**, 363–373.

Riemann, B. and Søndergaard, M. (1984). Measurement of diel rates of bacterial secondary production in aquatic environments. *Applied and Environmental Microbiology*, **47**, 632–638.

Staley, J. T. and Konopka, A. (1985). Measurement of *in situ* activities of nonphotosynthetic microorganisms in aquatic and terrestrial habitats. *Annual Review of Microbiology*, **39**, 321–346.

Stanley, P. M. and Staley, J. T. (1977). Acetate uptake by aquatic bacterial communities measured by autoradiography and filterable radioactivity. *Limnology and Oceanography*, **22**, 26–37.

Straskrabova, V. (1972). Microcolony method. In *Microbial Production and Decomposition in Fresh Waters*, IBP Handbook no. 23, Y. I. Sorokin and H. Kadota (Eds), IBP/Blackwell Scientific, Oxford, p. 77.

Tabor, P. S. and Neihof, R. A. (1982a). Improved method for determination of respiring individual microorganisms in natural waters. *Applied and Environmental Microbiology*, **43**, 1249–1255.

Tabor, P. S. and Neihof, R. A. (1982b). Improved microautographic method to determine individual microorganisms active in substrate uptake in natural waters. *Applied and Environmental Microbiology* **44**, 945–953.

Tabor, P. S. and Neihof, R. A. (1984). Direct determination of activities for microorganisms of Chesapeake Bay populations. *Applied and Environmental Microbiology*, **48**, 1012–1019.

Wright, R. T. (1973). Some difficulties in using ^{14}C-organic solutes to measure heterotrophic bacterial activity. In *Estuarine Microbial Ecology*, H. L. Stevenson and R. R. Colwell (Eds), University of South Carolina, pp. 199–217.

Wright, R. T. and Hobbie, J. E. (1966). Use of glucose and acetate by bacteria and algae in aquatic ecosystems. *Ecology*, **47**, 447–464.

Zimmermann, R., Iturriaga, R. and Becker-Birck, J. (1978). Simultaneous determination of the total number of aquatic bacteria and the number there of involved in respiration. *Applied and Environmental Microbiology*, **36**, 926–935.

Methods in Aquatic Bacteriology
Edited by B. Austin
© 1988 John Wiley & Sons Ltd.

14

Nitrate Metabolism by Aquatic Bacteria

C. M. Brown

Department of Brewing and Biological Sciences, Heriot-Watt University, Edinburgh EH1 1HX, Scotland

14.1 SIGNIFICANCE OF NITRATE IN AQUATIC ENVIRONMENTS

Nitrate is of central importance in the biological nitrogen cycle (Fig. 14.1), being the product of nitrification, the substrate of denitrification and nitrate dissimilation to ammonium and a nitrogen source for most bacteria and phytoplankton species in aquatic environments (nitrate assimilation). It is the most abundant fixed inorganic nitrogen source in most waters and occurs freely in oxidized sediments. Interest in nitrate and its metabolism is due also to the considerable leaching from land, especially agricultural land which has been treated extensively with ammonium fertilizers. This nitrate (produced by soil nitrification) passes into streams, aquifers, rivers, etc, from which industrial and domestic supplies are drawn. It has been calculated that in the UK, the nitrate concentration in river waters has risen by 50 to 400% in the last 20 years. The bulk of this nitrate is discharged into the sea. Again in the UK, the annual discharge from estuaries is estimated as 200 000 tonnes N. In addition, some 13 000–21 000 tonnes N are discharged annually as sewage sludge (The Royal Society, 1983).

Attention has been focused on the significance of nitrate in domestic supplies and the incidence of methaemoglobinaemia (the 'blue baby syndrome'). In addition, it has been suggested that ingestion of high concentrations of nitrate and nitrite might lead to the production of nitrosamines which are known animal carcinogens. It should be borne in mind, however, that many processed foods, traditionally, also contain high concentrations of nitrate and nitrite as preservatives. Nevertheless,

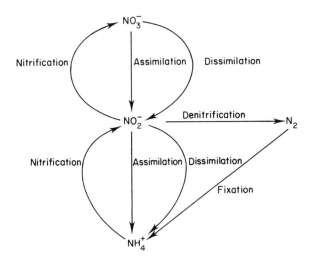

Fig. 14.1 The cycling of nitrate in aquatic environments

statutory limits of nitrate concentrations have been placed on drinking waters and the maximum acceptable concentration in the European Community is 11.3 mg nitrate $N l^{-1}$.

This chapter outlines both the principal methods for measuring nitrogen fluxes in aquatic environments and the main bacterial processes at a physiological level. Field and laboratory measurements of these processes are then discussed using selected examples.

14.2 MEASUREMENT OF NITROGEN FLUXES

In many studies of nitrogen cycling, it is necessary to estimate the concentrations and fluxes of a large number of N-containing materials including both inorganic and organic compounds. Only the most significant methods of following these processes are dealt with in this section. Quantitatively, the sediments are the most significant sites of nitrogen turnover in aquatic environments and researchers have used both sediment cores and slurries in their experiments. The measurement of any flux in an intact core is a difficult process but while work with slurries is easier experimentally the data produced will often reflect a potential rate rather than a true *in situ* rate. In only a few cases have true *in situ* experiments been carried out in the field.

14.2.1 Extraction of inorganic nitrogen compounds

In order to study their flux in aquatic environments, estimations of the concentrations of different forms of inorganic nitrogen must be performed on water samples, suspended solids and sediment materials. Nitrate, nitrite and ammonium are all highly soluble and may be extracted readily from sediments and suspended solids (after separation by filtration) with 1 or 2 M KCl or boiling water. The extraction of ammonium from clay minerals may require the use of an elevated pH. Centrifugation or squeezing methods are widely applied to extract "pore" water from sediments. For more details see Brown (1982).

14.2.2 Chemical estimations of inorganic nitrogen compounds

Nitrate may be estimated directly with a specific ion electrode, although this method often lacks sensitivity. An alternative direct method is to measure light absorption at 210 nm and this is often satisfactory for solutions which do not contain significant concentrations of organic materials which also absorb strongly in the UV. In many cases, however, nitrate is reduced chemically to nitrite (with Cd/Cu, hydrazine, etc.) since nitrite may be determined colorimetrically with great accuracy with, for example, the Griess Ilosvay reagents (sulphanilic acid and 8-amino-naphthalene-sulphonic-acid). This process has been adapted for automated analysis (Armstrong *et al.*, 1967). Nitrate and nitrite may also be determined accurately and with great sensitivity using high performance liquid chromatography (Thayer and Huffaker, 1980). This method also has the advantage of being non-destructive so allowing the collection of fractions for ^{13}N and ^{15}N analysis.

The colorimetric estimation of ammonium with, for example, a phenol/hyperchlorite or indophenol blue system (Solorzano, 1969; Koroleff, 1970), may also be carried out with accuracy and sensitivity. Many workers, however, prefer the fractional distillation procedure of Bremner and Keeney (1965). In this method, ammonia is distilled off after addition of MgO to increase the pH. Nitrate is then reduced to ammonia using Devardas alloy or titanous sulphate and the ammonia distilled off. The ammonia concentration in the distillate may be estimated colorimetrically or by titration with HCl. This method has the advantage that both ammonium and nitrate derived N may be retained for isotopic analysis if ^{15}N is employed or for radioisotope counting if ^{13}N is used.

14.2.3 The use of [15]N

[15]N is the only convenient nitrogen tracer available. Being non-radioactive it is easy to handle and stable in use but the sensitivity of detection is some

10^6 lower than that of a radioisotope. A range of substrates enriched in ^{15}N above the 'natural abundance' of 0.366% are available commercially containing N at up to 99% total N present. The usual methods of measuring N involve either mass spectrometry or optical emission spectroscopy. For a comprehensive review of general methodology the reader is referred to the article by Fiedler and Prokash (1975), while Leftley *et al.* (1983) have provided a useful summary of applications to aquatic systems. Both systems require the conversion of samples to nitrogen gas and operate under high vacuum. The mass spectrometer provides the greater accuracy while the emission spectrometer copes with a smaller sample size (1–10 μg compared with 0.5–5 mg). The costs of installing a mass spectrometer are high compared to an emission spectrometer although the running costs of the latter are often high due the use of UV-transparent discharge tubes.

14.2.4 The use of ^{13}N

^{13}N has a half-life of only 9.96 min and its use is, therefore, heavily restricted to those laboratories with facilities for its generation and to those applications which require only a short incubation time. The production of ^{13}N nitrate involves proton irradiation of water which produces ^{13}N radicals which themselves react with water to form ^{13}N nitrate, together with some nitrite and ammonium and the radioisotope ^{18}F. The nitrate is purified by HPLC and then used as any other radioisotope. Problems of contamination with ^{18}F are solved by counting at different time periods since ^{18}F has a half-life of 110 min. To date, ^{13}N has been used mainly for laboratory experimentation with pure culture systems although Tiedje *et al.* (1979) have used this isotope in studies of denitrification in soil. Focht (1982) has presented a useful summary of the potential of ^{13}N methods for estimating denitrification rates in natural systems.

14.2.5 The use of inhibitors and substrate analogues

A number of metabolic inhibitors may be used in studies of nitrogen fluxes in cultures and in environmental samples. For example, Knowles (1982) has provided a useful list of inhibitors of denitrification. The inhibition of the reduction of nitrous oxide by acetylene is well established and is detailed below. This is the basis of a well-established method for measuring denitrification. Acetylene does not appear to inhibit assimilatory nitrate or nitrite reductases. Respiratory poisons such as the uncoupler DNP are also inhibitors of denitrification, as are sulphide and a number of pesticides including dalapon and vapam. Inhibition of nitrification rates in soils is often used as a measure of the toxicity of pesticides and other 'environmental chemicals'. Inhibition of ammonia oxidation by nitropyrin has been used extensively in measurements of nitrification (see, for example, Goring, 1962; Billen, 1975) and is referred to below. Nitropyrin and carbon disulphide are such potent inhibitors of nitrification that they have been incorporated into ammonium fertilizers to slow down oxidation to the more soluble nitrates with consequent losses by leaching. Carbon disulphide (CD) application is usually by addition of potassium ethylxanthate which decays in soil to yield CD (see Prosser and Cox, 1982).

Chlorate is a structural analogue of nitrate and is a substrate for uptake and reduction by denitrifying bacteria. If reduced to chlorite then toxicity results and further nitrate reduction is inhibited. Acetylene is used widely as an analogue of nitrogen gas since it also contains a triple bond and is reduced by nitrogenase to produce ethylene. Acetylene and ethylene may be separated readily by gas chromatography and this forms the basis of the acetylene reduction test for estimating the rates of nitrogen fixation. Uptake studies on ammonium have often used the ^{14}C-labelled methylammonium as a substrate analogue in order to achieve sensitivity of assay. Assimilation of nitrate proceeds via ammonium to organic nitrogen compounds with glutamine being the first of these synthesized by bacteria under most circumstances. Analogues of glutamine, especially methionine sulphoximine are useful in that they inhibit ammonium assimilation and therefore cause a build up of earlier intermediates in nitrate reduction, thus aiding quantification of these processes.

14.3 PHYSIOLOGY OF BACTERIAL NITRATE METABOLISM

Physiological aspects of nitrate utilization and production are outlined as a background to a discussion on methods for studying nitrate metabolism in aquatic environments. Examples quoted are from aquatic bacteria whenever possible.

14.3.1 Nitrate uptake and assimilation

Nitrate uptake and assimilation clearly occur in all organisms able to utilize nitrate as a nitrogen source and these may be found in many bacterial genera. The overall process comprises at least three consecutive steps; uptake, reduction and incorporation into carbon skeletons and it is often difficult to resolve these experimentally.

A few workers have studied nitrate uptake into bacteria using ^{15}N nitrate. In *Pseudomonas fluorescens* nitrate is taken up against a concentration gradient in an active transport mechanism with a K_m for nitrate or $7\,\mu M$ (Betlach *et al.*, 1981). There were indications that this uptake represented a rate limiting step in nitrate assimilation. The presence of ammonium repressed uptake which was not sensitive to chlorate. Thayer and Huffaker (1982) also used ^{13}N, very short incubation times and nitrate concentrations in the range 1 μM to 1 mM in an elegant study of nitrate uptake in *Klebsiella pneumoniae*. They reported the presence of two transport processes with K_m values of 4.9 μM and 4.2 mM respectively. The low affinity system was the more active and both mechanisms were inhibited in the presence of ammonium. Nitrate uptake in *Azotobacter chroococcum* displayed saturation kinetics, was inhibited by nitrite and had a higher substrate affinity than did nitrate reductase for nitrate (Revella *et al.*, 1986). It was suggested that a proton electrochemical gradient was involved in nitrate uptake in this organism.

Nitrate has an oxidation state of $+5$ and an 8 electron reduction is required to reach the oxidation state of -3 typical of organic nitrogen. The assimilatory reduction of nitrate to ammonium occurs in two stages: first nitrate is reduced to nitrite (2-electron reduction catalysed by an assimilatory nitrate reductase NR) and nitrite is then reduced to ammonium in a 6 electron reduction catalysed by nitrite reductase (NiR).

The NR and NiR enzymes of a number of cyanobacteria have been isolated and characterized and both enzymes utilize photosynthetically reduced ferredoxin as physiological electron donor. NR synthesis is markedly influenced by the nitrogen source of the growth medium. In *Anacystis*, the regulation of NR and NiR occurs mainly through 'ammonium repression'. This requires the metabolism of ammonium via glutamine. In addition, NR activity is lowered in the presence of readily utilized sources of organic nitrogen. There is some evidence for the coordinated regulation of NR and NiR in *Anacystis nidulans* (see Herrero and Guerrero, 1986).

In *Thiobacillus neopolitanus*, the ability to reduce nitrate did not require the presence of substrate, was derepressed during nitrogen-limited growth on ammonium, nitrate and urea but was repressed during thiosulphate- or carbon dioxide-limitation with an excess of ammonium (Buedeker *et al.*, 1982).

Pseudomonas aeruginosa is able to synthesize separate NR enzymes for the assimilation (ammonium repressible and synthesized in aerobic cultures) and dissimilation (non-ammonium repressible and synthesized anaerobically) of nitrate. While these systems are not related functionally, they may share some gene products (Goldflam and Rowe, 1983). Several genes control assimilatory nitrate reduction. The *nasC* gene codes for the structural protein of NR while *nasA* and *nasB* code for products involved in the biosynthesis of a molybdenum-containing co-factor and the *nis* gene is required for NR synthesis (Jeter *et al.*, 1984). Cultures of *P. fluorescens* synthesized NR when grown on nitrate or under conditions of ammonium limitation. The presence of nitrate was not an obligatory requirement. In cultures with an active NR, ^{13}N nitrate was converted intracellularly to ammonium and then to amino acids via glutamine synthetase and glutamate synthase. These pathways of ammonium assimilation are well characterized in pseudomonads and many other bacteria (Brown, 1980).

14.3.2 Nitrification

Nitrification is the biological oxidation of reduced forms of nitrogen, chiefly ammonium, to nitrite and nitrate. While some nitrification may be carried out by chemoheterotrophs including bacteria and fungi, the bulk of the process involves a limited number of chemoautotrophic bacteria constituting the family Nitrobacteriaceae. These organisms utilize ammonium and nitrite oxidation for ATP generation. No single bacterium appears to be able to oxidize ammonium to nitrate and these organisms are classified into genera largely on the basis of inorganic electron donor and cell morphology. Ammonium (to nitrite) oxidizers belong to the genera *Nitrosomonas*, *Nitrosococcus*, *Nitrosospira*, *Nitrosolobus* and *Nitrosovibrio* although most physiological studies have been carried out on *Nitrosomonas* spp. Nitrite (to nitrate) oxidizers are contained in the genera *Nitrospina*, *Nitrococcus* and the most commonly isolated *Nitrobacter*.

The first step in ammonium oxidation is a cytochrome-dependent hydroxylation of ammonium to form hydroxylamine. This step is energy requiring rather than energy yielding. The further oxidation of hydroxylamine to nitrite is coupled to ATP generation via a membrane-bound electron transport system containing flavins and a number of cytochromes. The oxidation of nitrite to nitrate proceeds via hydration and dehydrogenation reactions and ATP generation from a cytochrome-electron transport system. In these nitrifiers, molecular oxygen serves as the terminal electron acceptor and nitrification is consequently an aerobic process.

There is evidence that nitrous oxide may be formed during ammonium oxidation. Ritchie and Nicholas (1972), working with *Nitrosomonas europaea*, reported that nitrous oxide arose from the oxidation of both ammonium and hydroxylamine in aerobic cultures and from nitrite anaerobically. Poth and Focht (1985) further reported that cultures of *N. europaea* produced nitrous oxide only under oxygen-limited conditions in experiments in which ^{15}N nitrate was converted to ^{15}N nitrous oxide in what appeared to be denitrification. Martikainen (1985) has presented data which indicated a close relationship between ammonium oxidation and nitrous oxide production in a coniferous forest soil. The significance of apparent denitrification by nitrifying bacteria in aquatic environments is not known.

While some *Nitrobacter* spp. are capable of mixotrophy and heterotrophy, the metabolism of nitrifying bacteria is predominantly autotrophic involving the reduction of carbon dioxide via a Calvin Cycle system. This places a particular constraint on these organisms since reducing power to reduce carbon dioxide must be generated via ATP-dependent reverse electron transport. Thus, while all autotrophs require 18 mol ATP and 12 mol reduced NAD(P) to fix 6 mol carbon dioxide and form 1 mol hexose, autotrophic nitrifiers also utilize about 5 mol ATP for each mol NAD (P) reduced. On this basis it may be calculated that the oxidation of 15 mol nitrite is required for the production of 1 mol hexose from carbon dioxide. This accounts for the relatively slow growth rates observed with these organisms (Prosser and Cox, 1982).

High concentrations of ammonium and nitrite inhibit both ammonium and nitrite oxidizers, with the latter being the most sensitive. Chlorate ions (analogues of nitrate ions) also inhibit nitrite oxidation. Nitrapyrin (N-serve or 2-chloro-6-trichloromethyl pyridine) is a potent inhibitor of ammonium oxidation.

14.3.3 Nitrate dissimilation and gas formation: denitrification

Denitrification refers to a series of energy generating reactions carried out by a wide range of bacteria able to utilize nitrate (or nitrite) as terminal electron acceptors in place of oxygen. The reduction of nitrate proceeds via nitrite to gaseous oxides of nitrogen, (N_2O and NO) and to molecular nitrogen itself. The overall process leads to the loss of fixed nitrogen from the environment. Such a loss may be considered to be disadvantageous in many environments that are poor in nitrogen, which must be replaced by the energetically expensive process of nitrogen fixation. In other situations, however, such as in water courses extracted to provide for domestic and industrial needs and which carry too high a concentration of nitrate, and in waste water disposal systems, active denitrification is of clear benefit.

Denitrifying bacteria have an essentially aerobic metabolism and nitrate/nitrite reduction occurs usually only when the oxygen supply is restricted. The taxonomic distribution of these bacteria is wide. Most are heterotrophic while some grow autotrophically on hydrogen and carbon dioxide or reduced sulphur compounds and photosynthetic denitrifiers have been characterized. From the frequency of isolation, *Pseudomonas* spp. and *Alcaligenes* spp. are probably of greatest significance in many environments. The reviews of Payne (1973) and Knowles (1982) should be consulted for details.

At least three distinct enzymes are involved in the reduction of nitrate to nitrogen gas. The best characterized is nitrate reductase (NR) which contains iron, labile sulphide and molybdenum. This enzyme is membrane bound and there is good evidence from studies with a number of organisms that NR exists on the inner membrane surface. In *Paracoccus denitrificans* the NR and membrane bound cytochromes appear to be co-regulated. The key factor in the induction of enzyme synthesis is the absence or near absence of molecular oxygen although maximal enzyme synthesis usually occurs only then in the presence of nitrate. A characteristic which distinguishes this enzyme from assimilatory NRs (even in the same organism) is the lack of repression by ammonium or amino acids. The dissimilatory reduction of nitrate is sensitive to the substrate analogue chlorate which is thought to be reduced by NR to form the toxic compound chlorite. There is some evidence that the assimilatory and dissimilatory processes share some gene products in *P. aeruginosa* including a Mo-containing cofactor (Goldflam and Rowe, 1983).

Nitrite reductase (NiR) catalyses the reduction of nitrite to gaseous products and has been purified from a number of denitrifying bacteria. Two types have been reported, haem-containing proteins as in *Alcaligenes faecalis* and *P. aeruginosa* and copper-containing flavoproteins as in *P. denitrificans* and the *Rhodopseudomonas sphaeroides*. NiR may be located on the inner side of the cytoplasmic membrane as is NR. The products of nitrite reduction are nitrous oxide and occasionally nitric oxide.

The obligatory presence of nitric oxide as an intermediate in denitrification and the presence of a nitric oxide reductase are debatable. What is not in doubt, however, is the involvement of nitrous oxide as an intermediate and the presence of a nitrous oxide reductase which catalyses the reduction of nitrous oxide to nitrogen gas. Nitrous oxide reductase is inhibited by acetylene and probably by cyanide, azide and sulphide.

14.3.4 Nitrate dissimilation and ammonium formation

This is the utilization of nitrate by a number of aerobes, facultative anaerobes and anaerobes as a terminal electron acceptor and occurs in the absence or near absence of molecular oxygen. Ammonium rather than nitrogen oxides and molecular nitrogen is the final product and is conserved within the environment. This process serves ecologically as the reverse of nitrification and may be considered to be a 'short circuit' of the nitrogen cycle (Cole and Brown, 1980). The range of organisms able to carry out this process is more restricted than those able to denitrify and includes members of the Enterobacteriaciae (including *Escherichia coli* and *Klebsiella (Enterobacter) aerogenes*) and a number of clostridia (see Cole and Brown, 1980; Dunn *et al.*, 1980; Hasan and Hall, 1975). There are also reports of this process occurring in sulphate-reducing bacteria (see below). Nitrate- and nitrite-reductases (NR and NiR) are involved in the dissimilatory reduction of nitrate to ammonia. The best characterized system is that of *E. coli*. In this organism the NR is formed anaerobically at maximum levels in the presence of nitrate. The enzyme is located on the cell membrane and consists of three components coded from three structural genes in the *nar* operon. The genes are transcribed in the order *narG, narH, narI* for the α and β subunits, respectively. The influence of nitrate on the *nar* operon is controlled via the *narL* gene product. The *E. coli fnr* gene product is involved as a positive regulator in the syntheses of several enzymes of anaerobic (metabolism including NR) and is inactive aerobically. The regulatory sequences of the *nar* operon have been mapped and cloned (Li *et al.*, 1985). The function of NR in *E. coli*, as in the denitrifying bacteria, is in the coupling of nitrate reduction to ATP-yielding electron transport. Three enzymes are involved in *E. coli* in the reduction of nitrite to ammonium (for a summary of these and their genetics see Macdonald *et al.*, 1985). An NADH-dependent, soluble, flavoprotein nitrite reductase contributes about 75% to the overall rate of nitrite reduction while a membrane bound formate-nitrite oxidoreductase contributes a further 20%. The synthesis of these enzymes is repressed during aerobic growth. The remaining nitrite reduction activity is accounted for by the presence of an NADPH-dependent nitrite reductase which functions physiologically as a sulphite reductase and is synthesized both aerobically and anaerobically. The anaerobic growth yield of *E. coli* on glucose is enhanced in the presence of nitrite. This increased growth yield occurs concomitantly with a decrease in ethanol and an increase in acetate excreted into the culture medium and is assumed to reflect an increased level of substrate-level phosphorylation at the level of acetyl phosphate. In addition, up to 70% nitrite N utilized may be accounted for as excreted ammonia.

Using pure cultures of a *Klebsiella aerogenes* isolated from estuarine sediments, Dunn *et al.* (1979) demonstrated that at dissolved oxygen tensions below 15 mmHg both NR and NiR were synthesized under growth conditions in which the nitrate supply was limiting. Anaerobically over 60% nitrate utilized was excreted as ammonium. Under conditions of carbon-limitation, however, no NiR was synthesized and nitrite was excreted. As in *E. coli*, growth yield was enhanced by the presence of nitrate and nitrite. In neither *E. coli* nor *K. aerogenes* is dissimilatory nitrate reduction subject to repression by ammonium or organic nitrogen sources.

In *Clostridium perfringens*, nitrite utilization is also associated with an increased growth yield and a shift in fermentation products from ethanol, butyrate and hydrogen to more oxidized products including acetate. Nitrate is therefore used as an electron acceptor and during this process is reduced quantitatively to ammonium (Hasan and Hall, 1975).

Membrane fractions prepared from the sulphate-reducing bacterium *Desulfovibrio gigas* growing in a lactate/sulphate medium were able to synthesis ATP by coupling the oxidation of hydrogen and the reduction of nitrite or hydroxylamine (Barton *et al.*, 1983). Keith and Herbert (1983) reported that strains of *D. desulfuricans* can utilize nitrate as well as sulphate as a terminal electron acceptor and demonstrated enhanced growth yields of this organism in the presence of nitrate in a model ecosystem (Keith and Herbert, 1985).

Dissimilatory nitrate reduction in *Pseudomonas putrefaciens* results in the production of nitrogen and nitrous oxide in addition to ammonium. The pathways to ammonium predominated at moderate Eh values but did not occur in cultures reduced with thioglycollate (Samuellson, 1985). Acetylene addition led to nitrous oxide accumulation as is usual in denitrifying organisms due to inhibition of nitrous oxide reductase. There are also reports of nitrous oxide production in non-denitrifying, nitrate and nitrite-dissimilating bacteria including a *Citrobacter* sp. (Smith, 1982), *Escherichia coli* (Smith, 1983), *Lactobacillus* spp. (Dodds and Collins-Thompson, 1985), propionibacteria (Kaspar, 1982) and a number of soil bacteria (Smith and Zimmerman, 1981). The significance of this production of nitrous oxide in a natural environment has not been assessed, nor is it known whether it influences to any degree the accuracy of data obtained by estimating denitrification by the acetylene blocking method.

14.4 ENVIRONMENTAL MEASUREMENTS

14.4.1 Nitrification

Nitrification rates may be estimated by both direct and indirect methods. Direct methods involve the chemical determination of increases in nitrite and nitrate and/or decreases in ammonium concentrations after incubation. This may be accomplished with or without amending with ammonium as substrate. Lack of sensitivity and the need to correct for other processes utilizing nitrate, nitrite and ammonium are the main drawbacks. Amendment with ammonium is likely to result in an overestimate of *in situ* rates in low ammonium environments. If amendment with ammonium is judged satisfactory, then the use of ^{15}N ammonium adds considerably to the potential of this method. Such amendments may totally distort the nitrogen regime of the sample, but are necessary due to the low sensitivity of ^{15}N measuring systems. For example, some 0.5 μmol nitrogen is required per sample for estimation of the $^{15}N/^{14}N$ ratio by emission spectroscopy. The lack of sensitivity of the measuring system demands an enrichment of ^{15}N over natural abundance of at least 0.2%. This may require, environmentally, highly significant additions of ^{15}N ammonium to achieve these concentrations and ratios. This is discussed in more detail below. The ^{15}N dilution method involves the addition of ^{15}N nitrate to water or sediment slurries and estimates the dilution of this label with ^{14}N nitrate (produced by nitrification) after incubation for a suitable time period (Koike and Hattori, 1978b). This method also suffers from the general lack of sensitivity of ^{15}N methods and relies on relatively long incubation times. While it can be used to estimate nitrifying activity at ambient ammonium concentrations it is likely to provide an underestimate of true activity due to the utilization of nitrate and nitrite in other processes. Clearly corrections might be applied to account for these losses.

Since nitrification is an oxidative process, activity may be estimated by monitoring the rate of oxygen uptake. The use of the nitrification inhibitor nitropyrin (N-serve, 2-chloro-6 (trichloromethyl) pyridine) to inhibit the metabolism of the nitrifying bacteria allows discrimination between oxygen uptake by these bacteria and that due to respiration by heterotrophs. Incubations should be carried out in the dark to avoid complications due to photosynthetic oxygen evolution.

The most popular method in use at the present time also employs nitropyrin as an inhibitor. As mentioned earlier, the bulk of nitrification in natural environments is due to the activities of a relatively small number of genera of autotrophic bacteria. The growth of these organisms may be monitored, therefore, by the rate of carbon dioxide fixation. The method of Billien (1975) has been applied widely to aquatic systems and involves the incorporation of ^{14}C from bicarbonate into biomass in the presence and absence of nitropyrin. This method has all the benefits of sensitivity of the use of radioisotopes and may be used without amending sediments with additional nitrogen. The equipment required is available in most laboratories and the methods most 'user friendly' in field applications. Assumptions have to made of the ratios of ammonium/nitrite oxidized to ^{14}C incorporated. Billen assumed ratios of 0.1 and 0.02 for ammonium and nitrite oxidation, respectively, and these have been widely applied. Belser (1984) measured the ratios of bicarbonate uptake to ammonium oxidized in cultures of *Nitrosomonas europaea* and *Nitrosospira* sp. over a range of growth rates (MGT 16 to 189 h) and pH values and obtained mean ratios of 0.086, essentially confirming Billen's data. In parallel experiments with *Nitrobacter* sp. a ratio of 0.023 was obtained for nitrite oxidation over a MGT range of 18 to 69 h, again confirming Billen's data. Belser also investigated the effects of high pH, high concentrations of ammonium and chlorate ions and reported that these did not inhibit bicarbonate uptake selectively from nitrite oxidation. He concluded that the nitropyrin inhibition system gave an accurate estimation of *in situ* nitrification in non amended systems. Since, however, no stoichiometric relationship could be established between (nitrogen) substrate concentration and bicarbonate incorporation, Belser suggested that this method was not applicable to kinetic studies or in amended systems. Nitropyrin has a half-life in water/sediments of only a few hours but the sensitivity of the ^{14}C methodology allows for incubation times of 24 h and less. Nitropyrin hydrolysis yields 6-chloro-picolinic acid which is usually considered ineffective as an inhibitor of nitrification. Recent work by Powell and Prosser (1985), however, suggests that 6-chloro-picolinic acid exerts an effect similar to nitropyrin but only after a lag period.

There have been few studies in which the different possible methods for estimating nitrification have been used side by side. Jones and Simon (1981) employed a nitropyrin method in intact cores of freshwater sediments by adding the inhibitor to the overlying water at a concentration of 10 mg l^{-1}. The nitropyrin was absorbed rapidly and 'inactivated' (perhaps hydrolysed and adsorbed on particulate organic material) and the dosing was repeated to retain a suitable inhibitory concentration. The cores were then sealed and changes in the oxygen concentration measured with an oxygen electrode over a 24 h incubation period. Control cores from the same sediment site lacked nitropyrin addition. In parallel, these authors used slurries prepared by diluting sediments 100 times in membrane-filtered core water amended with 10 µM ammonium. Nitrification was followed over periods up to 5 days by oxygen uptake (+/− nitropyrin) and chemical estimations of ammonium uptake and nitrate production. In samples from littoral sediments, there was good agreement between rates measured on slurries using the oxygen uptake and ammonium oxidation methods and these in turn were some 40–60% higher than rates obtained with oxygen uptake in intact cores and nitrate accumulation in slurries. In profundal sediments, however, differences were more marked; the nitrate accumulation assay gave results <20% those of the intact core system. Since these profundal sediments were highly reduced, this result may be due to nitrate loss in denitrification. Enoksson (1986) reported on simultaneous measurements with three isotope methods (^{15}N nitrate dilution, ^{15}N ammonium oxidation and ^{14}C/nitropyrin) of nitrification in the water column in the Baltic Sea. Rates from 1 to 280 nmol l^{-1} day^{-1} were recorded with highest activity beneath the halocline at depths near to an anoxic zone. This was probably due to a sufficiency of ammonium substrate under these conditions. Rates obtained with the ^{14}C/nitropyrin and ^{15}N dilution rates were similar but some 2–7 times less than those obtained with the ^{15}N oxidation method. This may merely be a reflection of the *in situ* ammonium concentration and since there was no attempt made to equalize the concentrations of nitrogen components (especially ammonium) the data from the three different methods are probably not strictly comparable. They are nevertheless an extremely useful indication of the pitfalls of placing too much emphasis on any single method of rate determination in a natural system. In Enoksson's opinion all three methods have disadvantages and none have been tested sufficiently to be used independently. He recommends the ^{15}N ammonium oxidation and ^{14}C/nitropyrin methods as being the least laborious to apply.

14.5 DENITRIFICATION

In situ measurements of denitrification may be carried out by trapping gases evolved from natural sediments and analyzing these by gas chromatography. This latter stage is essential since many sediments will also evolve methane and carbon dioxide. In addition, in sediments in which rapid methanogenesis is occurring, nitrogen gas may be released by gas stripping (of dissolved nitrogen) resulting in an overestimate of denitrification (Jones and Simon, 1981). Direct measurements from sediments may be carried out with relatively simple equipment such as an inverted plastic funnel and collecting cylinder but in all but lake sediments under a shallow water column, the servicing of this equipment is likely to require divers. This direct method may also be applied to sediment cores and slurries and to water samples incubated in the laboratory. In these latter cases, samples may be amended as required to give estimates of denitrification potential rather than *in situ* rate.

Denitrification may also be estimated by the rate of utilization of nitrous oxide, an intermediate in the reduction of nitrate to nitrogen gas. Jones and Simon (1981) used this method with slurries of freshwater sediments. The slurries were prepared by mixing sediments with equal volumes of membrane filtered core water. The slurries were placed into vials which were flushed with helium and sealed. Nitrous oxide was added as a substrate to a concentration in the gas phase of 0.17 mM and the rate of conversion of nitrous oxide to nitrogen gas estimated by gas chromatography. Autoclaved slurries were employed as controls.

The first enzyme of denitrification is nitrate reductase (NR) and since its synthesis is regulated by nitrate availability and anoxia it may be argued that estimation of enzyme activity would also estimate the rate of denitrification, assuming that NR activity was a rate limiting step in the overall process. Measurements of NR activity to estimate denitrification have been used by Packard *et al.* (1977) and Jones (1979) while measurements of algal assimilatory NR as an indication of primary production have been carried out by Eppley *et al.* (1969). In the experiments of Jones (1979), bacteria from water samples from a freshwater lake were concentrated on a glass fibre filter and sediment samples diluted x 10 before each were sonicated to disrupt the organisms present. The samples were then incubated for 1 h (at the non-*in situ* temperature of 35 °C) in the presence of methyl viologen reduced with sodium dithionite as electron donar and in the presence of nitrate. NR activity was measured as nitrite produced. NR activity was three to four times higher in sediments than in overlying water. Unfortunately, no parallel measurements of process rates were recorded and the potential of this method in estimating denitrification cannot be assessed. The use of chlorate enabled a comparison of assimilatory and dissimilatory NR enzymes to be made. In sediments where the Eh was greater than + 100mv, the dissimilatory enzyme accounted for some 60% activity. In more reduced sediments (Eh below + 50mv) then over 90% activity was ascribed to the dissimilatory enzyme.

During the development of gas exchange assays for the possible use in space probes searching for indications of extraterrestrial life, Federova *et al.* (1973, quoted by Balderstone *et al.*, 1976) noted that acetylene blocked the reduction of nitrous oxide to nitrogen gas in soils capable of denitrification. This observation was followed up initially in laboratory studies (e.g. Balderstone *et al.* 1976; Yoshinari and Knowles, 1976) and then applied to field estimations. Balderstone *et al.* (1976) used double side arm Warburg flasks as incubation vessels with one arm closed with a rubber stopper to allow for gas sampling. Experiments were carried out with the marine bacterium *Pseudomonas perfectomarinus* in which a reversible inhibition of nitrous oxide reduction was reported with acetylene partial pressures as low as 0.01 atm. These authors also carried out some preliminary experiments with marine sediment slurries, treated with acetylene, and reported nitrous oxide production from glucose-and acetate-dependent reduction of added nitrate. Sorensen (1987 a,b) applied this 'acetylene blocking method' to marine sediments using both intact cores and sediment slurries. Cores, from one experimental site, were taken in Plexiglas tubes 20 cm long and 2.6 cm diameter. Some cores were frozen immediately in liquid nitrogen for the subsequent assay of nitrate, nitrite and nitrous oxide. Other cores were used for the denitrification assay. These tubes were fitted with silicone rubber inserts for the 'sideward' injection of acetylene. The inserts were spaced at 1 cm intervals and 100 µl acetylene-saturated water samples were injected into the sediment behind each holes. The injections were carried out during the slow withdrawal of the needles and carried out horizontally and in five directions from each insert. The purpose of this was to introduce a sufficient quantity of acetylene to ensure blocking with a minimum of disturbance of the sediment. Such an injection procedure resulted in about 3% saturation of the core interstitial water. Preliminary experiments had shown that inhibition of nitrous oxide reduction was essentially complete in non-amended sediments with 1–2% acetylene saturation. The efficiency of acetylene blocking was also verified by measuring nitrous oxide consumption in slurries. After incubation at *in situ* temperatures the cores were frozen in liquid nitrogen. For assay purposes, core segments were thawed in an enclosed chamber in a stream of helium. Nitrous oxide and other gases were condensed from the gas stream in a liquid nitrogen trap and subsequently assayed by gas chromatography. Sorensen (1978a) also used a sediment slurry method for the simultaneous estimation of nitrate reduction to ammonium and denitrification. This is described in the next section. Results obtained for denitrification on these slurries by the acetylene blocking method and by measuring ^{15}N nitrogen produced from ^{15}N nitrate were in agreement.

Chan and Knowles (1979) used the acetylene blocking method to estimate denitrification *in situ*. A Plexiglas box system was inserted some 10 cm into sediment and enclosed some 12 l water and a sediment surface area of $0.04 \, m^2$. The incubation chamber was connected via tubing to a peristaltic pump operating from a boat or platform on the water surface. This pump was used to circulate the water overlying the sediment. All experiments were carried out within 18 h of inserting this system into the sediment and concluded after a further 22 h in order to minimize changes occurring due to this enclosure of a sediment/water system. Acetylene was inserted into the circulating water via a syringe and valve arrangement and used as a 5–10% solution after mixing. The sediments could be supplemented with nitrate by addition to the circulating water and ethane was used as an internal standard. The authors reported satisfactory performance of this system in sediments from a drainage pond and a small eutrophic lake.

Not all authors appear to be satisfied with the acetylene blocking technique. Jones and Simon (1981) reported inhibition of nitrous oxide reduction of between 7 and 54% in slurries from freshwater sediments and consequently estimated denitrification by an alternative method. Incomplete blockage of nitrous oxide reduction by acetylene may be due to too low a concentration of acetylene (Oremland *et al.*, 1984). Ronner and Sorensen (1985) employed both ^{15}N and acetylene blocking methods to estimate denitrification in low-oxygen waters of stratified areas of the Baltic Sea. Direct evidence of denitrification was shown by an oversaturation of nitrogen gas dissolved in the water column. This oversaturation was observed to develop at depths below the halocline concomitantly with a decrease in nitrate concentration. At the site of maximum denitrification, maximal concentrations of nitrite and nitrous oxide were also present. The nitrous oxide was removed by gas stripping and trapped in a 'Supelco' molecular sieve prior to assay by gas chromatography. At increased depths when conditions were anoxic then nitrate, nitrite and nitrous oxide were all consumed in denitrification.

Christensen and Sorensen (1986) studied denitrification in a plant-covered littoral sediment in a freshwater lake using the acetylene blocking method and measurements of nitrate uptake. They concluded that the nitrate uptake method would result in an overestimate due to nitrate

dissimilatory reduction and assimilation and that only the direct determination of nitrous oxide and nitrogen gas production were a true measure
of denitrification. They also noted that in midsummer the denitrification
rate was limited by nitrate availability and amended the samples
accordingly, so measuring potential rather than estimating an *in situ* rate.
This distinction is also clear from the paper of Oremland *et al.* (1984). In
their study of denitrification in San Francisco Bay intertidal sediments,
denitrification could be detected (by acetylene blocking) only in the upper
3 cm sediment but the potential of the deeper sediments to carry out this
process were apparent if these sediments were amended by nitrate
addition.

In a study of nitrous oxide reductase activity in sediments, Miller *et al.*
(1986) observed enzyme inhibition by oxygen, acetylene, heat treatment,
sulphide (10 mM) and nitrate at concentrations above 1 mM when added to
sediment slurries. Ambient nitrate concentrations of $<100 \mu M$ did not
influence enzyme activity while at $150 \mu M$ there was a lag before enzyme
activity was measurable. For an '*in situ*' assay of nitrous oxide reductase
activity, a solution of nitrous oxide in water was injected into sealed glass
core tubes containing intact sediment. Samples were removed and assayed
for nitrous oxide. Natural nitrous oxide production from nitrate was
inhibited by chlorate addition. It is of interest to note that Miller *et al.* also
reported that the rate of denitrification was often limited by nitrate
availability, a comment of obvious importance when interpreting data
obtained with amended sediments.

14.5.1 Dissimilatory nitrate reduction

Dissimilatory nitrate reduction to nitrite and ammonium may be demonstrated by incubating sediment slurries in the presence of added nitrate.
Ammonium formation in the absence of added nitrate (or the absence of
[15]N tracer methods to prove its origin) can be confused with ammonifica-
tion on the breakdown of organic nitrogen-containing substrates. Dunn *et
al.* (1980) amended estuarine and marine sediments with nitrate and
showed that some 30% added nitrogen could be recovered as nitrite and
ammonium. Similar chemical methods were used by Jones and Simon
(1981) in experiments with freshwater sediments from Blelham Tarn and
by Herbert (1982) with a large number of estuarine sediments. While all
these results indicate the clear potential for dissimilatory nitrate production, the data could not be considered meaningful for rate determinations
since no attempt was made to carry out a nitrogen balance on the systems
used.

Koike and Hattori (1978a) used ^{15}N as a tracer in studies of denitrification
and ammonium production in the coastal sediments of Mangoku-Ura,

Simoda Bay and Tokyo Bay. They demonstrated that in organically poor sediments most nitrate was denitrified while organically rich sediments converted nitrate to ammonium and organic nitrogen. Their method involved taking sediment cores with a corer or with a diver using core tubes from the top 3 cm. 'Within a few hours of sampling', about 1g wet weight of sediment was added to incubation flasks and 30 ml sterile seawater (low nitrogen, $<0.5\mu g$ at/l nitrate and nitrite) containing $15-30\mu g$ at/l nitrate or nitrite (50% N tracer) was added to form a slurry. The flasks were then flushed with nitrogen for 7 min to remove oxygen, filled with sterile seawater saturated with nitrogen and incubated at *in situ* temperatures. At timed intervals the reaction was stopped by addition of 1 ml mercuric chloride via a syringe. Samples treated with mercuric chloride at time zero acted as controls. After this treatment the incubation flasks were degassed *in vacuo* and the ^{15}N content of these gases analysed with a mass spectrometer. After gas extraction the contents of the incubation flasks were filtered through a 0.4 μm filter and the sediment washed with 2M KCl. The filtrate and washings were combined and ammonia removed by steam distillation and after oxidation to nitrogen with KOBr the ^{15}N content estimated by mass spectrometry. The residual sediment was washed with 3% NaCl to remove any residual nitrate and nitrite and the residue dried and ^{15}N content determined after conversion of nitrogen present to nitrogen by the Dumas method (heating with CuO *in vacuo*). This procedure allowed a nitrogen balance to be constructed. The overall rate of nitrate reduction was common at $10^{-2}\mu g$ at $N g^{-1}/h^{-1}$ for all three sediments studied but the detailed balance indicated wide differences in activity between assimilation, denitrification and dissimilatory reduction to ammonium.

Sorensen (1978a) estimated the rates of nitrate reduction to ammonium simultaneously with rates of denitrification in sediment slurries. Segments of sediments were mixed in beakers under a flow of nitrogen gas and 5 g samples placed into serum bottles, purged with nitrogen and stoppered with butyl rubber caps. For denitrification assays, acetylene was injected and nitrous oxide production monitored. For assays of ammonium production, 100 μl quantities of ^{15}N sodium nitrate were added to a concentration of 1.5 μmol nitrate/cm sediment. ^{15}N nitrogen released was estimated in an emission spectrometer—for a second measure of denitrification. Inorganic nitrogen compounds in the slurries were extracted with KCl and ammonium separated by the Conway microdiffusion method before assay for ^{15}N content by the method of Fiedler and Prokash (1975). ^{15}N methods of the type described above have also been applied to cultures of dissimilatory nitrate reducers isolated from the Baltic Sea (Samuelsson and Ronner, 1982) and estuarine sediments (Macfarlane and Herbert 1984). In the latter report, results were presented in agreement with those of

Koike and Hattori (1978b) in that ammonium production from nitrate was most prominant in organically rich sediments.

14.6 REFERENCES

Armstrong, T. A. S., Stern., C. R. and Strickland, J. H. D. (1967). The measurement of upwelling and subsequent biological processes. *Deep Sea Research*, **14**, 381–389.

Balderstone, W. L., Sherr, B. and Payne, W. J. (1976). Blockage by acetylene of nitrous oxide reduction by *Pseudomonas perfectomarinus*. *Applied and Environmental Microbiology*, **31**, 504–508.

Barton, L.L., Le Gell, J., Odom, J.M. and Peck, H.D. (1983). Energy coupling to nitrite respiration the sulfate-reducing bacterium *Desulfovibrio gigas*. *Journal of Bacteriology*, **153**, 867–871.

Belser, L. W. (1984). Bicarbonate uptake by nitrifiers: effects of growth rate, pH, substrate concentration and inhibitors. *Applied and Environmental Microbiology*, **48**, 1100–1104.

Betlach, M. R., Tiedje, J. M. and Firestone, R. B. (1981). Assimilatory nitrate uptake in *Pseudomonas fluorescens* studied using nitrogen-13. *Archives for Microbiology*, **129**, 135–140.

Billen, G. (1975). Evaluation of nitrifying activity in sediments by dark ^{14}C-bicarbonate incorporation. *Water Research*, **10**, 51–57.

Bremner, J. M. and Keeney, D. R. (1965). Steam distillation methods for determination of ammonium, nitrate and nitrite. *Analytica Chimica Acta* **32**, 485–496.

Brown, C. M. (1980). Ammonia assimilation and utilization in bacteria and fungi. In *Microorganisms and Nitrogen Sources*, J. W. Payne (Ed.), Wiley, Chichester and New York, pp. 511–535.

Brown, C. M. (1982). Nitrogen mineralisation in soils and sediments, in *Experimental Microbial Ecology*, R. G. Burns and J. H. Slater (Eds), Blackwell, Oxford, pp. 154–163.

Buedeker, R. F., Riegman, R. and Kuenen, J. G. (1982). Regulation of nitrogen assimilation by the obligative chemolithotroph *Thiobacillus neapolitanus*. *Journal of General Microbiology*, **128**, 39–47.

Chan, Y. K. and Knowles, R. (1979). Measurement of denitrification in two freshwater sediments by an *in situ* acetylene inhibition method. *Applied and Environmental Microbiology*, **37**, 1067–1072.

Christensen, P. B. and Sorensen, J. (1986). Temporal variation of denitrification activity in plant-covered littoral sediment from Lake Hampen, Denmark, *Applied and Environmental Microbiology*, **51**, 1174–1179.

Cole, J. A. and Brown, C. M. (1980). Nitrite reduction to ammonia by fermentative bacteria: a shortcut in the biological nitrogen cycle. *FEMS Microbiology Letters*, **7**, 65–72.

Dodds, K.L. and Collins-Thompson, D.L. (1985). Production of nitrous oxide during the reduction of nitrite by *Lactobacillus lactis* TS4. *Applied and Environmental Microbiology*, **50**, 1550–1552.

Dunn, G. M., Herbert, R. A. and Brown, C. M. (1979). Influence of oxygen tension on nitrate reduction by a *Klebsiella* sp. growing in a chemostat culture. *Journal of General Microbiology*, **112**, 379–383.

Dunn, G. M., Wardell, J. N., Herbert, R. A. and Brown, C. M. (1980). Enrichment, enumeration and characterisation of nitrate-reducing bacteria of the Tay Estuary. *Proceedings of the Royal Society of Edinburgh*, **78B**, 47–56.

Enokkson, V. (1986). Nitrification rates in the Baltic Sea: comparison of three isotope techniques. *Applied and Environmental Microbiology*, **51**, 244–250.

Eppley, R. W., Coatsworth, J. L. and Solorzano, L. (1969). Studies of nitrate reductase in marine phytoplankton. *Limnology and Oceanography*, **14**, 194–205.

Fiedler, R. and Prokash, G. (1975). The application of ^{15}N methods to biological systems. *Analytica Chimica Acta*, **78**, 1–62.

Focht, D. D. (1982). Denitrification. In *Experimental Microbial Ecology*, R. G. Burns and J. H. Slater (Eds), Blackwell, Oxford, pp. 194–211.

Goldflam, M. and Rowe, J. R. (1983). Evidence for gene sharing in the nitrite reductase systems of *Pseudomonas aeruginosa*. *Journal of Bacteriology*, **155**, 1446–1449.

Goring, C. A. (1962). Control of nitrification by 2-chloro-6(trichloromethyl) pyridine. *Soil Science*, **93**, 211–218.

Hasan, S. M. and Hall, J. B. (1975). The physiological functions of nitrate reduction in *Clostridium perfringens*. *Journal of General Microbiology*, **87**, 120–128.

Herbert, R. A. (1982). Nitrate dissimilation in marine and estuarine sediments. In *Sediment Microbiology*, D. Nedwell and C. M. Brown (Eds), Academic Press, London, pp. 53–71.

Herrero, A. and Guerrero, M. G. (1986). Regulation of nitrite reductase in the cyanobacterium *Anacystis nidulans*. *Journal of General Microbiology*, **132**, 2463–2468.

Jeter, R. M., Sias, S. and Ingraham, J. L. (1984). Chromosomal location and function of genes affecting *Pseudomonas aeruginosa* nitrate assimilation. *Journal of Bacteriology*, **157**, 673–677.

Jones, J. G. (1979). Microbial nitrate reduction in freshwater sediments. *Journal of General Microbiology*, **115**, 27–35.

Jones, J. G. and Simon, B. M. (1981). Differences in microbial decomposition processes in profundal and littoral sediments. *Journal of General Microbiology*, **123**, 297–312.

Jones, J. G., Simon, B. M. and Horsley, R. W. (1982). Microbial sources on ammonia in freshwater lake sediments. *Journal of General Microbiology*, **128**, 2813–2831.

Kaspar, H. F. (1982), Nitrite reduction by propionibacteria: detoxification mechanism. *Archives for Microbiology*, **133**, 126–130.

Keith, S. M. and Herbert, R. A. (1983). Dissimilatory nitrate reduction by a strain of *Desulfovibrio desulfuricans*. *FEMS Microbiology Letters*, **18**, 55–59.

Keith, S. M. and Herbert, R. A. (1985). The application of compound bi-directional flow diffusion chemostats to microbial interactions. *FEMS Microbiology Ecology*, **31**, 239–247.

Knowles, R. (1982). Denitrification. *Microbiological Reviews*, **46**, 43–70.

Koike, I. and Hattori, A. (1978a). Denitrification and ammonia formation in anaerobic coastal sediments. *Applied and Environmental Microbiology*, **35**, 278–282.

Koike, I. and Hattori, A. (1978b). Simultaneous determinations of nitrification and nitrate reduction in coastal sediments. *Applied and Environmental Microbiology*, **35**, 853–857.

Koroleff, F. (1970). Direct determination of ammonium as indophenol blue. *ICES Interlaboratory Report*, **3**, 19–22.

Leftley, J. W., Bonin, D. J. and Maestrini, S. Y. (1983). Problems in estimating marine phytoplankton growth, productivity and metabolic activity in nature: an overview of methodology. *Oceanography and Marine Biology Annual Review*, **21**, 23–66.

Li, S., Rabi, T. and Demoss, J. (1985). Delineation of two distinct regulatory domains in the 5' region of the *nar* operon of *E. coli*. *Journal of Bacteriology*, **164**, 25–32.

Macdonald, H., Pope, N. R. and Cole, J. A. (1985). Isolation, characterisation and complementation analysis of *nirB* mutants in *Escherichia coli*. *Journal of General Microbiology*, **131**, 2771–2782.

Macfarlane, G. T. and Herbert, R. A. (1984). Dissimilatory nitrate reduction and nitrification in estuarine sediments. *Journal of General Microbiology*, **130**, 2301–2308.

Martikainen, P. J. (1985). Nitrous oxide emission associated with autotrophic ammonium oxidation in acid coniferous forest soil. *Applied and Environmental Microbiology*, **50**, 1519–1525.

Miller, L. G., Oremland, R. S. and Paulsen, S. (1986). Measurement of nitrous oxide reductase activity in aquatic sediments. *Applied and Environmental Microbiology*, **51**, 18–24.

Oremland, R. S., Umberger, C., Culbertson, C. W. and Smith, R. L. (1984). Denitrification in San Francisco intertidal sediments. *Applied and Environmental Microbiology*, **47**, 1106–1112.

Packard, T. T., Dugdale, R. C., Goering, J. J. and Barber, R. T. (1977). Nitrate reductase activity in the subsurface waters of the Peru Current, *Journal of Marine Research*, **36**, 59–76.

Payne, W. J. (1973). Reduction of nitrogenous oxides by microorganisms, *Bacteriological Reviews*, 409–452.

Poth, M. and Focht, D. D. (1985). ^{15}N Kinetic analysis of nitrous oxide production by *Nitrosomonas europa*. *Applied and Environmental Microbiology*, **49**, 1134–1141.

Powell, S. J. and Prosser, J. I. (1985). The effect of nitrapyrin and chloropicolinic acid on ammonium oxidation by *Nitrosomonas europaea*. *FEMS Microbiology Letters*, **28**, 51–54.

Prosser, J. I. and Cox, D. J. (1982). Nitrification. In *Experimental Microbial Ecology*, R. G. Burns and J. H. Slater (Eds), Blackwell, Oxford, pp. 178–193.

Revilla, E., Llobell, A. and Paneque, A. (1986). Energy-dependence of the assimilatory nitrate uptake in *Azotobacter chroococcum*. *Journal of General Microbiology*, **132**, 917–923.

Ritchie, G. A. F. and Nicholas, D. J. D. (1972). Identification of the source of nitrous oxide produced by *Nitrosomonas eurupaea*. *Biochemical Journal*, **126**, 1181–1191.

Ronner, U. and Sorensen, F. (1985). Denitrification rates in the low-oxygen waters of the stratified Baltic proper. *Applied and Environmental Microbiology*, **50**, 801–806.

Samuelsson, M-O. (1985). Dissimilatory nitrate reduction to nitrite, nitrous oxide and ammonium by *Pseudomonas putrefaciens*. *Applied and Environmental Microbiology*, **50**, 812–815.

Samuelsson, M-O. and Ronner, U. (1982). Ammonium production by dissimilatory nitrate reducers isolated from Baltic Sea. *Applied and Environmental Microbiology*, **44**, 1241–1243.

Smith, M. S. (1982). Dissimilatory reduction of nitrate to ammonium and nitrous oxide by a soil *Citrobacter* sp. *Applied and Environmental Microbiology*, **43**, 854–860.

Smith, M. S. (1983). Nitrous oxide production by *Escherichia coli* is correlated with nitrate reductase activity. *Applied and Environmental Microbiology*, **45**, 1545–1547.

Smith, M. S. and Zimmerman, K. (1981). Nitrous oxide production by non-denitrifying soil nitrate reducers. *Soil Science Society of America Journal*, **45**, 865–871.

Solorzano, L. (1969). Determination of ammonia in natural waters by the phenol hypochlorite. *Limmnology and Oceanography*, **14**, 799–801.

Sorensen, J. (1978a). Capacity for denitrification and reduction of nitrate to ammonia in coastal marine sediment. *Applied and Environmental Microbiology*, **35**, 301–305.

Sorensen, J. (1978b). Denitrification rates in marine sediment as measured by the acetylene inhibition technique. *Applied and Environmental Microbiology*, **36**, 139–143.

Thayer, J. R. and Huffaker, R. C. (1980). Determination of nitrate and nitrite by high pressure liquid chromatography. *Analytical Biochemistry*, **102**, 110–119.

Thayer, J. R. and Huffaker, R. C. (1982). Kinetic evaluation using ^{13}N reveals two assimilatory nitrate transport systems in *K. aerogenes*. *Journal of Bacteriology*, **149**, 198–202.

The Royal Society (1983). *The Nitrogen Cycle of the United Kingdom*. The Royal Society, London.

Tiedje, J. M., Firestone, R. B., Betlach, M. R., Smith, M. S. and Casky, W. H. (1979). Methods for the production and use of nitrogen-13 in studies of denitrification. *Soil Science Society of America Journal*, **43**, 709–715.

Yoshinari, T. and Knowles, R. (1976). Acetylene inhibition of nitrous oxide reduction by denitrifying bacteria. *Biochemical Biophysical Research Communication*, **69**, 705–710.

Methods in Aquatic Bacteriology
Edited by B. Austin
© 1988 John Wiley & Sons Ltd.

15

Methods for the Study of Bacterial Attachment

J. N. Wardell

Department of Brewing and Biological Sciences, Heriot-Watt University, Edinburgh EH1 1HX, Scotland

15.1 INTRODUCTION

Bacterial adhesion in aquatic systems continues to be an area of increasing significance as shown by the output of research papers and symposium volumes within the last ten years. The interest is the result of a growing awareness of the impact, both in economic and environmental terms, of bacterial attachment. Microbial populations associated with surfaces feature in all natural environments, and it is arguably the activity of such populations which is of paramount importance in many instances. One only has to consider the processes of biofouling and corrosion, the action of saprophytes and the initiation of infection to begin to understand the scale of the problem. There are, however, more positive aspects such as the exploitation of surface-associated populations in biological filters and sewage treatment processes. Moreover, microbial films and immobilized cells are current development areas in the biotechnology industry where film-fermenters together with fluidized and fixed bed reactors form an important component of the manufacturing process. Naturally, therefore, the subject of bacterial attachment has been investigated by an ever increasing number of research workers from quite diverse backgrounds and the experimental approaches and techniques employed reflect this diversity.

It is beyond the scope of this chapter to give an overall review of bacterial attachment (the reader should refer to the reviews by Corpe, 1970b; Costerton *et al.*, 1981; Daniels, 1972; Meadows and Anderson, 1979; Fletcher, 1980; Sutherland, 1983; Wardell *et al.*, 1983 and to recent edited volumes by Berkeley *et al.*, 1980, Bitton and Marshall, 1980; and Savage and Fletcher, 1985) rather, the aim is to introduce the methodology which has been employed to investigate the process. In particular, the techniques used to enumerate attached populations; to investigate the mechanisms of attachment; factors affecting attachment; and to assess the metabolic activity of periphytic populations will be considered.

15.2 MEASUREMENT OF ATTACHED POPULATIONS: ENUMERATION

To be able to enumerate populations of bacteria attached to surfaces is obviously of fundamental importance; however, there are problems to overcome. Standard laboratory methods, or some variation of these, can often be applied to populations attached to experimental surfaces, especially in laboratory systems. On the other hand, it is quite a problem trying to enumerate organisms within biofilms *in situ* in natural environments. In many cases, the method used will be determined, at least in part, by the experimental design.

15.2.1 Dilution and plating

A straightforward and technically simple method consists of swabbing experimental surfaces followed by shaking the swab in a suitable diluent, serial dilution and plating for viable counts. This method has been used successfully by a number of workers (e.g. Corpe, 1974) despite the obvious drawbacks of the technique (Buck, 1979). The author has used the technique to enumerate bacteria attached to experimental glass surfaces within chemostat cultures (Wardell *et al.*, 1980, 1984) and found that, with care, the results obtained were comparable with direct microscope counts; at least of the same order of magnitude (Wardell, 1982). However, other workers have found significant differences (Corpe, 1970b, 1974). Discrepancies will arise from a number of sources. It is obviously important to rinse off loosely adhered organisms and residual culture before enumeration and the choice of rinsing fluid can be crucial; the use of distilled water or perhaps too great an ionic strength fluid may remove many of the attached organisms through ionic shock (Corpe, 1970b; Meadows, 1965) so it is important to treat surfaces for microscopy and swabbing in the same manner. Similarly, a suitable diluent (e.g. saline of the appropriate molarity, preferably buffered) should be used to reduce loss of viability during dilution steps. It should, perhaps, be noted at this point that using a simple rinsing apparatus and a tracer organism, Carson and Allsopp (1980) reported that the volume of rinsing fluid required to render sample glass slides free of unattached cells was 31, indicating that the method of rinsing and flow rate of fluid over the surface will be important.

Underestimations of microbial numbers will also occur, due to organisms being retained either on the surface itself or on the swab, and not being released into the suspending fluid on shaking. Gentle sonication or the use of alginate swabs, which dissolve, may be appropriate options to consider here. The technique is only to be recommended where relatively large numbers of bacteria are involved, in order to minimize errors, and for surfaces which are smooth. Moreover, some surfaces may not be amenable to direct microscopy, for instance where there has been development of a multilayered biofilm or where the surfaces are opaque such as metals and some plastics. For these situations, swabbing and plating for viable counts may be worth consideration.

A slightly different approach was taken by Molin *et al.* (1982) who examined the build-up of biofilm on glass rods immersed in chemostat cultures. In this case 'o'-rings, pushed along the glass rods, were used to scrape off the attached organisms. The 'o'-rings were then shaken in diluent, serially diluted and plated for viable counts. Many of the deficiencies attributed to swabbing also apply to this technique; perhaps more so since the 'o'-rings can only realistically be pushed along the glass rod once and would inevitably leave a film of micro-organisms behind. As the errors in enumeration are likely to be quite large the techniques could only be applied where thick, macroscopic, bacterial films were under investigation.

Direct or 'contact' plating of a surface onto solidified media is also a possibility. This technique was used during the early studies of bacterial attachment by ZoBell and Allen (1935) and in its simplest form involves rinsing the test surface followed by placing the surface into direct contact with a suitable solidified growth medium. The surface is then removed and the plate incubated for a subsequent colony count. The procedure was recently revived by Kjelleberg *et al.* (1985), who tried two variations of the techniques. In the first, glass microscope slides were placed in contact with the agar surface and turned through 360° before being removed; and in the second, slides were left in contact with the agar for 4 min before being removed. However, because of the difficulties in ensuring full contact between the slides and the agar, and of ensuring that attached organisms would be removed and enumerated after contact with the agar, the glass slides were washed in 2% (w/v) Tween-80 and the number of cfus in the washings also determined.

The basic technique will grossly underestimate the attached population but may prove valuable in determining, on a qualitative basis, the different types of organism which attach to a given surface and those organisms which occur in close proximity to each other.

15.2.2 Microscopy

A more reliable technique for enumeration of surface populations is microscopy. In the first instance, for experimental glass surfaces (slides, coverslips) the standard procedure of heat fixation followed by staining with crystal violet or dilute carbol fuchsin may be adequate. An even quicker procedure would be to use phase contrast microscopy, this would be particularly suitable for the rapid enumeration of organisms (Geesey et al., 1977), perhaps attached to easily manipulated glass surfaces (Hermansson and Marshall, 1985). Alternatively, the Gram stain (Cowan, 1974) may be of more value if enumerating organisms from a natural environment, since it may yield more information about the types of bacteria present.

Adequate precautions must be taken to ensure that unattached populations are rinsed from surfaces before fixation, and it may be that a chemical fixative such as Bouins fixative [Picric acid saturated aqueous solution 75 ml, formalin 25 ml, glacial acetic acid 5 ml (Gurr, 1979)] may be more appropriate than heat fixation, especially for plastics and other heat labile surfaces (Fletcher, 1976; 1977b). Meadows and Anderson (1968) used 2% (w/v) osmic acid (3 min) followed by Bouins fixative (2 min) for fixing bacteria attached to sand grains. Schofield and Locci (1985 a,b) used fixation in methanol (10 min) when examining the bacterial colonization of a variety of materials including glass, rubber, copper and stainless steel. The choice of fixative depends on the sample, and another possibility for aquatic bacteria (Corpe, 1974) is acetic acid (2% v/v, 2 min). As might be expected, a number of different stains have also been employed including: dilute carbol fuchsin (Meadows and Anderson, 1968), crystal violet (Corpe, 1974), Gram's stain (Wardell et al., 1980), erythrocin-B (Paerl and Merkel, 1982) and phenolic aniline-blue (Baker, 1984).

The problem of visualizing and enumerating organisms on surfaces has recently been addressed by Allison and Sutherland (1984). It has long been recognized (Corpe, 1970a; Costerton et al., 1978; Fletcher and Floodgate, 1973; Marshall et al., 1971 a,b; Sutherland, 1983) that extracellular polysaccharides are involved in the attachment process between cells and solid substrata. In the past it has been difficult to stain these structures adequately while enabling the organisms to be enumerated. Allison and Sutherland (1984) developed a staining technique, where colonized glass slides were treated with 10 mM cetyl pyridinium chloride, air dried and heat fixed. This procedure precipitates the polysaccharide and fixes the cells. The polysaccharide is then stained by flooding the slide with a 2:1 mixture of Congo red solution and 10% (v/v) Tween-80. After careful rinsing the organisms can be stained with dilute carbol fuchsin (10% v/v). The technique has been shown to be applicable to experimental surfaces recovered from natural environments as well as those from laboratory cultures.

Simple stains, such as those described above, suffer from the drawback that, by staining cytoplasm, both viable and non-viable cells will be enumerated. Therefore many workers have turned to fluorescent microscopy. This has a number of advantages in that, unless fixed, only viable cells will be stained (the stain is specific for nucleic acids which are rapidly degraded in dead cells); the specificity of the technique enables organisms to be enumerated against a background of detritus or polymeric material; the technique can be applied to opaque materials and if used in conjunction with monoclonal antibodies can be rendered highly specific. In addition, samples from the liquid phase of the culture or environment can be enumerated in the same way (Goulder, 1977) giving further standardization within the experimental regime. In the latter case the method would be particularly suitable where the liquid phase population is sparse and required trapping the organisms on a membrane filter. Typically, the method involves staining the cells with acridine orange ($100 \, \text{mg} \, l^{-1}$ in $0.1 \, \text{M}$ potassium phosphate buffer, pH 7.5) for several minutes (Geesey and Costerton, 1979; Geesey *et al.*, 1978). Alternatively, a period of 20 min using a concentration of $1:2500$ in citrate buffer, pH 6.6 may be used (Meyer-Reil, 1978). Detached or free-living bacteria may be similarly stained on membrane filters, which have been pre-stained with Irgalan black (Hobbie *et al.*, 1977).

Acridine orange has been successfully used to enumerate organisms on a variety of surfaces (Bright and Fletcher, 1983; Geesey *et al.*, 1978) and would seem to be the most commonly used fluorochrome (Harris and Kell, 1985). However, problems do arise with certain substrata which bind acridine orange, for example polystyrene, resulting in poor contrast between the background and the micro-organisms. The problem can be overcome by judicious choice of the fluorochrome (Paul, 1982). Where colonization by a known specific organism is under investigation, fluorescein-labelled monoclonal antibodies are a potentially useful tool for comparative enumeration. For example, the technique has been employed by Schofield and Locci (1985a,b) to investigate the colonization of components of hot water systems by *Legionella pneumophila*.

Finally, the increased availability of microcomputers and image analysers means that much information regarding cell size, volume, population density and area covered by surface colonizing organisms can be readily obtained from photomicrographs and electron micrographs (Krambeck *et al.*, 1981). Of particular importance is the application, in conjunction with scanning electron micrographs (see later), to enumerate organisms on three-dimensional structures such as plastic foam, glass beads and sand grains where the only realistic alternative may be some chemical assay. At the same time, it should be noted that the area covered in one field is very small compared with the surface as a whole and many representative fields would need to be examined to take into account edge effects, effects due to uneven surface charge (Floodgate, 1972) and other factors, which may give rise to local variations in the population density (Fletcher, 1977a).

15.2.3 Spectrophotometry

Photometric methods, which are routinely used in laboratories to estimate bacterial numbers in liquids, may also be used to determine surface population densities of some substrata. After fixing and suitable staining, the optical density of transparent materials can be determined and then used to estimate the surface population. For example, if oxalate-crystal violet is used the OD can be determined at 590 nm, the absorption maximum for crystal violet (Fletcher, 1976). To extrapolate to actual bacterial numbers, i.e. a quantitative rather than qualitative determination, a calibration curve of OD against population density determined by direct counts would be required. Nonetheless, the method does offer the attractive advantage that a large number of determinations could be carried out quickly and conveniently. Moreover, providing there is an even distribution of organisms across the surface, the method could be extended over a range of population densities where microscopic counting is no longer a reasonable proposition. Balanced against that is the restricted number of materials, glass, polystyrene, for example, which could be used.

15.2.4 ATP-assay

Since the method was developed, extractable adenosine triphosphate (ATP) has been used extensively to determine bacterial biomass in a variety of environmental samples (Karl, 1980) (See Chapter 2). ATP is rapidly lost from dead cells, consequently, any delay between sampling the population and performing the assay must be kept to a minimum. Therefore, the application of the method to surface associated populations is difficult unless the ATP can in some way be extracted *in situ* or if the experimental system is specifically designed around the ATP assay, for example using glass coverslips for colonization which can be subsequently crushed and ground to facilitate ATP extraction (see also Harber *et al.*, 1983).

The ATP assay has been used to estimate the biomass of a number of surface populations such as dental plaque (Robrish *et al.*, 1978), for *E. coli* adhering to polystyrene (Harber *et al.*, 1983), and for detecting microbial biomass in a range of different sediments and natural surfaces (see review by Karl, 1980). Like all methods, this particular technique is not without its limitations. Perhaps the most notable criticism is that the C:ATP ratio is often regarded as constant for all micro-organisms. This is not so (Karl, 1980): further investigations by Stuart (1982) showed that large overestimates in biomass were obtained from ATP × 250, and, in addition, C:ATP ratios varied with time. Harris and Kell (1985) listed the possible sources of error in the technique as:

1. incomplete extraction of ATP;
2. quenching of the reaction by extraction chemicals, etc.;
3. use of impure luciferin–luciferase reagents;
4. stress on cells which may cause loss of ATP (see also Karl, 1980);
5. activity of ATP-ases and other kinases, especially if there is any delay in processing the sample;
6. variation in intracellular ATP levels with growth and environmental conditions;
7. presence of free ATP of non-microbiological origin;
8. degradation of ATP by extraction procedure or other substances.

Moreover, as with any indirect technique the results obtained in terms of C:ATP must be extrapolated to biomass dry weight or cell numbers. Despite these possible disadvantages, for certain applications ATP assay may prove to be the most convenient and sensitive means of estimating surface-associated microbial populations.

15.2.5 Radiolabelling techniques

The use of radioisotopes purely for enumeration of attached populations does not seem to have attracted much attention. The reasons for this are probably self-evident. Dispersing labelled compounds into the natural environment is simply not acceptable which means that the technique can only be applied to closed or laboratory systems. The approach would be to measure the increase in radioactivity of clean surfaces exposed to a labelled population or to measure the decrease in radioactivity of the liquid phase population as organisms are removed by adsorption onto suitable substrata. The choice of nutrient substrate would seem to be important. While there is a number of labelled compounds available, usually incorporating ^{14}C or ^{3}H, different bacterial species will have diverse nutrient requirements and it may be that a compound which labels one species will prove unsuitable for another, perhaps because of a deficient uptake transport mechanism. While there may be particular circumstances where the approach described above may have potential, on the whole the requirements for containment facilities and the obvious need for accurate calibration would seem to render the technique too specialized for general use.

On the same theme the use of microautoradiographs to enumerate bacterial cells on surfaces also has potential. In addition to suffering from the same drawbacks as other radiolabelling techniques, there is the problem of grain size on the autoradiographs to consider. If the radioisotope used is too strong an emitter the number of silver grains exposed may be too large and mask the number of cells responsible. For this reason Paerl and Merkel (1982) have shown a preference for ^{33}P rather than the more usual ^{32}P. The ^{33}P being a 'softer beta emitter' gives better resolution when trying to enumerate bacterium sized particles.

15.2.6 Biochemical assay

With the improvements in analytical instrumentation which have taken place in recent years a number of reports have appeared advocating the use of chemical methods for estimating microbial biomass. The rationale being that if one can identify a key component for a particular group of organisms then, providing a suitable assay is available, it should be possible to use the component as a means of assaying the biomass. For example, microbial biomass may be estimated by analysing the phospholipid content of an environmental sample or more particularly prokaryotic biomass may be estimated by assaying for muramic acid (See Chapter 2). Since the muramic acids present in cyanophytes, Gram-positive and Gram-negative cells are different, assay methods capable of detecting the differences are required to resolve mixed populations. Palmitic acid, present in most microbial lipids can be assayed to a sensitivity equivalent to 5×10^5 cells of *E. coli* (Bobbie and White, 1980) and muramic acid to a sensitivity equivalent to 10^8 *E. coli* cells (see White, 1983). Judicious choice of the 'signature' compound enables the composition of mixed populations to be determined, e.g. lipopolysaccharide components for Gram-negative bacteria and teichoic acid components for Gram-positive organisms with sensitivities corresponding to 10^6–10^7 and 5×10^6 cells, respectively (Gehron *et al.*, 1982; Parker *et al.*, 1982).

The advantages of this type of approach to biomass estimation have been outlined by White (1983) as: estimation of microbial biomass without the need for further microbial growth; as chemical extractions are performed, there is no requirement for the quantitative recovery of micro-organisms from colonized surfaces; and, by analysing 'signature' compounds, some analysis of the community structure as it was *in situ* is possible. It is also possible that by correlating measurements of nutritional status with metabolic activity, biochemical analyses may be of value in determining true *in situ* microbial activity (White, 1983). Against all this are the disadvantages: the sensitivities of the techniques, although these will undoubtedly improve; the chemical extractions, so far applied to environmental samples such as sand and sediment particles, may be unsuitable for many substrata; and the necessary capital and running costs of the instrumentation may be prohibitive. It is beyond the scope of this chapter to give details of the chemical extractions, GLC, HPLC and mass-spectrophotometric methods. Reference should be made to the review by White (1983) and thence to the individual experimental papers.

15.3 THE ADHESION PROCESS

One of the fundamental aspects of the attachment of micro-organisms to surfaces is the mechanism of the adhesion. Leaving aside specialized mechanisms (fimbriae (pili), stalks and holdfasts), for the majority of bacteria involved in the primary 'conditioning film' it seems that polymeric adhesives are involved. It is widely accepted that in aquatic systems bacteria behave as colloidal particles (Lips and Jessup, 1979; Marshall, 1976) and that the initial interaction between the bacterial cell and surface is the result of physicochemical forces (Daniels, 1980). The most important of which seem to be the Van der Waals forces of attraction and the electrostatic repulsive forces (Marshall, 1976). Factors which will affect the initial interaction are therefore the nature and charge of the surface together with the surface components and charge of the cell.

15.3.1 Surface characteristics

The effect of surface charge on adhesion has been investigated by examining the colonization of different materials by the same organism under otherwise similar conditions. For example, Fletcher and Loeb (1979) examined the attachment of a marine pseudomonad to a variety of materials with differing surface charge: Teflon, polyethylene, polystyrene, which have little surface charge; germanium with neutral surface charge; and platinum with positive charge. The latter two were polished to mirror finish to provide a smooth surface comparable with the others. In addition, glass, mica and some oxidized plastics were examined as examples of hydrophilic, negatively charged surfaces. Glass and mica are also useful substrata in experimental systems since glass has an uneven surface charge, demonstrating 'edge effects' whilst the crystalline structure of mica imparts a uniform surface charge (Floodgate, 1972).

The surface charge of bacterial cells can be determined by electrophoretic mobility (Harden and Harris, 1953) and at pH values encountered in natural environments is usually negative (Burns, 1979; Ward and Berkeley, 1980). By manipulating the pH of the experimental system, it should be possible, within limits, to manipulate the charge on the bacterial cell since the cell surface charge is due to surface components of the cell envelope, for instance the teichoic acids of Gram-positive cells and the lipopolysaccharides of Gram-negative cells. Varying the culture conditions in order to change the surface components of the cell envelope is another method of manipulating the cell surface charge. For example, growth of bacteria in high carbon media will encourage the production of extracellular polymer (Brown *et al.*, 1977) and the phenomenon of 'turnover' in wall components is well documented. Ellwood and Tempest (1972) have shown that growing Gram-positive organisms (*B. subtilis*) under K -limited conditions, the cell walls had a high teichoic acid content and, therefore, high phosphate content; switching over to a phosphate—limited medium brought about a loss of teichoic acids from the cell wall and replacement with teichuronic acids. Such changes are bound to effect the surface charge of the bacterial cells. Many of the experimental methods in this area employ the ability of microbiologists to manipulate the cultural/ environmental conditions in order to bring about change in the physiological state of the bacterial cell and to investigate whether this has any effect on structural or physiological aspects of adhesion (Pringle *et al.*, 1983). Measuring attachment under different concentrations of divalent or trivalent cations is one way of examining the effect of ionic interactions during attachment (Fletcher, 1979; Marshall *et al.*, 1971a). Another approach is to coat surfaces with macromolecules or polymers which alter the surface charge characteristics. Fletcher (1976) investigated the effect of a number of proteins (bovine serum albumen, gelatin, fibrinogen, pepsin, protamine and histone) on the attachment of a marine pseudomonad to polystyrene. Any effect on attachment must be either by affecting the surface or the bacterial cell or both. Aspects of the effect of proteins on bacterial attachment to glass were also investigated by Meadows (1971) and Feldner *et al.* (1983).

An important feature of surfaces which become colonized in aquatic environments is the 'wettability', i.e. the hydrophilic/hydrophobic nature of the surface. The effect of this on adhesion can be determined for a particular system by employing test substrata of known different hydrophilic/hydrophobic character. Commonly employed materials would be plastic petri dishes as an example of a hydrophobic surface and plastic tissue culture dishes as an example of a hydrophilic surface; in addition, coating surfaces with polymer as described above will affect the wettability of the surface. Measuring the critical surface tension (γc) (Fletcher and Loeb, 1976) or, more recently, determining the contact angle (θ) and then calculating the work of adhesion (W_A) have been advocated to enable the hydrophilic/hydrophobic nature of a surface to be defined numerically (Pringle and Fletcher, 1983).

15.3.2 Polymeric adhesives

The permanent attachment of many bacteria to surfaces is mediated through the formation of polysaccharide adhesives. Initially only discrete amounts of specific polymer are involved (Fletcher and Floodgate, 1973; Marshall, 1976; Marshall *et al.*, 1971a; ZoBell, 1943) but later more extensive amounts of secondary polymer are involved (Corpe, 1964, 1970a, 1980; Fletcher and Floodgate, 1973; Geesey *et al.*, 1977; Jones *et al.*, 1969; Marshall, 1976; Marshall *et al.*, 1971a,b; Sutherland, 1983) associated with colonization of the surface. In aquatic environments, these structures are highly hydrated and therefore difficult to stabilize for microscopic examination (Costerton *et al.*, 1985). Once removed from the natural environment, these structures collapse perhaps giving a misleading impression of their natural appearance (Costerton *et al.*, 1981).

The problem of visualizing the structural components involved in the adhesion process brings one naturally to the application of electron microscopy. Both transmission electron microscopy (TEM) and scanning electron microscopy (SEM) have been applied successfully to a wide variety of surfaces from quite diverse environments. Due perhaps to the more complex procedure involved of fixing and sectioning, TEM has been less widely applied than SEM. However, much of the current understanding of the role played by acidic polysaccharides in the adhesion process is the result of the ability to stain the structures with alcian blue and ruthenium red, which defines the polysaccharide in the electron microscope. These stains are usually mixed in with the aldehyde fixative and must be maintained during subsequent washing stages due to their solubility. Fixed organisms and substrates which are penetrated by embedding resins can be prepared as a plastic block which can be subsequently sectioned. A problem occurs with metals and, for example, stone surfaces since these cannot be prepared in the same way. The appropriate treatment would be to embed the organisms and surface, then to shear away the surface by a sharp temperature change, e.g. dipping into liquid nitrogen. A useful protocol is given by Costerton (1980). Such techniques allow sections of bacteria in contact with surfaces to be examined in great detail and the structures involved in attachment to be elucidated (Costerton, 1980; Costerton *et al.*, 1978; Fletcher and Floodgate, 1973; Marshall, 1976). SEM has proved very popular in adhesion studies, the attractions being: the ability to view whole organisms *in situ*; to see the arrangement of cells and of any surface structures with minimal disturbance; and the ability to examine colonization of surfaces which are not amenable to TEM. Many excellent scanning electron micrographs of colonized surfaces have been published covering a variety of environments and surfaces (see edited volumes mentioned in section 15.1 and Dempsey 1981a,b).

The preparative procedure used in our laboratory is fairly standard: after careful rinsing with sterile saline of the appropriate ionic strength (0.2M for freshwater organisms, 0.4M for marine samples) to remove loosely attached cells, the adhered film is fixed by immersion in 2.5% (v/v) glutaraldehyde in cacodylate buffer. After washing in cacodylate buffer, the specimens are dehydrated by passage through a graded acetone series (10, 25, 50, 75, 100% (v/v) analytical grade acetone) with 15 min at each stage and then critical point dried from liquid CO_2. Acetone is an unsuitable dehydrating agent for some specimens, e.g. membrane filters; for these a suitable alternative is a graded ethanol series followed by final passage into amyl acetate since ethanol is immiscible with liquid CO_2. Alternatively, a graded ethanol series followed by a graded Freon-113 series can be used with critical point drying from liquid CO_2 or Freon-13. A protocol for this is given by Costerton (1980). Once dried, the specimens are coated with gold using a Polaron 'sputter coater' before viewing in the electron microscope. An example of the sort of detail which can be expected is shown in Fig. 15.1.

One of the criticisms often levelled at electron microscopy, and SEM in particular, is the generation of artefacts during sample preparation. For example, the fibril structures seen in many scanning electron micrographs (Dempsey, 1981a,b; Paerl, 1975; 1980; Wardell et al., 1984) are thought by some authors to be strands of condensed polymer resulting from the dehydration process rather than genuine structures. However, doubt remains since some authors (Dempsey, 1981a,b) have been able to demonstrate condensed polymer lying as a sheet or film overlying fibril structures. Moreover, Wardell et al. (1984) have been able to demonstrate extensive fibril structures in carbon-limited systems whereas in nitrogen-limited cultures (C-excess) such fibrils were less well developed and were absent from cells from the liquid phase of the culture, despite the presence of condensed loose polymer (Wardell, 1982). Therefore care must be exercised as always in interpreting electron micrographs.

Attempts have been made to overcome the problem of artefacts by stabilizing the structure with immunoglobulins or lectins before the dehydration steps and this has been successful in TEM studies (Bayer and Thurow, 1977; Birdsell et al., 1975; Mackie et al., 1979). As far as SEM is concerned any procedure which maintains a three-dimensional structure is preferable to air drying where the bacterial cells and surface structures collapse. In this latter case, it is often impossible to distinguish the bacterial cells under an overlying sheet of polymer.

15.4 ACTIVITY OF ATTACHED ORGANISMS

The metabolic activity of attached populations continues to be an area of intense interest. Many authors have made statements, or at least implied, that organisms growing as biofilms or in association with surfaces have an elevated metabolic activity compared to planktonic bacterial populations,

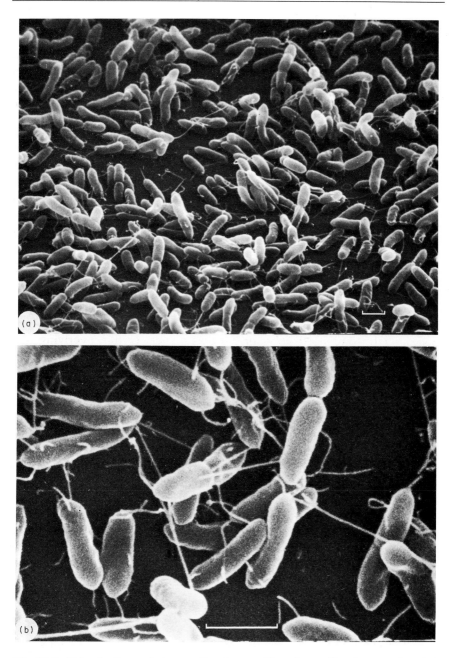

Fig. 15.1 Colonization of a smooth glass surface by a *Pseudomonas* sp. in chemostat culture under carbon limitation. Scanning electron micrographs: (a) ×6250, (b) ×19700 bar = 1 μm.

and for this reason the surface associated populations may be the most important in aquatic environments, responsible for nutrient cycling, degradation of xenobiotics, and corrosion processes. If this is so, then there are many implications, not only for microbial ecology but for the whole subject area of microbiology. It is, therefore, crucial that investigators are able to assess the metabolic activities of surface associated bacteria. To date, a number of different approaches have been tried, depending upon, as with enumeration, the environmental situation or the laboratory system under investigation. It must also be stated that the results do not always confirm the view, generally stated, earlier. The reasons for this are that there are a number of factors which may affect the activity of cells at surfaces:

1. Concentration effects of the substrate at the surface.
2. Change in pH or Eh at the surface.
3. Increase or decrease in inhibitor concentrations.
4. Masking of cell surface (and therefore transport sites etc) by a solid surface.
5. Change in the apparent activation energy.
6. Release of metabolically active molecules or ions from cells. (from Hattori and Hattori, 1976).

With so many variables involved, it is not surprising that experimental results are often conflicting, reporting both stimulatory and inhibitory effects on microbial activity. Moreover, there are inherent difficulties in trying to determine the metabolic activities of attached populations *in situ* compared to determinations carried out in the more closely controlled environment, possible within the laboratory. Initially, however, perhaps the most pertinent question is which particular kind of activity do we wish to measure and can this be extrapolated to assess 'metabolic activity' in general? The question is further complicated since some techniques detect only activity *per se* and do not necessarily give a *rate* of activity. Broadly speaking the approaches used can be described as:

1. Those techniques which measure changes in and/or growth rates of the population.
2. Assays of enzyme activities.
3. Measurement of respiration rates.
4. Measurement of specific substrate uptake rates.
5. Techniques which demonstrate non-specific metabolic activity.

15.4.1 Changes in the surface population

Assessment of microbial populations has been discussed at length earlier and will not be considered further, except to make the point that few studies have tried to assess the rate of growth at a surface. The main problem is trying to separate *growth* from the observed increase in the population, which is due to growth + adsorption (− detachment and predation (Baker, 1984). Some attempts have been made to separate these factors by mathematical modelling but the problem remains to be solved (Caldwell *et al.*, 1981; 1983).

Taking an experimental approach, Wardell *et al.* (1984) were able to demonstrate different rates of colonization of glass surfaces within chemostat cultures. At low concentrations of the limiting substrate (C or N), the rate of increase of the surface population was similar to that of the liquid culture, whereas at higher values the rate of increase of the surface population was undoubtedly greater than the growth rate of the liquid population. While this observation must, in part, be attributed to increased adsorption/deposition of cells onto the surface, a result of the increased liquid population density, the microscopic evidence seemed to suggest that growth, in the form of microcolonies, was also largely responsible. This aspect of attachment could be resolved using time-lapse video-microscopy; some progress has already been made in this area (Caldwell and Lawrence, 1986).

The rates of deposition of streptococci with time have been examined using flat glass capillaries and phase contrast microscopy (Rutter and Leech, 1980) but perhaps, more importantly, the rates of adsorption and removal, due to fluid shear, of *Bacillus* sp. cells have been examined using similar flat glass capillaries and video-microscopy (Powell and Slater, 1982, 1983). This latter technique seems to offer tremendous potential for observing the attachment to and colonization of glass surfaces over prolonged periods of time. It should be possible by using freeze-frame and image analysis techniques to distinguish between growth and adsorption; by noting the time element, the rate of growth at a surface could be determined. This approach has recently been employed by Caldwell and Lawrence (1986) to examine the growth kinetics of microcolonies of *Pseudomonas fluorescens* within surface microenvironments.

15.4.2 Enzyme activity

Direct measurement of cellular enzyme activity does not seem to have found much favour in adhesion studies. The reason for this is probably the delay involved in recovering and processing cells before the enzyme activity can be determined. As discussed earlier, such delays cast doubts on the relevance of the data obtained, especially with respect to activities *in situ*. However, Hendricks (1974) has compared the alkaline phosphatase, β-galactosidase and aryl sulfatase activities of both attached and planktonic bacterial populations from continuous cultures of river water. Glass coverslips were used as a convenient substratum for colonization, which were then crushed and ground, and the cells released harvested for enzyme assay. One millilitre volumes of cell suspension were incubated with a suitable nitrophenyl substrate (p-nitrophenyl phosphate, o-nitrophenyl, β-D-galactopyranoside or p-nitrophenyl sulphate) in an appropriate buffer and the enzyme activity determined as μg nitrophenol liberated per 24h per mg of protein. The nitrophenol was detected spectrophotometrically at 400 nm. The results from these assays showed that, in terms of activity of attached populations compared with suspended populations, only for alkaline phosphatase activity was there any signi-ficant difference—a 2.6 fold increase in activity in favour of the attached population (Hendricks, 1974). Similarly, p-nitrophenyl derivatives of glucose, mannose, galactose, phosphate, and sulfate have been used by White *et al.* (1979) to assay for esterase activity in detrital microflora. In this case, activity was determined by measuring release of p-nitrophenol at 25 °C in 25 mM sodium bicarbonate buffer at pH 9.0 spectrophotometrically at 410 nm.

15.4.3 Respiration rates

Measurement of respiration rate, as oxygen consumption, was one of the earliest indications that attached bacteria might be metabolically more active than free-living forms (ZoBell and Anderson, 1936). More recently, this aspect of bacterial activity with regard to attached populations has been re-examined. The most convenient approaches for determining respiration rates would seem to be either measuring the reduction in dissolved oxygen (O_2) or carbon dioxide evolution.

Reduction in dissolved oxygen can be determined quite easily with dissolved-O_2 electrodes. These require standardization which can be as % dissolved-O_2, partial pressure in mmHg or as moles O_2 l^{-1}, the choice depending upon the experimental regime. For example, Hendricks (1974) compared the respiration rates of suspended and attached bacteria from river water and found that the rate for attached bacteria, determined as nmoles O_2 consumed per hour per μg of protein using a commercial biological oxygen monitor, was 10% greater than the rate for suspended bacteria. Recent advances in instrumentation and the manufacture of probes now enable quite small environmental samples to be examined.

Perhaps more common in attachment studies is the estimation of CO_2 evolution using radiolabelled substrates. The advantages of the technique are sensitivity (nmole quantities) and the ability to correlate mineralization rates with uptake rates. The working volume of culture or cell suspension and the trapping system employed vary between different research groups and because radioisotopes are used, *in situ* environmental experiments are not possible in the absence of containment facilities. Nevertheless, a number of workers have employed labelled substrates for respiration studies.

Sessile populations taken from cobble surfaces in subalpine streams have been examined by Ladd *et al.* (1979). Typically an area of 4 cm² was scraped with a sterile scalpel to remove sessile bacteria which were dispersed by blending in sterile stream water, and then suitably diluted. Ten-millilitre volumes were dispersed into 50 ml sterile vials containing different concentrations of ^{14}C-glutamic acid. After incubation for 6–12 h the reaction (growth) was stopped by the addition of 0.2 ml of 2 M sulphuric acid. This procedure also releases the dissolved, labelled CO_2 which was trapped on filter paper wicks suspended from the seal of the container and soaked with 0.2 ml of β-phenylethylamine. Radioactivity was then determined by transferring the wicks to a toluene-based scintillation cocktail and counting in a scintillation counter. This method is based on that described by Harrison *et al.* (1971) for determining the mineralization rates in sediments from lakes. Harrison and his co-workers recommended a level of radiolabel of 0.1 μCi per 2 ml volume of substrate sample in 50 ml serum vials. One of the advantages of their method is that a large number of vials can be prepared and frozen prior to field work. Other experimental details to note are that the phenylethylamine should not be added to the filter wicks until all the CO_2 has been allowed to evolve from the culture suspensions (up to 2 h) and that up to 1 h should be allowed for the complete adsorption of the $^{14}CO_2$ by the phenylethylamine (Harrison *et al.*, 1971). The same experimental procedure has also been employed to measure the mineralization rate of ^{14}C-glucose by bacteria attached to suspended solid particles in river water (Goulder 1977).

Similarly, Bright and Fletcher (1983) used essentially the same technique to evaluate the respiration of a number of amino acids in artificial seawater by a marine pseudomonad attached to a variety of plastic substrata. Their control experiments suggested that a 2 h period should be allowed for the complete adsorbtion of the ^{14}C-CO_2 by the phenylethylamine.

Radio respirometry of ^{14}C-glucose and ^{14}C-glutamic acid by bacteria attached to hydroxyapatite has been described by Gordon *et al.* (1983). In their experimental protocol, 10 ml respirometry flasks with a centre well were employed and 0.1 ml of 10% w/v sodium hydroxide, added to the well, used to trap the $^{14}CO_2$ evolved. In this case, 85% phosphoric acid was used to stop the incubation and release labelled CO_2 from the hydroxyapatite.

15.4.4 Specific substrate uptake

This naturally follows on from the previous section, the difference being that, instead of trapping labelled CO_2 evolved, the organisms are removed from the culture by filtration for suspended organisms or those attached to suspended solids or small particles. For larger particles, pieces of crushed coverslip, etc., sedimentation or low-speed centrifugation may suffice. In all cases it is important to rinse carefully but thoroughly to remove traces of label in the culture fluid without removing attached cells. For this reason filtration, through a 0.45 μm (Gordon et al., 1983) or 0.2 μm membrane (Bright and Fletcher, 1983; Ladd et al., 1979) is preferable. The technique employed by Bright and Fletcher (1983) is applicable to larger experimental surfaces. Pieces of colonized substratum, which had been exposed to labelled substrate, were incubated in the presence of 'Lumasolve' (LKB, Croydon, England) for 12 h at 50 °C to solubilize the cells. The scintillation cocktail was subsequently added to the digest prior to determining the activity of the sample in a scintillation counter.

Such techniques have been used to investigate the rate of assimilation of a variety of substrates: glucose, arginine, glutamic acid, glycine and leucine (see references cited earlier). As previously mentioned, the results differ depending on the organism, substrate and interface examined, and will no doubt also depend on the sensitivity of the assay system. However, the more recent study suggests that while respiration rates may not be dissimilar for attached and free-living populations, the assimilation rate of attached populations is likely to be greater (Bright and Fletcher, 1983).

An interesting alternative to ^{14}C labelling is to use ^{13}C substrates and look for enrichment of ^{13}C in cellular components by GLC analysis coupled to mass-spectrometry. White (1983) has outlined the advantages of this approach as:

1. No radioactivity is involved so experiments can be done more readily in the field.
2. The specific activity is high (>99%) so precursor concentrations can be used which are nearer to actual ambient values and there is less likelihood of distortion in experimental results. This is an important point since providing labelled substrate often means elevating the substrate concentration above that normally encountered in the environment. This itself may stimulate metabolic activity, thereby providing misleading data.
3. Enrichment of ^{13}C can be easily detected using the selected ion detection mode in the mass spectrometer.

Finally White makes the point that while these (labelling) techniques may be excellent when applied to thin films of micro-organisms on easily manipulated surfaces such as detritus, glass coverslips or fouling coupons, in sediments the disturbances following the exposure of the system to labelled substrates may in itself induce metabolic activity.

15.4.5 Methods to detect non-specific metabolic activity

15.4.5.1 Microcalorimetry

Although microcalorimetry has been available as a potential tool for measuring microbial activity for some 15 years, the number of reports, especially with regard to attached populations, are limited. Reasons for this are undoubtedly the level of sensitivity required of the apparatus and the subsequent cost. The main expense being the provision of an ultra-thermostable water/air bath with a specification of $\pm 0.01\,°C$ (Gordon *et al.*, 1982), or $\pm 0.005\,°C$ (Lock and Ford, 1983). However, the technique has been used in a variety of applications, and an introduction to its use in sediment ecology is given by Parmatmat (1982). Until fairly recently, the technique had only been applied to populations of free-living bacteria at cell densities of ca. 10^7 cfu ml^{-1} but there are now a few reports describing the application of the technique to attached populations of bacteria. The construction and operation of a microcalorimeter for measuring the heat production of attached and sedimentary micro-organisms has been described by Lock and Ford (1983). The limit of detection of the apparatus was $3\,\mu W$ ($\pm 2\%$), low enough to measure the heat output of river epilithon growing on black glass beads, river sand and marine sand; the outputs detected being $0.8–6.8\,\mu W$ cm^{-2}, $9.8\,\mu W$ cm^{-3} and $15.3\,\mu W$ cm^{-3}, respectively. The output detected from the experimental chambers ($32–284\,\mu W$) being at least 10-fold higher than the limit of detection. Similarly, Gordon *et al.* (1982, 1983) have employed a micro-mixing calorimeter to investigate the effect of hydroxyapatite particles on the heat output of *Vibrio alginolyticus* growing on glucose or glutamate as carbon source. Although the bacterial population employed was quite high (6×10^7 cfu ml^{-1}) in terms of natural aquatic populations, the sensitivity of the instrument allowed submicromolar concentrations of substrate, in the range $0.35\,\mu M–8\,mM$, to be employed; that is, concentrates of substrate more akin to those encountered in natural waters (Gordon *et al.*, 1982, 1983).

It seems likely that, providing capital costs can be met, microcalorimetry has potential for detecting metabolism activity of organisms attached to a range of substrata providing the substratum and colonizing population can be manipulated with minimum interference.

15.4.5.2 Chemical indicators of metabolic activity

Actively respiring micro-organisms will utilize tetrazolium salts (triphenyl tetrazolium chloride (TTC), 2-(paraiodophenyl)-3-(*p*-nitrophenyl)-5-phenyl tetrazolium chloride (INT)), as alternative terminal electron acceptors in the electron transport chain and in doing so the tetrazolium salt is reduced to a deep red formazan dye. Incorporated into plates this shows as red pigmented colonies, however, in individual cells this results in optically dense red intracellular bodies which can be detected microscopically (Zimmermann et al., 1978). This property has been exploited to detect metabolically active microbes in natural water and attached to the surfaces of particles (Harvey and Young, 1980).

Early attempts to detect active microbial cells by this technique suffered from the drawback that the formazan 'spots' within the cells do not delineate the cell, some confusion with optically dense particles of non-microbial origin was therefore inevitable. Zimmermann et al. (1978) devised a method which combined epifluorescence microscopy with exposure to INT. Epifluorescence enabled enumeration of the micro-organisms and alternate bright field illumination allowed enumeration of those cells with active electron transport chains. The procedure was to allow INT, at a final concentration of 0.02% (w/v), to react with samples for 20 min in the dark at the sampling temperature. This 20-min period was chosen to ensure all respiring cells had sufficient time to accumulate a maximum quantity of the INT-formazan. Samples were then fixed, and further reaction stopped by the addition of formaldehyde. This was followed by filtering the samples onto polycarbonate membrane filters and staining with acridine orange for epifluorescent microscopy.

While the method outlined above was only applied to free-living organisms, the technique has been modified by Harvey and Young (1980) to enable organisms attached to small particles to be similarly enumerated. The essential modification to the technique of Zimmermann et al. (1978) was to arrange a simultaneous bright-field-fluorescent image by reducing the bright field image to 50% intensity. By altering the plane of focus and adjusting the intensity of the transmitted light organisms attached to particles were enumerated and whether or not they were actively respiring noted. Using this method, Harvey and Young (1980) estimated that in the salt marsh estuary sampled, 95% of the actively respiring bacteria were attached to particles. More recently the technique has been further modified by Tabor and Neihof (1982) to try and eliminate the reduced counts encountered with time due to dissolution of the formazan in immersion oil. In addition, it is often difficult to detect very small cells due to the small amount of INT-formazan formed.

While the use of INT is obviously of value in detecting microbial activity for populations on dispersed particles, its role in the evaluation of sediment samples or microbial films, for example on glass surfaces, remains to be proven. However, there does seem to be some interest in this approach. Similar studies employing fluorescein diacetate have been reported by Chrzanowski *et al.* (1984). In this technique, dissociated microbial cells were incubated for 20 min at the *in situ* temperature in the presence of fluorescein diacetate (FDA) at a final concentration of 5 µg ml^{-1}. FDA, normally non-fluorescent, can be transported across cell membranes and then deacetylated by non-specific esterases to fluorescein which fluoresces green when irradiated with blue light. The fluorescein formed resists export from the cell and accumulates, delineating the cells for microscopy. For bacteria, fluorescein diacetate transport appears to be passive and since accumulation depends on an intact cell membrane and active metabolism (to carry out deacetylation) only metabolically active cells should fluoresce (Chrzanowski *et al.* 1984). However, while application of the technique to several freshwater habitats has resulted in estimates of active cells as 6–24% of the total population, these estimates were low compared to estimates based on electron transport chain activity. One reason postulated for this is that Gram-negative cells are less permeable to FDA than Gram-positive cells, presumably due to the more complex structure of the cell envelope and, in particular, the presence of the outer membrane (Chrzanowski *et al.*, 1984). This very clearly limits the usefulness of the technique; moreover, the applicability to surface associated populations remains unproven. Nevertheless, it is a technique which may be worth considering for specific experimental systems.

As with most experimental procedures, including attached populations, the experimental protocols selected for determining activity will need to be modified to suit the particular experimental design and the environment under investigation.

In conclusion, the techniques described will hopefully indicate the range of possible experimental approaches which can be applied to the study of bacterial attachment in aquatic environments, and introduce the reader to the relevant literature. Perhaps more than anything else this chapter illustrates the complexity and importance of what may at first sight appear to be a relatively simple phenomenon.

15.5 REFERENCES

Allison, D. G. and Sutherland, I. W. (1984). A staining technique for attached bacteria and its correlation to extracellular carbohydrate production. *Journal of Microbiological Methods*, **2**, 93–99.

Baker, J. H. (1984). Factors affecting the bacterial colonization of various surfaces in a river. *Canadian Journal of Microbiology*, **30**, 511–515.

Bayer, M. E. and Thurow, H. (1977). Polysaccharide capsule of *Escherichia coli*: microscope study of its size, structure and site of synthesis. *Journal of Bacteriology*, **130**, 911–936.

Berkeley, R. C. W., Lynch, J. M., Melling, J., Rutter, P. R. and Vincent, B. (Eds) (1980). *Microbial Adhesion to Surfaces*, Ellis Horwood, Chichester.

Birdsell, D. C., Doyle, R. J. and Morgenstein, M. (1975). Organisation of teichoic acid in the cell wall of *Bacillus subtilis*. *Journal of Bacteriology*, **121**, 726–734.

Bitton, G. and Marshall, K. C. (Eds) (1980). *Adsorption of Micro-organisms to Surfaces*, Wiley, New York.

Bobbie, R. J. and White, D. C. (1980). Characterization of benthic microbial community structure by high resolution gas chromatography of fatty acid methyl esters. *Applied and Environmental Microbiology*, **39**, 1212–1222.

Bright, J. J. and Fletcher, M. (1983). Amino acid assimilation and respiration by attached and free-living populations of a marine *Pseudomonas*. *Microbial Ecology*, **9**, 215–226.

Brown, C. M., Ellwood, D. C. and Hunter, J. R. (1977). Growth of bacteria at surfaces: influence of nutrient limitation. *FEMS Microbiology Letters*, **1**, 163–166.

Buck, J. D. (1979). The plate count in aquatic microbiology. In *Native Aquatic Bacteria: Enumeration, Activity and Ecology*, ASTM STP 695, J. W. Costerton and R. R. Colwell (Eds), American Society for Testing and Materials, pp. 19–28.

Burns, R. G. (1979). Interaction of microorganisms, their substrates and their products with soil surfaces. In *Adhesion of Microorganisms to Surfaces*, D. C. Ellwood, J. Melling and P. Rutter (Eds), Special publication of Society for General Microbiology, Academic Press, London, pp. 109–138.

Caldwell, D. E., Brannan, D. K., Morris, M. E. and Betlach, M. R. (1981). Quantitation of microbial growth on surfaces. *Microbial Ecology*, **7**, 1–11.

Caldwell, D. E., Malone, J. A. and Kieft, T. L. (1983). Derivation of a growth rate equation describing microbial surface colonization. *Microbial Ecology*, **9**, 1–6.

Caldwell, D. E. and Lawrence, J. R. (1986). Growth kinetics of *Pseudomonas fluorescens* microcolonies within the hydrodynamic boundary layers of surface microenvironments. *Microbial Ecology*, **12**, 299–312.

Carson, J. and Allsopp, D. (1980). The enumeration of marine periphytic bacteria from a temporal sampling series. In *Biodeterioration: Proceedings of Fourth International Symposium, Berlin*, D. Allsopp, T. A. Oxley and G. Becker (Eds), Pitman, London, pp. 193–198.

Chrzanowski, T. H., Crotty, R. D., Hubbard, J. G. and Welch, R. P. (1984). Applicability of the fluorescein diacetate method of detecting active bacteria in freshwater. *Microbial Ecology*, **10**, 179–185.

Corpe, W. A. (1964). Factors influencing growth and polysaccharide formation by stains of *Chromobacterium violaceum*. *Journal of Bacteriology*, **88**, 1433–1441.

Corpe, W. A. (1970a). An acidic polysaccharide produced by a primary film-forming marine bacterium. In *Developments in Industrial Microbiology II*, E. D. Murray (Ed.),. *Proceedings of the Society for Industrial Microbiology*, pp. 402–412.

Corpe, W. A. (1970b). Attachment of marine bacteria to solid surfaces. In *Adhesion in Biological Systems*, R. S. Manley, (Ed.) Academic Press, New York, pp. 73–87.

Corpe, W. A. (1974). Periphytic marine bacteria and the formation of microbial films on solid surfaces. In *Effect of Ocean Environment on Microbial Activity*, R. R. Colwell and R. Y. Morita (Eds), University Park Press, Baltimore, pp. 397–417.

Corpe, W. A. (1980). Microbial surface components involved in adsorption of microorganisms onto surfaces. In *Adsorption of Microorganisms to Surfaces*, G. Bitton and K. C. Marshall (Eds), Wiley, New York, pp. 105–144.

Costerton, J. W. (1980). Some techniques involved in study of adsorption of microorganisms to surfaces. In *Adsorption of Microorganisms to Surfaces*, G. Bitton and K. C. Marshall (Eds), Wiley, New York, pp. 403–423.

Costerton, J. W., Geesey, G. G. and Cheng, K-J. (1978). How Bacteria Stick. *Scientific American*, **238** (i), 86–95.

Costerton J. W., Irvin, R. T. and Cheng, K-J. (1981). The bacterial glycocalyx in nature and disease. *Annual Review of Microbiology*, **35**, 299–324.

Costerton, J. W., Marrie, T. J. and Cheng, K-J. (1985). Phenomena of bacterial adhesion. In *Bacterial Adhesion: Mechnisms and Physiological Significance*, D. C. Savage and M. Fletcher (Eds), Plenum Press, New York, pp. 1–43.

Cowan, S. T. (1974). *Cowan and Steel's Manual for the Identification of Medical Bacteria*, 2nd edn revised, Cambridge University Press, Cambridge.

Daniels, S. L. (1972). The adsorption of microorganisms onto solid surfaces: A review. In *Developments in Industrial Microbiology*, Vol. 13, E. D. Murray (Ed.) *Proceedings of the Society for Industrial Microbiology*, pp. 211–253.

Daniels, S. L. (1980). Mechanisms involved in sorption of microorganisms to solid surfaces. In *Adsorption of Microorganisms to Surfaces*, G. Bitton and K. C. Marshall (Eds), Wiley, New York, pp. 7–58.

Dempsey, M. J. (1981a). Marine bacterial fouling: a scanning electron microscope study. *Marine Biology*, **61**, 305–315.

Dempsey, M. J. (1981b). Colonization of antifouling paints by marine bacteria. *Botanica Marina*, **24**, 185–191.

Ellwood, D. C. and Tempest, D. W. (1972). Effects of environment on bacterial wall content and composition. *Advances in Microbial Physiology*, **7**, 83–117.

Feldner, J., Bredt, W. and Kahane, I. (1983). Influence of cell shape and surface charge on attachment of *Mycoplasma pneumoniae* to glass surfaces. *Journal of Bacteriology*, **153**, 1–5.

Fletcher, M. (1976). The effects of proteins on bacterial attachment to polystyrene. *Journal of General Microbiology*, **94**, 400–404.

Fletcher, M. (1977a). Attachment of marine bacteria to surfaces. In *Microbiology 1977*, D. Schlessinger (Ed.), *American Society for Microbiology*, Washington DC, pp. 407–410.

Fletcher, M. (1977b). The effects of culture concentration and age, time and temperature on bacterial attachment to polystyrene. *Canadian Journal of Microbiology*, **23**, 1–6.

Fletcher, M. (1979). The attachment of bacteria to surfaces in aquatic environments. In *Adhesion of Microorganisms to Surfaces*, D. C. Ellwood, J. Melling and P. Rutter (Eds), Special publication of Society for General Microbiology, Academic Press, London, pp. 88–108.

Fletcher, M. (1980). Adherence of marine micro-organisms to smooth surfaces. In *Bacterial Adherence*, E. H. Beachey (Ed.), Receptors and Recognition, Series B, vol. 6, Chapman and Hall, London, pp. 354–374.

Fletcher, M. and Floodgate, G. D. (1973). An electron-microscope demonstration of an acidic polysaccharide involved in the adhesion of a marine bacterium to solid surfaces. *Journal of General Microbiology*, **74**, 325–334.

Fletcher, M. and Loeb, G. I. (1976). The influence of substratum surface properties on the attachment of a marine bacterium. In *Colloid and Interface Science*, vol. III, M. Kerker (Ed.), Academic Press, London, pp. 459–469.

Fletcher, M. and Loeb, G. I. (1979). Influence of substratum characteristics on the attachment of a marine pseudomonad to solid surfaces. *Applied and Environmental Microbiology*, **37**, 67–72.

Floodgate, G. D. (1972). The mechanism of bacterial attachment to detritus in aquatic systems. In *Proceedings of the IBP–UNESCO Symposium on Detritus and its Role in Aquatic Ecosystems, Memorie dell'Istituto italiano di idrobiologia Dott. Marco de Marchi*, **29**, suppl. 309–323.

Geesey, G. G. and Costerton, J. W. (1979). Microbiology of a northern river:

bacterial distribution and relationship to suspended sediment and organic carbon. *Canadian Journal of Microbiology*, **25**, 1058–1062.

Geesey, G. G., Mutch, R., Costerton, J. W. and Green, R. B. (1978). Sessile bacteria: an important component of the microbial population in small mountain streams, *Limnology and Oceanography*, **23**, 1214–1223.

Geesey, G. G., Richardson, W. T., Yeomans, H. G., Irvin, R. T. and Costerton, J. W. (1977). Microscopic examination of natural sessile bacterial populations from an alpine stream. *Canadian Journal of Microbiology*, **23**, 1733–1736.

Gehron, M. J., Moriarty, D. J. W., Smith, G. A. and White, D. C. (1982). Determination of the Gram-negative Gram-positive bacterial content of sediments, cited by D. C. White (1983).

Gordon, A. S., Millero, F. J. and Gerchakov, S. M. (1982). Microcalorimetric measurements of glucose metabolism by marine bacterium *Vibrio alginolyticus*. *Applied and Environmental Microbiology*, **44**, 1102–1109.

Gordon, A. S., Gerchakov, S. M. and Millero, F. J. (1983). Effects of inorganic particles on metabolism by a periphytic marine bacterium. *Applied and Environmental Microbiology*, **45**, 411–417.

Goulder, R. (1977). Attached and free bacteria in an estuary with abundant suspended solids. *Journal of Applied Bacteriology*, **43**, 399–405.

Gurr, (1979). *Gurr's Biological Staining Methods*, Hopkin and Williams, Chadwell Heath, England.

Harber, M. J., Mackenzie, R. and Asscher, A. W. (1983). A rapid bioluminescence method for quantifying bacterial adhesion to polystyrene. *Journal of General Microbiology*, **129**, 621–632.

Harden, V. P. and Harris, J. O. (1953). The isoelectric point of bacterial cells. *Journal of Bacteriology*, **65**, 198–202.

Harris, C. M. and Kell, D. B. (1985). The estimation of microbial biomass. *Biosensors*, **1**, 17–84.

Harrison, M. J., Wright, R. T. and Morita, R. Y. (1971). Method for measuring mineralization in lake sediments. *Applied Microbiology*, **21**, 698–702.

Hattori, T. and Hattori, R. (1976). The physical environment in soil microbiology: an attempt to extend principles of microbiology to soil organisms. *CRC Critical Reviews in Microbiology*, **4**, 423–461.

Harvey, R. W. and Young, L. Y. (1980). Enumeration of particle-bound and unattached respiring bacteria in the salt marsh environment. *Applied and Environmental Microbiology*, **40**, 156–160.

Hendricks, C. W. (1974). Sorption of heterotrophic and enteric bacteria to glass surfaces in the continuous culture of river water. *Applied Microbiology*, **28**, 572–578.

Hermansson, M. and Marshall, K. C. (1985). Utilization of surface localized substrate by non-adhesive marine bacteria. *Microbial Ecology*, **11**, 91–105.

Hobbie, J. E., Daley, R. and Jasper, S. (1977). Use of Nucleopore filters for counting bacteria for fluorescence microscopy. *Applied and Environmental Microbiology*, **33**, 1225–1228.

Jones, H. C., Roth, I. L. and Sanders, W. M. (1969). Electron microscope study of a slime layer. *Journal of Bacteriology*, **99**, 316–325.

Karl, D. M. (1980). Cellular nucleotide measurements and applications in microbial ecology. *Microbiological Reviews*, **44**, 739–796.

Kjelleberg, S., Marshall, K. C. and Hermansson, M. (1985). Oligotrophic and copiotrophic marine bacteria—observations related to attachment. *FEMS Microbiology Ecology*, **31**, 89–96.

Krambeck, C., Krambeck, H-J. and Overbeck, J. (1981). Microcomputer-assisted biomass determination of plankton bacteria on scanning electron micrographs. *Applied and Environmental Microbiology*, **42**, 142–149.

Ladd, T. I., Costerton, J. W. and Geesey, G. G. (1979). Determination of the heterotrophic activity of epilithic microbial populations. In *Native Aquatic Bacteria: Enumeration, Activity and Ecology*, J. W. Costerton and R. R. Colwell (Eds), ASTM STP 695, *American Society for Testing and Materials*, pp. 180–195.

Lips, A. and Jessup, N. E. (1979). Colloidal aspects of bacterial adhesion. In *Adhesion of Microorganisms to Surfaces*, D. C. Ellwood, J. Melling and P. Rutter (Eds), Special publication of the Society for General Microbiology, Academic Press, London, pp. 5–27.

Lock, M. A. and Ford, T. E. (1983). Inexpensive flow microcalorimeter for measuring heat production of attached and sedimentary aquatic microorganisms. *Applied and Environmental Microbiology*, **46**, 463–467.

Mackie, E. B., Brown, K. W. Lam, J. and Costerton, J. W. (1979). Morphological stabilization of capsules of group B Streptococci, Types Ia, Ib, II and III with specific antibody. *Journal of Bacteriology*, **138**, 609–617.

Marshall, K. C. (1976). *Interfaces in Microbiology Ecology*, Harvard University Press, Cambridge, Massachusetts.

Marshall, K. C., Stout, R. and Mitchell, R. (1971a). Mechanism of the initial events in the sorption of marine bacteria to surfaces. *Journal of General Microbiology*, **68**, 337–348.

Marshall, K. C., Stout, R. and Mitchell, R. (1971b). Selective sorption of bacteria from seawater. *Canadian Journal of Microbiology*, **17**, 1413–1416.

Meadows, P. S. (1965). Attachment of marine and fresh-water bacteria to solid surfaces. *Nature (London)*, **207**, 1108.

Meadows, P. S. (1971). The attachment of bacteria to solid surfaces. *Archiv für Mikrobiologie*, **75**, 374–381.

Meadows, P. S. and Anderson, J. G. (1968). Microorganisms attached to marine sand grains. *Journal of Marine Biological Association, U. K.*, **48**, 161–175.

Meadows, P. S. and Anderson, J. G. (1979). The microbiology of interfaces in the marine environment. *Progress in Industrial Microbiology*, **15**, 207–265.

Meyer-Reil, L. A. (1978). Autoradiography and epifluorescence microscopy combined for the determination of number and spectrum of actively metabolizing bacteria in natural waters. *Applied and Environmental Microbiology*, **36**, 506–512.

Molin, G., Nilsson, I. and Stenson-Holst, L. (1982). Biofilm build-up of *Pseudomonas putida* in a chemostat at different dilution rates. *European Journal of Applied Microbiology and Biotechnology*, **15**, 218–222.

Paerl, H. W. and Merkel, S. M. (1982). Differential phosphorous assimilation in attached *vs* unattached microorganisms. *Archiv für Hydrobiologie*, **93**, 125–134.

Paerl, H. W. (1975). Microbial attachment to particles in marine and freshwater ecosystems. *Microbial Ecology*, **2**, 73–83.

Paerl, H. W. (1980). Attachment of microorganisms to living and detrital surfaces in freshwater systems. In *Adsorption of microorganisms to surfaces*. G. Bitton and K. C. Marshall (Eds), Wiley, New York, pp. 375–402.

Parmatmat, M. (1982). Heat production by sediment: ecological significance. *Science*, **215**, 395–397.

Parker, J. H., Smith, G. A., Fredrickson, H. L., Vestal, J. R. and White, D. C. (1982). Sensitive assay, based on hydroxy fatty acids from lipopolysaccharide lipid A, for Gram-negative bacteria in sediments. *Applied and Environmental Microbiology*, **44**, 1170–1177.

Paul, J. H. (1982). Use of Hoechst Dye 33258 and 33342 for enumeration of attached and planktonic bacteria. *Applied and Environmental Microbiology*, **43**, 939–944.

Powell, M. S. and Slater, N. K. H. (1982). Removal rates of bacterial cells from glass surfaces by fluid shear. *Biotechnology and Bioengineering*, **24**, 2527–2537.

Powell, M. S. and Slater, N. K. H. (1983). The deposition of bacterial cells from laminar flow onto solid surfaces. *Biotechnology and Bioengineering*, **25**, 891–900.

Pringle, J. H. and Fletcher, M. (1983). Influence of substratum wettability on attachment of freshwater bacteria to solid surfaces. *Applied and Environmental Microbiology*, **45**, 811–817.

Pringle, J. H., Fletcher, M. and Ellwood, D. C. (1983). Selection of attachment mutants during the continuous culture of *Pseudomonas fluorescens* and relationship between attachment ability and surface composition. *Journal of General Microbiology*, **129**, 2557–2569.

Robrish, S. A., Kemp, C. W. and Bowen, W. H. (1978). Use of extractable adenosine triphosphate to estimate the viable cell mass in dental plaque samples obtained from monkeys. *Applied and Environmental Microbiology*, **35**, 743–749.

Rutter, P. and Leech, R. (1980). The deposition of *Streptococcus sanguis* NCTC 7868 from a flowing suspension. *Journal of General Microbiology*, **120**, 301–307.

Savage, D. C. and Fletcher, M. (Eds) (1985). *Bacterial Adhesion—Mechanisms and Physiological Significance*, Plenum Press, New York.

Schofield, G. M. and Locci, R. (1985a). Colonization of components of a model hot water system by *Legionella pneumophila*. *Journal of Applied Bacteriology*, **58**, 151–162.

Schofield, G. M. and Locci, R. (1985b). The persistance of *Legionella pneumophila* in non-sterile, sterile and artificial hard waters and their growth pattern on tap washer fittings. *Journal of Applied Bacteriology*, **59**, 519–527.

Stuart, V. (1982). Limitations of ATP as a measure of microbial biomass. *South African Journal of Zoology*, **17**, 93–95.

Sutherland, I. W. (1983). Microbial exopolysaccharides—their role in microbial adhesion in aqueous systems. *CRC Critical Reviews in Microbiology*, **10**, 173–201.

Tabor, P. S. and Neihof, R. A. (1982). Improved method for determination of respiring individual microorganisms in natural waters. *Applied and Environmental Microbiology*, **43**, 1249–1255.

Ward, J. B. and Berkeley, R. C. W. (1980). The microbial cell surface. In *Microbial Adhesion to Surfaces*, R. C. W. Berkeley, J. M. Lynch, J. Melling, P. R. Rutter and B. Vincent (Eds), Society of Chemical Industry, Ellis Horwood, Chichester, pp. 47–66.

Wardell, J. N. (1982). Bacterial communities and surface associated growth, Ph.D. thesis, Heriot-Watt University, Edinburgh.

Wardell, J. N., Brown, C. M. and Ellwood, D. C. (1980). A continuous culture study of the attachment of bacteria to surfaces. In *Microbial Adhesion to Surfaces*, R. C. W. Berkeley, J. M. Lynch, J. Melling, P. R. Rutter and B. Vincent (Eds), *Society of Chemical Industry*, Ellis Horwood, Chichester, pp. 221–230.

Wardell, J. N., Brown, C. M., Ellwood, D. C. and Williams, A. E. (1984). Bacterial growth on inert surfaces. In *Continuous Culture 8: Biotechnology, Medicine and the Environment*, A. C. R. Dean, D. C. Ellwood and C. G. T. Evans (Eds), Ellis Horwood, Chichester, pp. 159–168.

Wardell, J. N., Brown, C. M. and Flannigan, B. (1983). Microbes and surfaces. In *Microbes in their Natural Environments*, J. H. Slater, R. Wittenbury and J. W. T. Wimpenny (Eds), Symposium 34, Society for General Microbiology, Cambridge University Press, Cambridge, pp. 351–378.

White, D. C. (1983). Analysis of microorganisms in terms of quantity and activity in natural environments. In *Microbes in their Natural Environments*, J. H. Salter, R. Whittenbury and J. W. T. Wimpenny (Eds), Symposium 34, Society for General Microbiology, Cambridge University Press, Cambridge, pp. 37–66.

White, D. C., Bobbie, R. J., Herron, J. S., King, J. D. and Morrison, S. J. (1979). Biochemical measurements of microbial mass and activity from environment samples. In *Native Aquatic Bacteria: Enumeration, Activity and Ecology*, ASTM STP 695 J. W. Costerton and R. R. Colwell (Eds), American Society for Testing and Materials, pp. 69–81.

Zimmermann, R., Iturriaga, R. and Becker-Birck, J. (1978). Simultaneous determination of the total number of aquatic bacteria and the number thereof involved in respiration. *Applied and Environmental Microbiology*, **36,** 926–935.

ZoBell, C. E. (1943). The effect of solid surfaces upon bacterial activity. *Journal of Bacteriology*, **46,** 39–56.

ZoBell, C. E. and Allen, E. C. (1935). The significance of marine bacteria in the fouling of submerged surfaces. *Journal of Bacteriology*, **29,** 239–251.

ZoBell, C. E. and Anderson, D. Q. (1936). Observations on the multiplication of bacteria in different volumes of stored sea water and the influence of oxygen tension and solid surfaces. *Biological Bulletin*, **71,** 324–342.

Index of Organism Names

General Index